Understanding Globalization

Understanding Globalization

The Social Consequences of Political, Economic, and Environmental Change

4th Edition

Robert K. Schaeffer

ROWMAN & LITTLEFIELD PUBLISHERS, INC.
Lanham • Boulder • New York • Toronto • Plymouth, UK

ROWMAN & LITTLEFIELD PUBLISHERS, INC.

Published in the United States of America
by Rowman & Littlefield Publishers, Inc.
A wholly owned subsidary of The Rowman & Littlefield Publishing Group, Inc.
4501 Forbes Boulevard, Suite 200, Lanham, Maryland 20706
www.rowmanlittlefield.com

Estover Road, Plymouth PL6 7PY, United Kingdom

British Library Cataloguing in Publication Information Available

Library of Congress Cataloging-in-Publication Data

Schaeffer, Robert K.
    Understanding globalization : the social consequences of political, economic,
and environmental change / Robert K. Schaeffer. — 4th ed.
        p.   cm.
    Includes bibliographical references and index.
    ISBN-13: 978-0-7425-6179-3 (cloth : alk. paper)
    ISBN-10: 0-7425-6179-8 (cloth : alk. paper)
    ISBN-13: 978-0-7425-6180-9 (pbk. : alk. paper)
    ISBN-10: 0-7425-6180-1 (pbk. : alk. paper)
    ISBN-13: 978-0-7425-6519-7 (electronic)
    ISBN-10: 0-7425-6519-X (electronic)
    1. Globalization. 2. Social history—1945– 3. World politics—1945–1989.
4. Economic history—1945– 5. Global environmental change. I. Title.
    JZ1318.S33 2009
    303.48'2—dc22
                                                                2008041867

Printed in the United States of America

♾™ The paper used in this publication meets the minimum requirements of
American National Standard for Information Sciences—Permanence of Paper
for Printed Library Materials, ANSI/NISO Z39.48-1992.

Contents

The "redistribution" and "reorganization" of production in
the United States, Western Europe, and Japan during the
postwar period has contributed to the "globalization" of
production. This chapter analyzes the role that government
policies, business strategies, and stock markets played in
these developments and discusses how they contributed to
changing work roles for women and men.

In 1971 and again in 1985, U.S. presidents devalued the dollar
to improve U.S. competitiveness. But this monetary policy
failed to reduce trade deficits with our principal competitors
or persuade consumers to "Buy American." This policy
contributed to domestic problems, resulting in the sale and
export of timber and the subsequent loss of jobs in the Pacific
Northwest. And it created a host of problems for countries
around the world. This chapter examines global problems
associated with changing exchange rates and explains why

v

monetary policy is gendered: why it is designed to protect jobs
for men, not women.

In 1971, U.S. officials began battling inflation, which they
regarded as a problem because inflation is a discriminatory
economic process. The Federal Reserve eventually used high
interest rates to curb inflation. But this policy triggered the
collapse of the domestic savings and loan industry and
contributed to rising homelessness in U.S. cities. These
developments affected men and women in different ways. The
U.S. battle against inflation also triggered a massive debt crisis
for countries around the world, a crisis that persists today.

Many poor, "developing" countries borrowed heavily in the
1970s, tapping the financial pools in Eurocurrency markets to
promote economic growth. But rising U.S. interest rates and
falling commodity prices forced them into bankruptcy and
triggered a massive debt crisis. Despite an arduous, two-
decade effort to repay their loans, many poor countries
remain mired in debt. Moreover, the imposition of structural
adjustment programs on indebted governments has
contributed to poverty, environmental destruction, and a
decline in government services, developments that have been
particularly hard on women.

Globalization has triggered different kinds of migration
around the world. This chapter examines the causes and
consequences of contemporary economic, environmental, and
gender-based migrations, which includes a discussion of
migrating orphans, brides, and trafficked women. This is
followed by a look at the cross-border migrations triggered by
political developments such as partition, conflict, and genocide.

Since 1974, dictators in more than thirty countries have been
replaced by civilian democrats. This remarkable political
change was largely the product of regional economic crises,
which were compounded, in individual states, by assorted
political problems. The future of democratizing states
depends on their ability to address the problems that

first triggered the collapse of capitalist and communist dictators alike.

to the rise of the Taliban. This chapter explores the origins of the Taliban and its ally, Al Qaeda, which organized the attacks of September 11, 2001.

The attacks of September 11, 2001, led to U.S. invasions in Afghanistan and Iraq. This chapter examines the problems associated with the "War on Terror," its impact on Indo-Pakistani and Arab-Israeli conflicts, and its contribution to the current global recession.

In the mid-1980s, scientists warned that the release of heat-trapping gases could trigger global warming, with disastrous results. This chapter explores the debate about contemporary climate change, examines the roles that different heat-trapping gases play, and discusses some of the political and economic strategies that might reduce the threat of global warming.

Acknowledgments

A number of people and organizations contributed to the development of this book. My colleagues at Friends of the Earth and Greenpeace supported much of the initial work. My colleagues at Pugwash Conferences on Science and World Affairs, particularly Metta Spencer and Joseph Rotblat, encouraged my work on democratization. Participants in meetings sponsored by the American Sociological Association and the Political Economy of the World-System reviewed and commented on papers that addressed many of the problems discussed in this book. Immanuel Wallerstein and Giovanni Arrighi encouraged me to think globally and historically. Torry Dickinson, my friend, colleague, and collaborator, deserves special mention for her intellectual contribution to all of my work. Her interest in history and social change helped shape my ideas and direct my research. To all of you, my sincere thanks.

Introduction

We live in a time of global change. But people experience change in different ways. Global change is not a universal or uniform process. It affects some people more than others, and it can have very different consequences—good and bad—for people in different settings.

On the Weather Channel, journalists use satellite images to identify the global pressure systems that shape the weather in different regions, and use Doppler radar to track the movement of weather-system fronts across the landscape. They might note that a low-pressure system that moves off the Pacific and tracks east across North America can have very different consequences for people living along its path. The storm may bring fog to people living along the coast, rain to people living in inland valleys, snow to people in the mountains, and clear skies to people living in the plains beyond. As a result, people living along the path of the storm may experience changing weather in very different ways.

Like a meteorologist following a series of storms, I will narrate the history of some important global developments—inflation, debt crisis, democratization, cross-border migration, the rise of China, conflicts in the Middle East and 9/11, and global warming—that have recently swept across the landscape. These accounts will pay particular attention to the different and sometimes unexpected consequences of change for people affected by global change.

Many of the global changes examined here are interconnected. But the stories about change will be told separately. The idea is to provide readers with accounts of change that can be studied individually or read collectively. Where storylines intersect, reference will be made to related developments in other chapters. Some of the early chapters focus on changes that originated in the United States. This is because the United States is a central actor in the world and its initiatives are felt widely, and because it will help U.S. readers appreciate from the outset their relation to global change. As we examine contemporary global processes, keep in mind that shared global developments may have a different meaning for people living in different places.

As this book went to press in 2008, economic problems plunged the United States into a recession and triggered a global economic downturn. The problems, which originated in the housing industry, created a crisis for large-scale financial institutions, triggered a collapse of stock market prices, dried up credit, forced U.S. automakers into near bankruptcy, convinced consumers to cut back on their spending, and increased unemployment, which deepened the ensuing recession. These events are examined in chapter 12 ("Aftermath of 9/11"). Of course, it is too soon to provide a comprehensive and detailed account of these developments, and their consequences for people in the United States and around the world. Still, they can be understood by examining developments in past years that shaped or resembled current ones: the impact of Federal Reserve policies in the 1980s (chapters 3 and 4) and the Fed's role as both the architect and crisis-manager of economic problems today; the problems experienced by indebted countries and farmers in the 1980s (chapter 4) and by indebted homeowners today; the role that rising energy prices played in the 1970s (chapter 3), and the importance of soaring gas prices in recent years; the collapse of the savings and loans in the 1980s (chapter 3), and the crisis of large financial institutions in 2008; the emergence of the stock market as a force for economic change in the 1980s (chapter 1), and its prominent role in the current crisis; and the decline of U.S. automakers in the 1970s (see chapter 1) and their near-death experience in 2008. An examination of these developments contributes to an understanding of contemporary events and the likely or possible global consequences of rapid social change in the years ahead.

Acronyms

AFM	Armed Forces Movement (Portugal)
AIDS	Acquired Immunodeficiency Syndrome
ANC	African National Congress
CFCs	chlorofluorocarbons
EC	European Community
EU	European Union
FDI	foreign direct investment
GATT	General Agreement on Tariffs and Trade
GDP	gross domestic product
GNP	gross national product
GPCR	Great Proletarian Cultural Revolution
G-5	Group of Five
G-7	Group of Seven
IMF	International Monetary Fund
IPCC	Intergovernmental Panel on Climate Change
IRA	individual retirement account
LIBOR	London Interbank Offered Rate
NASA	National Aeronautics and Space Administration
NATO	North Atlantic Treaty Organization
OPEC	Organization of Petroleum-Exporting Countries
PLO	Palestine Liberation Organization
PRI	Institutional Revolutionary Party (Mexico)
S&L	savings and loan
SAPs	structural adjustment programs

TNC   transnational corporation
UN   United Nations
WTO   World Trade Organization
XBCs   cross-border corporations

1

✿

Globalizing Production in the United States, Western Europe, and Japan

Since 1945, two economic developments have altered the way people work in the United States, Western Europe, and Japan. First, the production of goods and services in these three regions has been redistributed. Second, the production of goods and services in these regions has been reorganized. The redistribution and reorganization of production has contributed to the globalization of production.

In the period between 1945 and 1970, the redistribution of production generally promoted economic development in the United States, Western Europe, and Japan, the group of countries described for many years as the "first world" or the "West," and now referred to as the "Triad" or the "North." But after 1970, the redistribution of production generally came at U.S. expense, a process economists in the 1970s called "deindustrialization" because U.S. industries lost business and jobs to firms based in Western Europe and Japan. This development contributed to changing gender relations in the United States because men lost jobs in manufacturing industries at a time when women found work in increasing numbers in the service sector.

Then, in the 1980s and 1990s, U.S. firms began reorganizing production. Rising stock prices on Wall Street encouraged corporations to buy up other firms and introduce new technologies in the newly merged firms. The resulting reorganization of production made U.S. firms more competitive with businesses in Western Europe and Japan in the 1990s. But this reorganization also led to downsizing, or job loss for many workers in the United States.

To appreciate the causes and consequences of these two developments, we will return to the 1950s, when the process of redistributing production among businesses in the United States, Western Europe, and Japan began.

REDISTRIBUTING WORK, 1945–1970

After World War II, male wage workers in the United States produced most of the manufactured goods consumed in the United States, Western Europe, and Japan. Male and female farm households in the United States also produced most of the food consumed in the United States, Western Europe, and Japan. But that changed during the next twenty-five years. By 1973, the United States produced only one-fifth (21.9 percent) of the world's manufactured goods, down from more than half (56.7 percent) in 1948. Meanwhile, businesses in Western Europe and Japan doubled their share of world manufacturing, from 15 to more than 30 percent.[1] By the mid-1970s, Western Europe had also become self-sufficient in food production, so their consumption of food grown in the United States declined substantially. Essentially, a significant share of manufacturing and agricultural production had shifted from the United States to Western Europe and Japan. This redistribution of production, what may be called the globalization of production, began early in the postwar period. Significantly, production was redistributed primarily from the United States to Western Europe and Japan, not to other poor countries. So the globalization of production was a partial and limited process, not a universal development.

Why was production redistributed during this period? Because governments, businesses, and consumers all adopted policies and practices that shifted the location of jobs in manufacturing and agriculture from the United States to other locations.

U.S. policies played a crucial role in helping redistribute production during the postwar period. Although U.S. policies were designed to promote political and military cooperation and foster economic growth within the core, they also contributed to the redistribution of wage work. Here's how.

First, the U.S. government provided public aid worth billions of dollars to its allies through the Marshall Plan and related programs.[2] It also directed vast quantities of military aid to Western Europe and Japan, aid amounting to as much as $2 trillion between 1950 and 1970.[3] Public aid provided capital that allied governments used to rebuild wrecked infrastructure and industries destroyed by war. This created jobs for demobilized servicemen in construction and manufacturing. Without this aid, Western Europeans would have been unable to rebuild.[4] U.S. military spending in Western Europe and Japan provided numerous economic benefits (it also caused some social problems). Because the military purchased goods for overseas U.S. bases from local suppliers, U.S. defense spending generated jobs in defense-related industries. U.S. purchases of French aircraft for NATO, for example, created jobs in an industry that would later compete with U.S. aircraft manufacturers.[5] The hundreds of thousands of U.S. servicemen stationed in Western Europe and Japan spent their wages there, resulting in jobs for local businesses and injecting dollars, a scarce and important commodity, into local economies. These practices provided capital and cash that was used to create national and local manufacturing and service industries in Western Europe and Japan. And by serving abroad, U.S. soldiers released young men in Western Europe and Japan from military obligations, so they could take jobs producing goods rather than standing guard.

Second, the U.S. government allowed its allies to levy high tariffs (taxes on goods they imported from the United States) and establish strict controls on capital movements. These policies encouraged U.S. firms to invest heavily in Western Europe. General Electric, for example, quadrupled the number of factories it operated in Western Europe between 1949 and 1969.[6] The $78 billion that U.S. firms invested in Western Europe during the 1950s and 1960s was used to create jobs and produce goods there, so that European consumers could purchase goods made by Europeans rather than buying imported goods made by Americans.[7] Private U.S. investments in this period may have resulted in the loss of two million jobs in the United States.[8] This practice, which was encouraged by government policies in the United States and Western Europe, contributed to the redistribution of production in manufacturing industries.

Third, U.S. officials established a global system of fixed exchange rates during the war. The Bretton Woods agreement, as it was called, allowed Western European and Japanese firms to compete as equals in U.S. markets, even though they were not yet competitive with U.S. firms (see chapter 2).[9] Generous postwar exchange rates, which made Western European and Japanese goods seem cheap in U.S. markets, and low U.S. tariffs on imported goods, encouraged worker-consumers in the United States to purchase toys, sewing machines, radios, and alcohol from Western Europe and Japan. Exchange rates also persuaded U.S. workers to travel abroad. The $4.8 billion that U.S. worker-tourists spent overseas, most of it in Western Europe, generated jobs in service and tourist industries and injected dollars into local economies.[10]

For their part, governments in Western Europe and Japan made the most of opportunities provided by U.S. policies, business practices, and consumer behavior. U.S. public aid, military assistance, private investment, and consumer spending provided them with capital, cash, and markets that they used to create jobs and rebuild industries. Governments in Western Europe and Japan also adopted policies that tapped another important resource: their own domestic workers.

Generally speaking, governments in Western Europe and Japan adopted monetary, trade, and tax policies designed to discourage consumption by domestic workers, encouraging them instead to save. By making it difficult for them to purchase imported goods or buy big-ticket items such as houses or cars, they forced workers to save a high percentage of their income. Japanese worker households, for example, put aside nearly 20 percent of their income in the 1950s and 1960s.[11] The money that workers deposited in banks and postal accounts was then collected by banks and the government and used to finance the growth of domestic manufacturing industries.[12]

To compensate workers for working hard and saving money, governments in Western Europe gave generous social welfare benefits to workers: pensions, health care, unemployment compensation, and vacations. The "welfare states" established in Western Europe after World War II created electoral support for conservative governments. In Japan, the government took a rather different approach, offering workers few social benefits. Instead, the government provided generous financing to industries, which then paid benefits to male workers in manufacturing, promising them lifetime employment (*shushin koyo*) and a seniority-based wage system (*nenko joretsu seido*).[13] Women employed by large firms in Japan were typically hired only on a temporary basis, so they were largely excluded from the benefits designed to compensate worker households for their thrift. The rewards offered male workers were nevertheless sufficient to persuade households to support conservative government throughout the postwar period.

Policymakers in the United States could encourage and permit a redistribution of production, a process that resulted in the distribution of U.S. jobs to manufacturing industries located elsewhere, because there was a considerable amount of work that needed to be done. Workers were needed to rebuild whole economies in Western Europe and Japan, wage wars in Korea and Vietnam, fashion weapons and vehicles for arms and space races with the Soviet Union, build houses and supply durable goods for baby-boom households that had scrimped during the Depression and saved during the war, and supply newly independent countries with goods financed by the World Bank and foreign-aid programs. There was so much work to be done that the United States could surrender a significant share of production to its allies and still provide work for most male workers in the United States. There was so much work available that industries could, for the first time, even offer jobs to large numbers of minorities, women, and immigrants.

Minorities

During World War II, the lure of paid work in the North and West, and the pain of institutional racism in the segregated, "Jim Crow" South, persuaded five million African American workers to leave southern farms for jobs in big cities in the North and West, where many found jobs in manufacturing and service industries.[14]

Women

At war's end, many women were forced out of manufacturing industries to make room for returning servicemen. But while the percentage of women in the labor force dropped from 34.7 percent in 1944 to 31.1 percent in 1954, the decline was small and women retained a claim on a significant share of the available jobs.[15]

Immigrants

U.S. industries even provided work for a large number of immigrants, one million in the late 1940s, 2.5 million more in the 1950s, and another 3.3 million in the 1960s. Agricultural firms also annually recruited another 300,000 to 445,000 workers from Mexico through the government's Bracero Program.[16]

In Western Europe, there was such a large demand for workers that industries could provide virtually full employment for domestic males, jobs to eight million ethnic Germans who were forced to emigrate from Eastern Europe and the Soviet Union after the war, and work for another three million Germans who fled East Germany before the Berlin Wall was built in 1961. There was so much work available that Western European countries could also recruit millions of other workers from Spain, Portugal, southern Italy, Yugoslavia, Greece, and Turkey (each donated about one million workers to the labor force in Western Europe) through various "guest worker" programs during the 1950s and 1960s.[17]

In Japan, meanwhile, industry provided full employment for men, jobs for many women, though on unequal terms, and jobs for another 2.6 million Japanese immigrants, who had emigrated from areas occupied by Japan during the war, much as ethnic Germans in Eastern Europe had done.[18]

Because the demand for workers was so strong, and because many workers belonged to trade unions, which used strikes to demand higher salaries, wages rose in all three regions, though at different rates. In the United States, wages doubled between 1950 and 1970. Wages rose at an even faster rate in Western Europe and Japan. By 1970, they had become comparable to wage levels in the United States.[19]

But while wages rose more rapidly for workers in Western Europe and Japan, their standards of living did not measure up to the living standard of U.S. workers. Policies that discouraged consumption and promoted savings in Western Europe and Japan forced workers to pay high taxes, spend more of their income on food, and made it difficult for them to purchase cars or homes that were comparable to those available, at a lower cost, to workers in the United States. One striking measure of different living standards is this: in 1970, 96 percent of U.S. worker households had flush toilets in their homes; but only 9.2 percent of worker households in Japan had flush toilets in their apartments.[20]

During the twenty-five years after the war, work was widely available, salaries rose, wage differentials narrowed, and standards of living improved for most workers in the United States, Western Europe, and Japan. Under these conditions, the redistribution of production in manufacturing industries was regarded as unproblematic, even beneficial. For U.S. policymakers, the provision of U.S. jobs to industries in Western Europe and Japan was a relatively small price to pay for military unity, political cooperation, and economic growth in the core. But this would change after 1970, when the price of redistributive policies became apparent to businesses and workers in the United States.

REDISTRIBUTION AND DEINDUSTRIALIZATION, 1970–1979

In 1971, the United States posted a modest trade deficit, its first since 1893. Though small, the $2.3 billion trade deficit signaled that the United States had already lost a significant share of production to industries located in Western Europe and Japan. During the next twenty years, U.S. job losses and trade deficits would mount, and much of the production previously performed in U.S. manufacturing industries would be redistributed abroad, a rapid globalization process known in the United States as "deindustrialization." The redistribution of production accelerated in the 1970s because economic conditions had changed. In the early 1970s, the demand for manufactured goods fell because the United States withdrew from the war in Vietnam and slowed the pace of the arms and space races with the Soviet Union. It fell, too, because the Organization of Petroleum-Exporting Countries (OPEC) oil embargo forced up energy prices and poor Soviet grain harvests raised food prices. As energy and food prices rose, consumers cut back and demand for manufactured and agricultural goods weakened, triggering a global recession.

Meanwhile, the global supply of manufactured goods had steadily increased. The recovery and growth of manufacturing industries in Western Europe and Japan increased supplies of goods from these regions. As a result, the battle for a share of global markets and a claim to a share of production in manufacturing industries intensified. As it did, many important manufacturing industries in the United States lost markets, and the jobs they provided were redistributed to industries located in Western Europe and Japan.

Why was production in U.S. manufacturing industries redistributed during the 1970s? There is no single answer. The reasons varied from one industry to the next. A brief look at government policies, industry practices, and consumer behavior in three important manufacturing industries—steel, autos, and aircraft—illustrates some of the different reasons why production was redistributed.

Steel

According to Benjamin Fairless, head of U.S. Steel in 1950, the U.S. steel industry was "bigger than those of all other nations on the earth put together."[21] But the steel industry declined slowly in the 1960s and then rapidly in the 1970s, victimized by U.S. government policy and its own business practices.

During the postwar period, successive U.S. presidents worked hard to keep steel prices low. They did so to prevent steel price increases from triggering inflation, and to ensure that the other U.S. industries that used steel—auto makers, appliance manufacturers—paid low prices for it. When U.S. steel companies announced price hikes, presidents Kennedy, Johnson, and Nixon attacked the steel industry, lobbied its leaders to rescind price increases, and ordered federal agencies to purchase steel from low-price competitors.[22] But government efforts to keep down prices lowered profit rates and made it difficult for the steel industry to use its earnings to modernize tired, aging plants.[23] The government also used antitrust suits to prevent mergers and promote competition. This helped keep prices low, though mergers might have helped the industry reorganize and increase its efficiency. Ironically, this policy led officials to reject proposed mergers among U.S. firms, but allowed them to be acquired by foreign firms.[24] When steel-industry firms asked the government to levy tariffs on steel imports or prosecute foreign firms that illegally "dumped" cheap steel in U.S. markets, officials repeatedly refused.[25]

For their part, industry leaders were slow to adopt new, energy-efficient technology, relying instead on aging plants and outmoded technologies because they wanted to pay for these before investing in new capacity. In 1978, 45 percent of U.S. plate mills were more than twenty-five years old, while only 5 percent of comparable Japanese mills were that old.[26] The industry's acrimonious relations with labor unions also triggered a series of long strikes in the 1950s, forcing the industry to raise worker pay. The industry might have afforded wage increases if it had invested in new technology and increased productivity, but the government's low-price policies made this difficult to do.[27]

Business customers also played a role in the steel industry's decline. U.S. businesses that used steel wanted cheap supplies, so they lobbied hard against steel-industry efforts to raise prices or secure government protection against unfair foreign competition. General Motors, for instance, argued that actions against countries that dumped low-price steel in the United States "will have a negative effect on the prices [General Motors pays] for finished products with high steel content."[28] Then, in the 1970s, as inflation pushed up steel prices, businesses began using plastic and aluminum materials to replace steel in cars and appliances.[29] This reduced the demand for steel, both foreign and domestic.

The U.S. steel industry was among the first U.S. industries to experience deindustrialization. As early as 1959, the United States imported more steel than it exported. By 1970, the U.S. share of world steel production had plummeted to 20 percent, down from 50 percent in 1945. Steel production then fell from 130 million tons in 1970 to 88 million tons in 1985. Today, the industry does not produce enough steel to meet domestic demand, and the United States is "the only major industrial nation that is not self-sufficient in steel."[30]

Autos

The decline of the U.S. auto industry in the 1970s was due less to government policy than to business practices and consumer habits. Its decline was significant because 7.5 million people build, sell, or repair cars and trucks in the United States.[31]

The Volkswagen Beetle was the first import to make inroads in the U.S. market in the 1960s, largely because its size, price, and durability was unmatched by models made in Detroit. By 1970, it had captured 15 percent of the U.S. market.[32] During the 1970s, the Beetle was superseded by Japanese models. Japanese cars captured U.S. markets for a variety of reasons. Car makers in Japan adopted new technologies like the system of just-in-time production, or *kanban*, which reduced inventory costs, and made cars that were stronger and used less steel. They developed amicable relations with workers in Japan, enabling them to increase production and improve quality. So when oil prices rose in the 1970s, cost-conscious consumers in the United States turned to high-mileage, inexpensive, durable cars made in Japan. They bought only four million Japanese cars in 1970, but purchased twelve million in 1980.

For their part, U.S. automakers were slow to adopt new technologies and develop cheap, high-quality, fuel-efficient models that could compete with imports. Their acrimonious relations with workers and their unions prevented the industry from significantly improving productivity or quality in the 1960s and 1970s.[33] In 1980, the four major U.S. automakers lost $4 billion and Chrysler was on the verge of bankruptcy.[34]

Still, unlike the steel industry, U.S. automakers were not without resources. In the 1950s and 1960s, Ford had opened factories in Western Europe, building cars and employing workers there. In the 1970s and 1980s, other U.S. car makers followed suit, opening factories overseas, many in Latin America. By 1980, the industry had itself moved 37.2 percent of its production abroad. Production in the U.S. auto industry was redistributed both because other core firms captured U.S. markets, and because U.S. firms themselves redistributed work to other settings.

Aircraft

Unlike steel or autos, U.S. policymakers gave massive aid to the aircraft industry, which they viewed as essential to national defense. In the 1940s, the government built dams that provided cheap electricity to smelt aluminum, the essential raw material for modern aircraft, and purchased hundreds of thousands of planes from private manufacturers during the war.[35] After the war, the military poured billions of dollars into the industry, financing new technology and designs and providing demand for the development of new military and commercial aircraft. The government's purchase of a transport plane from Boeing enabled it to launch its first successful commercial aircraft, the 707.[36] As a result, the industry captured 90 percent of the world market, a position it held well into the 1970s.

But U.S. dominance did not go unchallenged. In 1965, aircraft firms in Western Europe organized Airbus, a consortium that used government aid to develop commercial aircraft. Government subsidies and private investment from European and American banks enabled Airbus to develop its first plane (with wings from Britain, cockpit from France, tail from Spain, edge flaps from Belgium, body from West Germany, and, importantly, engines from the United States).[37] Unlike its U.S. competitors, the plane ran on two engines rather than three and required two pilots instead of three, which saved fuel and lowered operating costs. These were important considerations for Eastern Airlines, which made the first significant purchases of the new plane.[38] By 1988, Airbus had captured 23 percent of the world market. It wrested markets and jobs first from weak U.S. firms like Lockheed and McDonnell Douglas. During the next decade, Airbus challenged Boeing, the world leader. In 1999, for the first time ever, Airbus received more orders for new planes than Boeing.[39] And in 2000, Airbus began developing a new behemoth jet that will compete with Boeing for the lucrative long-haul business. The Boeing 747, which has long monopolized this business, has been Boeing's most successful plane, accounting for roughly half of its annual profits.[40] The development of a successful challenge by Airbus could have a huge impact on the distribution of aircraft production.

Although the deindustrialization of the U.S. aircraft industry came later than it did to the steel and auto industries, the effect was much the same. "Every time a $50 million airplane is sold by Airbus instead of Boeing," one expert observed, "America loses about 3,500 high-paying jobs for one year."[41]

THE REDISTRIBUTION OF PRODUCTION AND CHANGING GENDER ROLES

The redistribution of production generally resulted in job loss and falling wages for males in U.S. manufacturing industries. By eliminating jobs long reserved for men, the redistribution of production helped transform gender roles in the United States. When the steel, auto, and aircraft industries surrendered markets and ceded jobs to overseas competitors, they laid off the men who smelted steel, assembled cars, and fabricated aircraft. Except for a brief time during World War II, few women worked in these industries. During the postwar period, indeed for most of this century, work in manufacturing had given men economic power in the labor force (largely through labor unions), political power in public life (primarily through the Democratic Party), and social authority in households (based largely on their role as breadwinners). The loss of wage work in manufacturing undermined male power in public life and male authority in private life.

Job loss has not always resulted in the erosion of male power and authority. During the Great Depression, men lost manufacturing jobs in droves. But because few women were employed in manufacturing or service industries, and those who were lost their jobs, too, male job loss did not significantly alter gender roles. In the 1970s, however, male job loss was accompanied by the entry of women into service industries. It was the combination of these two, simultaneous developments—the exit of men from manufacturing and the entry of women into service industries—that transformed gender relations.

During the 1970s, a growing number of women secured work in service industries. This is somewhat surprising. One might think that the end of the war in Vietnam, the demobilization of servicemen, and the recession triggered by rising oil and food prices would have resulted in the expulsion of women from the labor force, just as they had done after World War II. But women were not expelled because the number of returning servicemen was small, few women worked in manufacturing industries, and the service industries where women were employed in large number were actually growing in this period.

Women entered the labor force in large numbers during the 1970s for two reasons. First, women needed to secure wage work to maintain household incomes in an inflationary-recessionary, job-and-income-loss environment. In a sense, deindustrialization pushed women into the labor force. Second, the service industry, which historically had reserved jobs for women, needed workers as the demand for its goods and services increased. Growing demand was the product of several related developments. Massive advertising and widely available credit encouraged U.S. workers to spend, not save. Workers spent an increasing percentage of their disposable income on consumer goods and services, increasing the demand for services from the private sector. Increased government spending on social service–welfare programs also increased the demand for public service-sector workers, and women found jobs as teachers, health care workers, and social service administrators.[42] Moreover, as women left home to take private- and public-sector service jobs, worker households began buying services that women could no longer or easily provide as housewives. This further stimulated the demand for women workers and also teenagers in service industries.[43] In 1964, for example, only 1.7 million Americans worked in restaurants and bars. But 7.1 million did so in 1994.[44] So the expansion of the service industry helped pull women into the labor force.

As a result of economic push and pull, the percentage of women in the paid work force increased from 38 percent in 1970 to 43 percent in 1980. This increase was comparable to the gains made by women during World War II. After 1980, women continued to enter the labor force, though at a slower rate, rising to 45 percent by 1990.[45]

Of course, women who took paid jobs did not stop working at home. They still shouldered substantial workloads as housewives.[46] So women's total work (unpaid household work plus wage work) increased substantially in this period, from an average of about 1,400 hours in 1969 to 1,700 hours in 1987.[47]

As women assumed more prominent economic roles, they also began playing a larger role in public and private life. The emergence of the women's movement in the 1970s encouraged women to play more visible roles in politics and public life. The social status women gained by wage income, and the autonomy provided by new reproductive technologies and legal rights (the birth control pill, divorce law reform, abortion rights), made it possible for many women to assume new roles in worker households. At the same time, changing economic, political, and social roles for women and men frequently increased tensions between women and men, resulting in high divorce rates and the rise of female-headed households.

This development was perhaps first apparent for poor African American worker households. During the 1950s and 1960s, black men and women had migrated from the South to northern cities, and men found wage work in heavy industries. Because black men were heavily "concentrated in industries like steel," and in cities where manufacturing industries made their home, deindustrialization in the 1970s had a catastrophic impact on jobs and employment for African American males.[48]

The exit of black men from manufacturing, and the entry of black women into service industries and government assistance programs such as Aid to Families with Dependent Children, transformed gender relations and contributed to the rise of female-headed households. But while this development has often been portrayed as symptomatic of problems unique to African American households, it can be more usefully understood as the early expression of problems common to many white and Hispanic households in the United States. The problems evident in African American households were not an aberration, but a harbinger. As it turned out, the exit of men from manufacturing and the entry of women into service industries transformed gender relations not only for African American households, but for other ethnic groups as well.

The ongoing redistribution of production in the United States, Western Europe, and Japan was joined in the 1980s and 1990s by another development: the reorganization of production.

REORGANIZING PRODUCTION
IN THE UNITED STATES, 1980–2000

The redistribution of production, which generally came at the expense of male manufacturing workers in the United States, did not go unnoticed or unchallenged. In the 1980s, U.S. officials, alarmed about the loss of U.S. hegemony, adopted monetary and trade policies designed to reassert U.S. control over the redistributive process, stem manufacturing losses, and reclaim a share of production.

As a first step, U.S. policymakers devalued the dollar, first in 1971 and again in 1985 (a more extensive discussion of this development follows in chapter 2). As a second step in 1979, they raised interest rates (this is discussed at greater length in chapter 4). And third, in 1986 they initiated a series of trade negotiations with members of the General Agreement on Tariffs and Trade (GATT) and with neighboring states in North America.[49]

U.S. dollar devaluations encouraged investors from Western Europe and Japan to buy U.S. assets and open factories of their own in the United States. For example, in 1987, Japanese firms built or acquired 239 factories in the United States, up from only 43 in 1984. Total foreign investment in the United States, most of it from Western Europe and Japan, increased from $184 billion in 1985 to $304 billion in 1988.[50]

High U.S. interest rates in the early 1980s persuaded investors in Western Europe, Japan, and Latin America to purchase U.S. treasury bonds. In 1980, for example, foreign investors bought $71 billion worth of U.S. bonds, with two-thirds of the money coming from Western Europe and Japan, and most of the rest from Latin America.[51] These developments encouraged European and Japanese investors to invest, for the first time, in the United States.

Prior to 1980, public resources and private investment had generally traveled in one direction, from the United States to Western Europe and Japan. But U.S. monetary policies in the 1980s altered investment traffic patterns. As Western European and Japanese investment in the United States increased (first buying public and then purchasing private assets), investment became multilateral, not unilateral. The emergence of multilateral or "globalized" investments, however, was generally restricted to these three central regions.[52]

Much the same was true of trade. U.S. trade negotiations in the late 1980s and early 1990s were designed to reduce core barriers to U.S. exports. By persuading Western Europe and Japan to reduce trade and other barriers, U.S. officials hoped to increase U.S. exports and change the direction of trade flows. In the 1970s and 1980s, trade goods had been moving from Western Europe and Japan to the United States. The trade agreement adopted by GATT members in 1994 helped stimulate the flow of U.S. goods to its main trading partners, thereby multilateralizing or globalizing trade, along with investment.

Taken together, U.S. monetary and trade policies helped attract Western European and Japanese investment to the United States and open their doors to some U.S. goods. Essentially, these measures rescinded the economic advantages that U.S. policymakers had given Western Europe and Japan in the late 1940s, when generous monetary and trade policies were used to promote economic recovery and political cooperation during the Cold War. But while these steps helped level the economic playing field, they did not greatly improve U.S. performance on the field. The redistribution of production continued in the 1980s, though at a slower pace than it had in the 1970s.

THE STOCK MARKET AND THE REORGANIZATION OF PRODUCTION IN THE UNITED STATES

Although U.S. monetary and trade policies helped level the playing field, they did not improve U.S. performance on it. But when U.S. officials amended Social Security, income tax, and antitrust programs in the early

1980s, they made it possible for U.S. corporations to undertake a vast re-organization of production in manufacturing and service industries. This reorganization helped U.S. industries regain markets in the 1990s, when industries in Japan and Western Europe slumped. For U.S. workers, how-ever, this reorganization also resulted in job loss or downsizing and de-clining incomes.

The massive reorganization of production in the United States during the 1990s was propelled by obscure but important policy changes in the early 1980s. In 1981, the Reagan administration passed legislation to re-form Social Security. As part of the package, officials made Individual Re-tirement Accounts (IRAs) more widely available to worker households. At the time, officials regarded this as a minor change to the Social Secu-rity program. They had little idea that it would have a huge impact on the stock market or trigger a massive reorganization of production.

As a result of changes to IRAs, workers rushed to take advantage of the tax breaks given to IRA accounts, and deposits in IRAs increased from $30 billion in 1980 to $370 billion in 1990. Much of the money deposited in IRA and 401(k) accounts, which also grew rapidly as a result of legislation adopted in 1978, was invested in the stock market, typically through mu-tual funds.[53] In 1982, less than 10 percent of U.S. households owned stocks. But tax-free accounts encouraged millions to invest in the stock market, and by 1998, nearly 49 percent of U.S. households owned stocks.[54]

In the mid-1980s, wealthy households and foreign investors also began investing heavily in the stock market. Wealthy Americans used money given them by tax cuts (the Reagan administration cut taxes on wealthy households from 70 percent to 28 percent in the early 1980s) to invest in the stock market. Foreign investors rushed to purchase U.S. stocks after the dollar was devalued in 1985 (the devaluation cut the price of U.S. as-sets for foreign buyers), and foreign investment in the stock market to-taled $176 billion in 1986 (see chapter 2).

Investment from these sources poured rivers of money into the stock market. Think of them as downpours that filled the Mississippi (IRAs), the Tennessee (401[k]s), the Ohio (wealthy individuals), and the Missouri (foreign investors), which swelled the rivers and raised the barges and boats (stocks and bonds) that floated downstream. This flood of invest-ment into the stock market essentially bid up stock prices and lifted the market into a long bull market.

Although government policies pushed investors toward the stock mar-ket, Wall Street exerted its own pull. The market was able to attract in-vestors from worker, wealthy, and foreign households because stock prices were rising for the first time in a decade. The market's initial rise was given a jump start by two new government policies. First, the Rea-gan administration cut corporate taxes. Corporate income-tax cuts allowed businesses to increase their dividends to shareholders, making them more attractive to investors. Second, and more importantly, the Reagan administration stopped enforcing antitrust laws (the 1890 Sherman Act and the 1914 Clayton Act), which had long prevented corporations from merging and monopolizing the production of goods and services. This allowed businesses to merge with other firms, cut costs, increase profits, and raise dividends, making them even more attractive to investors.

As new investment was pushed and pulled into Wall Street, stock prices rose, bid up by growing demand. Rising stock prices, in turn, put enormous pressure on U.S. corporations to boost profits and increase their payouts to investors, who expected dividends to keep pace with rising stock prices. To keep up with the Dow Joneses, corporate managers reorganized production. They merged with other firms, rearranged production, introduced new technology, and laid off or downsized workers in an unrelenting effort to raise productivity, cut costs, and increase profits. Higher profits could then be used to increase shareholder dividends, and this in turn helped boost stock prices.

Between 1982 and 1987, the Dow Jones Industrial Average rose from 777 to 2,722, a threefold increase. This bull market came to an end on October 19, 1987, when prices fell 508 points, a 22 percent decline that resulted in a $1 trillion loss for investors. But the market soon recovered because the demand for stocks did not evaporate, as it had after the 1929 crash. Demand remained strong because worker-investors who held stocks in IRAs and 401(k)s could not easily withdraw their money without incurring stiff tax penalties, and because they had invested for the long term, using IRAs to provide for their retirement. Worker-investors were, in effect, forced by the tax code to stay in the market and prop up prices. When other investors realized that the government and worker households had built a floor under the stock market, below which prices could not easily fall, investment resumed. By 1990, investment had returned to pre-crash levels and prices began to rise again. Stock prices then surged upward, and the Dow climbed from 3,000 in 1990 to more than 11,000 in 1999, the longest bull market in U.S. history. As stock prices rose, the reorganization of production accelerated.

For businesses, rising stock prices put enormous pressure on managers to increase their profits so they could pay higher dividends to investors.[55] Firms that failed to do so were punished by investors, who sold off stock and drove down its price. When that happened, managers were fired and the firm became prey to others. To prevent this and survive in an inflationary stock-price environment, managers have adopted two strategies to increase profits, payouts, and share prices. Both strategies typically resulted in job loss for workers.

The first strategy has been for managers to reorganize production by merging with other firms, sometimes divesting parts to relieve themselves of unprofitable burdens or to raise cash for other parts. By merging with other firms, managers tried to create economies of scale or obtain control of markets that would enable them to increase profits. This strategy only became feasible because the federal government stopped enforcing antitrust law.[56]

In 1980, the year before the current merger wave began, corporations announced mergers worth $33 billion. Since then, businesses have merged and merged again. On just one day in 1998 (November 23), corporate managers announced mergers worth $40 billion, a sum greater than the value of all mergers in 1980.[57] Between 1981 and 1996, firms arranged mergers worth $2 trillion. And the pace accelerated, with mergers worth $1 trillion recorded in 1997 and $1.6 trillion in 1998.[58] All told, there were 151,374 mergers worth $13 trillion between 1980 and 2000. "We're in the greatest merger wave in history," said John Shepard Wiley, a professor of anti-trust law at U.C.L.A. "There has been a sea change in [public] attitudes toward large mergers."[59]

A second strategy has been for managers to introduce new technology, lay off workers, and cut costs to increase productivity. For example, managers at Caterpillar, a heavy equipment manufacturer, closed nine plants and spent $1.8 billion to modernize its remaining factories. As new technology was introduced, the firm cut its work force from 90,000 to 54,000 and increased production. "We've almost doubled our productivity since the mid-1980s," Caterpillar executive James Owens enthused.[60]

Business efforts to increase productivity have not been limited to manufacturing industries. Computer, phone, fax, and other electronic technologies—scanners, automatic tellers, and so on—have enabled managers to reorganize service-sector firms, where it had long been difficult to deploy technology as a way of increasing productivity. The demand for technology that can improve productivity has spawned the growth of the computer industry, which, in turn, has transformed service industries. In 1995, experts predicted that "half of the nation's 59,000 branch banks will close and 450,000 of the 2.8 million jobs in the banking industry will disappear [by the year 2000]" as a result of new bank technologies like automated tellers.[61] As Carl Thur, president of the American Productivity Center, put it, "The trick [for U.S. business] is to get more output without a surge in employment."[62]

In the 1970s and early 1980s, productivity in U.S. firms increased slowly. But by merging with other firms, reorganizing business, introducing new technologies, laying off workers, and cutting costs, managers were able to increase the productivity of their firms. Between 1982 and 1994, "productivity increased about 19.5 percent."[63] Since then, it has increased at high annual rates: 2.8 percent in 1996, 2.5 percent in 1997, and 3.0 percent in 1998. These rates are significantly higher than the 1.1 percent annual increases reported from 1973 to 1989.[64]

Increased productivity helped U.S. firms raise profits. Between 1983 and 1996, annual corporate profits nearly quadrupled, from less than $200 billion to $736 billion.[65] Higher profits made it possible to increase dividends to shareholders. This, in turn, has increased the value of corporate stock, drawn new money into the stock market, and driven stock prices higher, and higher still.

For workers, the reorganization of production, which was driven by the stock market, has resulted in massive job loss or, as it came to be known, "downsizing." Like other phrases used in the 1980s and 1990s—"involuntary force reductions," "right-sizing," "repositioning," "deselection," "reducing head count," "separated," "severed," "unassigned," and "reductions in force"—downsizing has meant one thing: "You're fired."

Mergers resulted in job loss because some jobs in combined firms overlapped. Merged banks did not need two branches on the same street; merged manufacturing firms did not need two sets of accountants to keep the books, much less two assembly lines making the same goods under different brand names.

Between 1981 and 1991, four million workers at Fortune 500 companies lost their jobs, and total employment in these large firms fell from sixteen to twelve million workers.[66] By 1998, the eight hundred largest U.S. firms employed only 17 percent of the workforce, down from nearly 26 percent in 1978.[67] Firms throughout the economy accelerated the pace of layoffs: 1.42 million workers were laid off in 1980, 3.26 million in 1995.[68] Of course, new jobs were also created, but many of them were on a part-time, temporary, or contractual basis. Some firms even "leased" their workers to other firms to cut costs and evade labor-law restrictions.[69] As many as thirty million workers are now employed on a part-time, temporary, or contractual basis.[70]

Previous waves of change affected male workers in manufacturing: steel, autos, aircraft, construction. But contemporary downsizing has affected men and women, in manufacturing and service industries. It has affected skilled workers and college graduates, not just workers with high-school diplomas.[71] It has affected white workers, not just blacks and Hispanics. It has affected managers in offices and assembly-line workers in factories. It has created two-tier workplaces, where permanent employees work alongside temporary or "permatemp" workers, who do the same jobs but receive very different salaries and benefits.[72] The only group of workers that has been relatively immune from downsizing has been government workers and public-sector employees like teachers.

Ongoing job loss and the rise of temporary and part-time employment has made it extremely difficult for workers to earn higher wages, even though their productivity has increased and profits have grown. This contrasts sharply with the early postwar period, when productivity gains enabled firms to increase profits.

In the earlier period, corporations raised wages both because they could afford to do so and because widespread union membership helped workers insist that productivity gains be shared. But this has changed. The reorganization of production has disorganized workers and weakened unions. Today, the percentage of unionized workers in private industry (only 9.4 percent) is what it was in 1929.[73] Unions would have to recruit fifteen million new members to regain their postwar strength.[74]

Although the reorganization of production has weakened worker claims on the profits created by productivity increases, it has strengthened investor claims on corporate profits. As a result, profits have been redistributed from workers to managers and shareholders. Because the reorganization of production is now being driven largely by the stock market, investors can now insist that any gains be shared with stockholders and managers, not with workers. As a result, workers have not been able to gain higher wages, despite the fact that corporations could afford to do so.

There is serious irony here. Workers who invested in the stock market to provide for their retirement helped fuel the stock-price inflation that forced corporations to reorganize and, in the process, downsize workers. As investors, many workers benefited from rising stock prices and the corporate distribution of profits to shareholders. But as workers, many investors experienced job and income loss, which resulted from the reorganization of production and redistribution of profits. One *New York Times* writer captured this irony for workers in a headline, which read, "You're Fired! (but Your Stock Is Way Up)."[75]

The reorganization of production led to job loss and declining wages. Labor's share of the national income declined, and the wealth it once claimed has been redistributed upward.[76] The richest 2.7 million Americans now claim as much wealth as the bottom 100 million Americans. The average income of the top 20 percent of American households increased from $109,500 in 1979 to $167,000 in 1997. The richest 1 percent saw their average income skyrocket from $420,000 in 1979 to $1.2 million in 1997, largely because their income from stock dividends and rising prices grew enormously.[77] But for the majority of workers during this period, wages actually declined, despite working harder, longer, and more productively. The average income for the bottom 20 percent of American households declined from $11,890 in 1979 to $11,400 in 1997.[78] In 1999, 215 million American workers took "home a thinner slice of the economic pie than [they did] in 1977."[79]

Moreover, men and women are working longer hours. Juliet Schor, author of *The Overworked American*, estimates that the hours worked by wage workers in the United States increased to 1,966 in 1999, surpassing the Japanese by seventy hours and Europeans by 320 hours (or almost nine full work weeks). "Excessive working time is a major problem," she concluded.[80]

Because men and women are working longer, many workers have less time for family, friends, or vacations. Working parents today spend "40 percent less time with their children than they did 30 years ago," MIT economist Lester Thurow reported.[81] Friendships also suffer from heavy workloads. Kim Sibley, who juggles two jobs in Flint, Michigan, was asked by a reporter whether her friends also had dual careers. "I don't have time for friends," she replied.[82]

Vacations are another disappearing entity. In 1996, 38 percent of all worker families in the United States did not take any vacation, an increase from the 34 percent that did not vacation in 1995. And the average length of vacations for those who do manage to enjoy time off has declined from five days to four days in the last ten years.[83]

Increased workloads can sometimes be fatal. The number of workers asked to work evening or night shifts has increased 30 percent since 1985, and fifteen million workers now work nondaytime shifts. As a result, the number of fatigue-related auto accidents has increased. Government highway safety officials estimate that 1,500 traffic deaths and 40,000 injuries are annually caused by fatigued workers, particularly those with late shifts.[84]

LONG-TERM PROBLEMS WITH THE STOCK MARKET

There are two long-term problems with the emergence of the stock market as the institution that drives economic change in the United States. First, the market, which has relied on a steady infusion of cash to push stock prices up, may have difficulty obtaining the money it needs to continue growing in the future. Second, the market may not be able to sustain stock prices when baby-boomers retire and begin withdrawing money from their IRAs and 401(k)s.

During the past twenty years, the stock market has relied on a steady flow of cash from different sources. The flood of cash from IRAs, 401(k)s, and pension funds has been critical to its success, helping drive up prices during the longest bull market in history. But the supply of cash from these sources is not unlimited, in part because the amount of money people can contribute to IRAs is restricted by law and by their ability to save.

To encourage a new flood of cash into the stock market, Congress in 2001 passed legislation to increase the amount that individuals can contribute to their IRAs, from $2,000 to $5,000 by 2008.[85] But while this may encourage a new wave of investment, it will come primarily from rich households who can afford to set aside $10,000 (per couple) each year for their retirement. Working- and middle-class households, who now have substantial debts and little savings, will not be able to contribute the full amount. This source of investment is largely tapped out. In the long run, this means that the level of investment that assisted the growth of the stock market in the 1980s and 1990s may no longer be forthcoming. And without new infusions of cash, the market may not be as robust as it has been.

Policymakers have proposed privatizing part of Social Security to make more money available to the stock market. They propose that workers be allowed to take one-half of the money withheld each month from their paychecks and invest it themselves for their retirement. This would vastly increase the amount of money flowing toward Wall Street. But there are several problems inherent with this plan. Because the stock market is volatile, it goes up and down; workers will assume greater risk than they now do under Social Security, which promises fixed benefits after retirement.[86] Some workers will invest more wisely than others, so workers earning the same amount may end up with very different amounts of money available to them when they retire. Women earn less than men (about 30 percent less on average), so men will have more money available to invest. This could reinforce gender inequality when women and men retire.

A second long-term problem is this: Many of the working- and middle-class workers who invested in the stock market did so to provide for their retirement. When stock prices declined in 1987 and 2001, they stayed in the market, helping prop up prices and avert a more serious collapse. They did not sell because they would have been penalized if they withdrew money from IRAs and because they were investing, long-term, for their retirement. Their steadfast support was good for the market and for the economy that depends on its health.

But eventually these worker investors will retire and begin withdrawing money from the market. They will do so in significant numbers because the baby-boom generation to which they belong is large. As they begin to pull cash out of the market, stock prices could weaken. Remember that money flowing into the market helps create a bull market, while money taken out of the market contributes to a bear market. "If demography has played a part in driving the market up," Niall Ferguson has argued, "it can only have the reverse effect as the 'Baby Boomers' retire and begin to live off their accumulated assets."[87]

The decline in stock prices during 2001, particularly after the September 11 disasters at the World Trade Center towers, the Pentagon, and Shanksville, Pennsylvania, illustrated the problems with relying on the stock market to provide money for people's retirement. People with 401(k) accounts saw the value of their investments decline, on average, from $46,740 to $41,919.[88] This was the first time in twenty years that "the average account lost money, even after thousands of dollars of new contributions."[89] People who retire during a periodic market downturn face a very different financial future than people who retire during an upturn. Jim Dellinger, a baker who has a 401(k) through his employer, Giant Foods, saw his two mutual funds decline by 12 and 16 percent during the first six months of 2001. "I think about it every day. I have basically my life savings in there," he said.[90]

BOOM IN THE UNITED STATES, BUST IN WESTERN EUROPE AND JAPAN, 1990–2000

The reorganization of production resulted in job and income loss for both women and men in the United States. But it also helped manufacturing and service industries increase productivity and profitability, which helped industry increase investment in research and development.[91] These developments improved the performance of U.S. industries in the 1990s in redistributive battles with industries based in Western Europe and Japan. The gains made by U.S. businesses in the 1990s promoted overall economic growth, providing jobs for downsized wage workers, at least for a time, though on a more casual and lower-paid basis.

U.S. businesses expanded and unemployment rates fell after 1992 as a result of several developments. Slumping economies in Western Europe and Japan weakened manufacturing and service industries based there, providing reorganized U.S. industries with the opportunity to make redistributive gains. U.S. businesses did well in the 1990s because businesses in Japan and Western Europe did not.

In the early 1990s, Japan and Western Europe both experienced economic crises. Although the crises had different origins in each area, both regions were confronted with their first real setbacks since World War II, problems that persisted throughout the decade.

Japan

In Japan, problems began in 1985. The devaluation of the dollar eventually doubled the value of the yen, increasing the value of assets held by

Japanese banks (see chapter 2). Banks and workers invested this new-found wealth in the stock market and in real estate. The flood of new investment bid up stock and real estate prices. The Nikkei Index (the Japanese stock market) rose from 12,000 in 1986 to 38,916 in 1990, a threefold increase.[92] The price of residential and commercial real estate quadrupled between 1985 and 1990.[93] It was said in 1990 that the value of land in Tokyo alone was worth more than all the land in the United States. Japanese consumers had so much money to burn that they even poured money into the market for pet insects, particularly for rare beetles called *ohkuwagata*. In 1990, single bugs sold for $7,000 at department stores, and one huge specimen sold for $30,000.[94]

But too much money can cause problems. The money pouring into the stock and real estate markets drove prices to unsustainably high levels. In the stock market, prices soared while dividend yields fell.[95] The "bubble" of high stock prices burst in 1990. Prices fell one-half by 1991 and continued falling. Between 1990 and 1992, the Nikkei Index registered a 61 percent decline.[96] Stock prices did not recover from the crash in Japan, as they had in the United States after the 1987 crash, because investors fled the market and did not return.

The Japanese real estate market soon followed. In 1990, many households in Japan found they would need to use the wages of a lifetime just to buy one *tsubo* (six feet by six feet) of land in Tokyo.[97] When owners discovered they could not sell high-priced property, the residential and commercial real estate markets collapsed and prices plummeted, bankrupting individuals, businesses, and banks that had used land as collateral for other loans. The pet insect market also collapsed. Bugs that sold for $7,000 during the beetle-mania of the 1980s were marked down to only $300 in 1999.[98]

For workers, the recession led to widespread layoffs and rising unemployment rates, which doubled in the 1990s. This came as a great shock to workers in a country where businesses routinely provided lifetime employment and regular wage increases to their male workers. Corporations began laying off workers, hiring temporary workers, eliminating seniority-based pay systems, and introducing merit pay.[99] "For years, everyone's pay increased as they got older," observed Shoji Hiraide, general manager of a Tokyo department store. "It made everyone think that we are all in the middle class. But lifetime employment is crumbling and salaries are based more on merit and performance. In seven or eight years, Japanese society will look much more like Western society, with gaps between rich and poor that can be clearly seen."[100]

The recession fell most heavily on female workers, who were long treated as temporary workers by corporations that guaranteed lifetime employment only to men. Women in temporary jobs were dismissed first.

The government disguised rising unemployment for women by recording them as "housewives," not "unemployed workers." This practice understated real unemployment rates. The fact that Japanese businesses downsized even men in lifetime positions meant that they had already laid off a great many women. "Somehow, although I've done nothing wrong, I feel like a criminal," Kimiko Kauda said after being fired from her job of thirty years. "I have never heard of people being fired in my neighborhood or among my friends."[101]

Western Europe

During the 1990s, Western Europe also became mired in recession, though for different reasons than Japan. Problems began in Germany. The 1989 collapse of communist government in East Germany led to German unification. The German government then spent $600 billion to rebuild the region's economic infrastructure, provide benefits to workers, purchase voter loyalty, and prevent a "widespread social explosion" by workers laid off as the East deindustrialized.[102] Because spending on this scale—$600 billion for a small region with the population of New York State—can trigger inflation, the government raised taxes, and the Bundesbank, which controls monetary policy, raised interest rates to reduce inflationary pressures. These measures triggered a sharp recession and widespread job loss.[103] In the early 1990s, unemployment rates doubled from 6 percent to 12 percent in Germany, and were twice this rate in the East, where deindustrialization and recession were joined. Faced with high unemployment rates, German unions agreed to substantial pay cuts (10 percent in 1997), benefit reductions, shorter vacations, and work-rule concessions. "Corporations want to abolish the social consensus in Germany," union negotiator Peter Blechschmidt said of the 1997 wage cuts. "They are trying to change this into a different country."[104] A country more like the United States.

The situation in Germany was not unique. Countries throughout Western Europe also experienced recession and rising unemployment, partly due to the cost of European unification. In 1991, most Western European states agreed at Maastricht, in the Netherlands, to adopt a common currency (the United Kingdom did not do so). To prepare for the introduction of a single currency in 1999, governments in the European Union (EU) set a number of common economic goals: reducing budget deficits, stabilizing exchange rates, and, most important, reducing inflation. To meet this last goal, member governments raised interest rates. As in Germany, high interest rates triggered a regional recession, and unemployment soared throughout Europe. Unemployment rose to 12 percent in France and Italy, 13 percent in Ireland, 22 percent in Spain, and even more in regions like southern Italy.[105]

Although recession and job loss affected workers throughout Western Europe, women were the big losers, particularly in eastern Germany. Nearly 60 percent of the four million who had been employed in 1989 lost their jobs during the next four years.[106] Only half as many men lost their jobs in the same period. Young people have also been affected in disproportionate numbers. Like Japan, Western European industries had strong seniority systems. So when recession hit, they laid off younger workers. In general, young people were unemployed at twice the rate (21.8 percent) as workers over twenty-five years old.[107]

By the late 1990s, governments and industries in Western Europe and Japan responded to renewed U.S. competitiveness by reorganizing production. They did so by adopting measures pioneered in the United States. For a start, governments and businesses tried to increase the role played by investors and stock markets. In Germany, for instance, worker-consumers are being encouraged to adopt an *Aktienkultur*, or "stock culture," and invest their substantial savings in the stock market.[108] To facilitate this, the government plans to cut capital gains taxes on German corporations, which would make it easier for them to reorganize and consolidate industry.[109] And business has increased advertising expenditures to encourage stock market investment. Deutsche Telekom recently spent $150 million on a campaign to advertise a $10 billion stock offering, which was then used to purchase Telecom Italia, one of the first big cross-border mergers in Western Europe.[110] These policies and practices are helping jump-start the *Aktienkultur*, not only in Germany but across Europe.[111]

Mergers have played a growing role in the reorganization of production in Europe. The value of annual mergers in Western Europe jumped dramatically in the second half of the 1990s, growing from about $150 billion in 1994 to more than $600 billion in 1999.[112] As businesses merged and modernized, they typically downsized workers, just like their corporate counterparts in the United States. This has helped keep unemployment rates high. Downsizing, together with efforts to curb seniority-based pay systems and exact wage concessions from unionized workers, has kept wages from rising.

The spreading merger wave in the United States and Western Europe has itself begun to alter the redistributive process and reorganize production in new ways. Initially, the redistribution of production was managed largely by government policies—exchange rates, tariff barriers, defense spending, and foreign aid. But today, the redistribution of production is directed increasingly by the cross-border corporations that formed when industries reorganized: Mercedes-Chrysler; Ford-Volvo; Renault-Nissan; Aegon TransAmerica–Deutsche Telekom–Telecom Italia; Volkswagen–Rolls Royce; and MCI–British Telecom. These cross-border corporations (XBCs) differ from their transnational corporation (TNC) predecessors. TNCs were firms based in one country, which conducted businesses through subsidiaries in other states. XBCs, by contrast, are firms with origins in more than one state, usually formed by mergers, which conduct business in other states. Because XBCs now redistribute production as a process internal to a corporation that spans multiple states, government policies that used to shape the redistributive process now play a less significant role in determining who gets what jobs.

It is important to note that the redistribution of wage work, managed first by governments and more recently by XBCs, generally resulted in the redistribution of production in the United States, Western Europe, and Japan. Jobs in U.S. industries were redistributed primarily to industries in Western Europe and Japan. Ford workers in Detroit lost jobs to Toyota workers in Yokohama and Volkswagen workers in Munich; Boeing workers in Seattle lost jobs to Airbus workers in London, Paris, Milan, and Hamburg. Some production was redistributed to industries in Latin America and East Asia, and in the late 1990s, to China (see chapter 7). Until recently, the redistribution and reorganization of production, two processes that define contemporary globalization, have been generally confined in the rich countries. As such, globalization should be understood as a "selective," not "ubiquitous," process.

As industries in Western Europe and Japan reorganize along U.S. lines (the process is more advanced in Western Europe than it is in Japan), their ability to compete in redistributive battles with the United States will increase, and they may reclaim some of the production obtained by U.S. industries in the 1990s.

The redistribution and reorganization of production has resulted in job and income loss for workers in all three regions. But while workers in the United States, Western Europe, and Japan face many of the same problems, they do so with different resources at their disposal.

In the United States, worker households are heavily indebted. To maintain their standards of living, they have spent down their savings. During the 1990s, workers in the United States used their savings and borrowed money to shop and buy. "The American consumer has taken the globe from deep contraction back to flatness to recovery," one investment analyst observed.[113] But 1998 was a turning point. This was the first year since the Great Depression that U.S. workers did not acquire any net savings.[114] As savings declined, household debts increased. The average consumer debt per household nearly doubled in the last decade, rising from nearly $39,000 in 1990 to $66,000 in 2000. And since 1973, debt has grown as a percentage of income from 58 percent to 85 percent.[115]

Although much of the $5.5 trillion in total household debt is in the form of home loans, worker households now owe $350 billion on their credit cards.[116] Not surprisingly, bankruptcies are at record levels: one in a hundred households annually declare bankruptcy.[117] "There is a lid on earnings, but meanwhile, people's cost of living and their desire for fancier lifestyles go unabated," observed A. Stevens Quigley, a Seattle bankruptcy lawyer.[118] Student debt also grew from $18 billion to $33 billion between 1991 and 1997, and graduates owe $18,000 on average when they leave college.[119]

Although worker households are up to their ears in debt, some can tap other resources that are not typically counted in savings-rate/debt-burden ledgers. The generation of workers who accumulated savings, pensions, and houses during the 1950s and 1960s has transferred important assets to their children, the heavily indebted baby-boomers. Some economists estimate that as much as 25 percent of worker-household income comes from parents and relatives.[120] Essentially, the postwar generation has helped the current generation of wage workers survive. In addition to income from this source, worker households who used IRAs to invest in the stock market have generally seen the value of their stocks rise, which would boost their real savings.[121] But even after adjusting for income from these two sources, which are not available to most households, U.S. workers are still heavily indebted.

Compare the condition of worker households in the United States with households in Japan and Western Europe. Although savings rates in Japan and Western Europe have recently declined, as workers used up some savings during the recession and retired workers spent their accumulated savings, they still save a large percentage of their income.[122] In Japan, households saved, on average, 12 percent of their disposable income in 1999 and had deposited $100,000 in the bank.[123]

In the United States, workers are heavily indebted and household account balances are in the red. But in Japan and Western Europe, where workers have substantial savings and few debts, household account balances are in the black. As a consequence, households in Japan and Western Europe are in a much better position to weather changes associated with the redistribution and reorganization of production than their peers in the United States. While workers in all three regions now have comparable incomes and face similar problems, they confront economic change with different resources. So when new economic storms emerge, their fortunes may diverge.

To understand some of these developments in greater detail, we will now return to the early 1970s, when U.S. policymakers first confronted two important problems: (1) rising competition with businesses in Western Europe and Japan; and (2) rising inflation. The solutions U.S. officials advanced to solve these two problems had a huge impact on people in the United States and around the world. And the consequences of decisions made then are still being felt today.

NOTES

1. Bertrand Bellon and Jorge Niosi, *The Decline of the American Economy* (Montreal: Black Rose Books, 1988), 29.

2. T. E. Vadney, *The World since 1945* (London: Penguin, 1992), 73.

3. Ruth Sivard, *World Military and Social Expenditures, 1987–88* (Washington, D.C.: World Priorities, 1987), 37.

4. Eric Helleiner, *States and the Reemergence of Global Finance: From Bretton Woods to the 1990s* (Ithaca, N.Y.: Cornell University Press, 1994), 58–69.

5. Henry C. Dethloff, *The United States and the Global Economy since 1945* (New York: Harcourt Brace, 1997), 72.

6. Richard J. Barnet and John Cavanah, *Global Dreams: Imperial Corporations and the New World Order* (New York: Touchstone, 1994), 113; Peter Dicken, *Global Shift: The Internationalization of Economic Activity* (New York: Guilford Press, 1992), 67.

7. Ikeda Satoshi, "World Production," in *The Age of Transition: Trajectory of the World-System, 1945–2025*, ed. Terence K. Hopkins and Immanuel Wallerstein (London: Zed Books, 1996), 48; A. G. Kenwood and A. L. Lougheed, *The Growth of the International Economy, 1820–1990* (London: Routledge, 1992), 250; Ernst Mandel, *Europe vs. America: Contradictions of Imperialism* (New York: Monthly Review Press, 1970), 13; Cynthia Day Wallace and John M. Kline, *EC 92 and Changing Global Investment Patterns: Implications for the U.S.-EC Relationship* (Washington, D.C.: Center for Strategic and International Studies, 1992), 2; Barnet and Cavanah, *Global Dreams*, 42.

8. This is a rough estimate. Economists calculate that $1 billion of U.S. investment overseas results in the loss of 26,500 domestic jobs. So investments worth $78 billion would result in the loss of more than two million jobs in the United States, the number of people living in Boston, Kansas City, Miami, and San Francisco.

9. Shigeto Tsuru, *Japan's Capitalism: Creative Defeat and Beyond* (Cambridge: Cambridge University Press, 1993), 49–51, 78.

10. Saskia Sassen, *Globalization and Its Discontents* (New York: Harcourt Brace, 1997), 39, 79.

11. Jon Halliday, *A Political History of Japanese Capitalism* (New York: Pantheon, 1975), 279.

12. Tsuru, *Japan's Capitalism*, 109; Halliday, *Political History of Japanese Capitalism*, 273; Michael J. Piore and Charles F. Sabel, *The Second Industrial Divide: Possibilities for Prosperity* (New York: Basic Books, 1984), 161.

13. Halliday, *Political History of Japanese Capitalism*, 224–27. In 2000, one woman successfully sued her employer for job discrimination because managers had denied her a pay raise for twenty-one years. Howard W. French, "Women Win a Battle, but Job Bias Still Rules Japan," *New York Times*, 26 February 2000.

14. Nicholas Lehman, *The Promised Land: The Great Black Migration and How It Changed America* (New York: Knopf, 1991), 6.

15. Ruth Milkman, "Union Responses to Workforce Feminization in the United States," in *The Challenge of Restructuring: North American Labor Movements Respond*, ed. Jane Jenson and Rianne Mahon (Philadelphia: Temple University Press, 1993), 229.

16. Lawrence Mishel, Jared Bernstein, and John Schmitt, *The State of Working America, 1998–99* (Ithaca, N.Y.: Economic Policy Institute, Cornell University

Press, 1999), 182; Paul Hirst and Grahame Thompson, *Globalization in Question: The International Economy and the Possibilities of Governance* (Cambridge: Polity Press, 1996), 26; Sassen, *Globalization and Its Discontents*, 35; Richard B. Craig, *The Bracero Program: Interest Groups and Foreign Policy* (Austin: University of Texas Press, 1971), 102–3.

17. Saskia Sassen, *Losing Control? Sovereignty in an Age of Globalization* (New York: Columbia University Press, 1996), 81; Faruk Tabak, "The World Labour Force," in *The Age of Transition: Trajectory of the World-System, 1945–2025*, ed. Terence K. Hopkins and Immanuel Wallerstein (London: Zed Books, 1996), 94; Robert K. Schaeffer, *Power to the People: Democratization around the World* (Boulder, Colo.: Westview Press, 1997), 66–67, 189.

18. Tsuru, *Japan's Capitalism*, 68.

19. B. J. McCormick, *The World Economy: Patterns of Growth and Change* (Oxford: Philip Allan, 1988), 188.

20. Juliet B. Schor, *The Overworked American: The Unexpected Decline of Leisure* (New York: Basic Books, 1991), 111; Lehman, *The Promised Land*, 111; Halliday, *Political History of Japanese Capitalism*, 231.

21. Judith Stein, *Running Steel, Running America: Race, Economic Policy, and the Decline of Liberalism* (Chapel Hill: University of North Carolina Press, 1998), 7.

22. Paul R. Lawrence and Davis Dyer, *Renewing American Industry* (New York: The Free Press, 1983), 72; Stein, *Running Steel*, 209, 223.

23. Stein, *Running Steel*, 210.

24. Stein, *Running Steel*, 290.

25. In 1967, for example, the Japanese steel industry charged Japanese customers $116 a ton for steel, but charged U.S. customers only $96 a ton, a clear case of illegal dumping. Stein, *Running Steel*, 218, 257.
Despite evidence that foreign producers were dumping steel and seizing U.S. markets, officials refused to take action, arguing that doing so would increase domestic prices or undermine political relations with its allies. President Clinton rejected tariffs on cheap steel imports from Russia, notwithstanding obvious dumping, because, he said, it would jeopardize U.S.-Soviet political relations and deprive them of the earnings they needed to repay their debts to the West. "It's extremely important that we not cut off one of the few sources of raising hard currency that the Russians have right now," one advisor said. David E. Sanger, "U.S. Says Japan, Brazil Dumped Steel," *New York Times*, 13 February 1999.

26. Ira C. Magaziner and Mark Patinkin, *The Silent War: Inside the Global Business Battles Shaping America's Future* (New York: Vintage Books, 1990), 309.

27. Stein, *Running Steel*, 16.

28. Leslie Wayne, "American Steel at the Barricades," *New York Times*, 10 December 1998.

29. Lawrence and Dyer, *Renewing American Industry*, 72.

30. Stein, *Running Steel*, 295, 303; Frank Levy, *Dollars and Dreams: The Changing American Income Distribution* (New York: Russell Sage Foundation, 1987), 91; Dethloff, *The United States and the Global Economy*, 124.

31. Tom Redburn, "A Revolution Built in Mr. Ford's Factory," *New York Times*, 2 January 2000.

32. William J. Abernathy, Kim B. Clark, and Alan M. Kantrow, *Industrial Renaissance: Producing a Competitive Future for America* (New York: Basic Books, 1983), 47, 54; Magaziner and Patinkin, *Silent War*, 5–7.

33. James A. Geschwender, *Racial Stratification in America* (Dubuque, Iowa: Wm. C. Brown, 1978), 224, 236–37.

34. Lawrence and Dyer, *Renewing American Industry*, 18; Kim Moody, "Labor Givebacks and Labor Fightbacks," in *The Imperiled Economy*, vol. 2, *Through the Safety Net*, ed. Robert Cherry (New York: Union for Radical Economics, 1988), 161.

35. Dethloff, *The United States and the Global Economy*, 44.

36. William Greider, *One World, Ready or Not: The Manic Logic of Global Capitalism* (New York: Simon & Schuster, 1997), 125; Ian McIntyre, *Dogfight: The Transatlantic Battle over Airbus* (Westport, Conn.: Praeger, 1992), 2.

37. McIntyre, *Dogfight*, 44–45; Magaziner and Patinkin, *Silent War*, 230, 244, 251–52.

38. McIntyre, *Dogfight*, xx, 44; Magaziner and Patinkin, *Silent War*, 255.

39. John Tagliabue, "A Yankee in Europe's Court," *New York Times*, 11 February 2000; Magaziner and Patinkin, *Silent War*, 232.

40. John Tagliabue, "Airbus Industry Is Considering a Very Big Jet," *New York Times*, 3 February 2001.

41. Magaziner and Patinkin, *Silent War*, 257.

42. Many of the women who found public sector jobs were organized by unions. The growth of public sector unions, representing predominantly female workers, prevented unions from declining more sharply than they did. Indeed, unions were much more successful organizing women than men after 1970. Milkman, "Union Responses," 237, 239.

43. Schor, *Overworked American*, 26.

44. David Cay Johnston, "The Servant Class Is at the Counter," *New York Times*, 27 August 1995.

45. Milkman, "Union Responses," 229.

46. Schor, *Overworked American*, 87–88.

47. Schor, *Overworked American*, 29.

48. Stein, *Running Steel*, 306, 316; Barnet and Cavanah, *Global Dreams*, 54.

49. Stock price inflation is a discriminatory economic process. See also the discussion in chapter 3.

50. Edward M. Graham and Paul R. Krugman, *Foreign Direct Investment in the United States* (Washington, D.C.: Institute for International Economics, 1991), 14, 21; Neil Reid, "Japanese Direct Investment in the U.S. Manufacturing Sector," in *The Internationalization of Japan*, ed. Glenn D. Hook and Michael A. Weiner (London: Routledge, 1992), 66; Alan Scott, ed., *The Limits of Globalization: Cases and Arguments* (London: Routledge, 1997), 141.

51. Bill Orr, *The Global Economy in the '90s: A User's Guide* (New York: New York University Press, 1992), 287.

52. Arthur A. Alderson, "Globalization and Deindustrialization: Direct Investment and the Decline of Manufacturing Employment in 17 OECD Nations," *Journal of World-Systems Research* 3, no. 1 (1997): 5.

53. Danny Hakim, "Controlling 401(k) Assets," *New York Times*, 17 November 2000.

54. Richard W. Stevenson, "Fed Says Economy Increased Net Worth of Most Families," *New York Times*, 19 January 2000; Mishel, Bernstein, and Schmitt, *State of Working America*, 268.

55. Floyd Norris, "Dividends Rise, but Not as Fast as Stocks," *New York Times*, 3 January 1997.

56. In the 1980s, Republican administrators abandoned antitrust because they believed mergers would increase business efficiency and make U.S. companies stronger and more competitive with mega-firms in Western Europe, where antitrust laws are weak, and in Japan, where they are virtually nonexistent.

57. Laura M. Holson, "A Day for Mergers, with $40 Billion in Play," *New York Times*, 24 November 1998.

58. Stephen Labaton, "Merger Wave Spurs a New Scrutiny," *New York Times*, 13 December 1998; Laura M. Holson, "The Deal Still Rules," *New York Times*, 14 February 1999. Bankers and lawyers received more than $2 billion in fees for arranging the Exxon-Mobil merger in 1998. *New York Times*, "Costs of Exxon-Mobil Deal to Top $2 Billion," 6 April 1999.

59. Stephen Labaton, "Despite a Tough Stance or Two, White House Is Still Consolidation Friendly," *New York Times*, 8 November 1999.

60. James Sterngold, "Facing the Next Recession without Fear," *New York Times*, 9 May 1995.

61. Saul Hansell, "Wave of Mergers Is Transforming American Banking," *New York Times*, 21 August 1995.

62. Stanley Aronowitz and William DiFazio, *The Jobless Future: Sci-Tech and the Dogma of Work* (Minneapolis: University of Minnesota Press, 1994), 3.

63. John Judis, "Should an Economist Be in Charge of the Economy?" *New Republic*, 7 June 1999.

64. Michael Wallace, "Downsizing the American Dream: Work and Family at Century's End," in *Challenges for Work and Family in the Twenty-First Century*, ed. Dana Vannoy and Paula J. Dubeck (New York: Aldine de Gruyter, 1998), 23; Mishel, Bernstein, and Schmitt, *State of Working America*, 29.

65. Robert J. Samuelson, "Economic Mythmaking," *Newsweek*, 8 September 1997; Mishel, Bernstein, and Schmitt, *State of Working America*, 69.

66. Robert D. Hershey Jr., "Survey Finds 6 Million, Fewer Than Thought, in Impermanent Jobs," *New York Times*, 19 August 1995.

67. Hershey, "Survey Finds 6 Million."

68. Louis Uchitelle and N. R. Kleinfield, "On the Battlefields of Business, Millions of Casualties," *New York Times*, 3 March 1996. See the critique of this in John Cassidy, "All Worked Up," *New Yorker*, 22 April 1996, 52–53; David L. Birch, "The Hidden Economy," *Wall Street Journal*, 10 June 1998; and the *Times*'s response: Louis Uchitelle, "Despite Drop, Rate of Layoffs Remains High," *New York Times*, 23 August 1996.

69. Christopher D. Cook, "Workers for Rent," *In These Times*, 22 July 1996; Barry Meier, "Some 'Worker Leasing' Programs Defraud Insurers and Employers," *New York Times*, 20 March 1992.

70. Chris Tilly, "Short Hours, Short Shrift: The Causes and Consequences of Part-Time Employment," in *New Policies for the Part-Time and Contingent Workforce*, ed. Virginia L. DuRivage (Armonk, N.Y.: M. E. Sharpe, 1992), 15, 17; Hershey, "Survey Finds 6 Million."

71. One in five college graduates now works in a low-wage job that does not require a college degree. Sylvia Nasar, "More College Graduates Taking Low-Wage Jobs," *New York Times*, 7 August 1992.

72. Steven Greenhouse, "Equal Work, Less-Equal Perks," *New York Times*, 30 March 1998.

73. Steven Greenhouse, "Growth in Unions' Membership in 1999 Was the Best in Two Decades," *New York Times*, 20 January 2000.

74. Andrew Hacker, "Who's Sticking to the Union?" *New York Review of Books*, 18 February 1999; Steven Greenhouse, "Union Membership Slides Despite Increased Organizing," *New York Times*, 22 March 1998.

75. Floyd Norris, "You're Fired! (but Your Stock Is Way Up)," *New York Times*, 3 September 1995.

76. Louis Uchitelle, "As Class Struggle Subsides, Less Pie for the Workers," *New York Times*, 5 December 1999.

77. Richard W. Stevenson, "Study Details Income Gap between Rich and Poor," *New York Times*, 31 May 2001.

78. Stevenson, "Study Details Income Gap"; Jeff Faux and Larry Mishel, "Inequality and the Global Economy," in *Global Capitalism*, ed. Will Hutton and Anthony Giddens (New York: New Press, 2000), 10.

79. David Cay Johnston, "Gap between Rich and Poor Found Substantially Wider," *New York Times*, 5 December 1999.

80. Schor, *The Overworked American*. See Steven Greenhouse, "So Much Work, So Little Time," *New York Times*, 5 September 1999. Median-income worker households with two wage earners together worked 20 percent more hours in 1997 than they did in 1979.

81. Lester Thurow, "Companies Merge; Families Break Up," *New York Times*, 3 September 1995.

82. John Foren, "Spotlight on Moonlighters," *San Francisco Chronicle*, 31 May 1992.

83. Edwin McDowell, "The Abbreviated Tourist," *New York Times*, 31 July 1997; Edwin McDowell, "More Work or Less Work Can Equal No Time Off," *New York Times*, 6 July 1996.

84. *New York Times*, "Sleep-Depriving Jobs Linked to Accidents," 4 June 1999.

85. David Cay Johnston, "Tax Bill Expands Limits on Retirement Savings," *New York Times*, 28 May 2001.

86. Hal R. Varian, "With Privatization, Market Risks Could Put a Hole in the Social Security Safety Net," *New York Times*, 31 May 2001.

87. Niall Ferguson, *The Cash Nexus: Money and Power in the Modern World, 1700–2000* (New York: Basic Books, 2001), 302.

88. Danny Hakim, "401(k) Accounts Are Losing Money for the First Time," *New York Times*, 9 July 2001.

89. Hakim, "401(k) Accounts."

90. Peter T. Kilborn, "Stock Slide Sinks Hopes in Industrial City," *New York Times*, 16 March 2001.

91. From 1994 to 1999, R&D spending increased from $97.1 billion to $166 billion, far more than R&D outlays in Japan, which totaled $95 billion in 1998. William J. Broad, "U.S. Back on Top in Industrial Research," *New York Times*, 28 December 1999; Louis Uchitelle, "The $1.2 Trillion Spigot," *New York Times*, 30 December 1999.

92. Sheryl WuDunn, "The Heavy Burden of Low Rates," *New York Times*, 11 October 1996; Gretchen Morgenson, "Beware of Japanese Bearing Promises," *New York Times*, 21 June 1998; Charles P. Kindleberger, *World Economic Primacy 1500 to 1900* (Oxford: Oxford University Press, 1996), 206–7.

93. Sheryl WuDunn and Nicholas D. Kristof, "Crisis in Banking Is Japanese, but Implications Are Global," *New York Times*, 27 June 1998; Kevin Phillips, *The Politics of Rich and Poor: Wealth and the American Electorate in the Reagan Aftermath* (New York: Random House, 1990), 151.

94. Nicholas D. Kristof, "Long Mandibles, Sleek Carapace: A Steal at $300," *New York Times*, 10 April 1999.

95. Kindleberger, *World Economic Primacy*, 207.

96. Kindleberger, *World Economic Primacy*, 207.

97. Tsuru, *Japan's Capitalism*, 169.

98. Kristof, "Long Mandibles, Sleek Carapace."

99. Stephanie Strom, "Japan's New 'Temp' Workers," *New York Times*, 17 June 1998; Andrew Pollack, "Japanese Starting to Link Pay to Performance, Not Tenure," *New York Times*, 2 October 1993; Sheryl WuDunn, "When Lifetime Jobs Die Prematurely," *New York Times*, 12 June 1996; Stephanie Strom, "Toyota Is Seeking to Stop Use of Seniority to Set Pay," *New York Times*, 8 July 1999; Howard W. French, "Economy's Ebb in Japan Spurs Temporary Jobs," *New York Times*, 12 August 1999; David E. Sanger, "Look Who's Carping about Capitalism," *New York Times*, 6 April 1997.

100. Stephanie Strom, "Tradition of Equality Fading in New Japan," *New York Times*, 4 January 2000.

101. Andrew Pollack, "Jobless in Japan: A Special Kind of Anguish," *New York Times*, 21 May 1993.

102. Stephen Kinzer, "Help Wanted: One Mayor, Please," *New York Times*, 13 March 1995.

103. John Judis, "Germany Dispatch: Middle of Nowhere," *New Republic*, 29 November 1999.

104. Alan Cowell, "German Workers Fear the Miracle Is Over," *New York Times*, 30 July 1997; Nathaniel C. Nash, "In Germany, Downsizing Means 10.3% Jobless," *New York Times*, 7 March 1996; Ferdinand Protzman, "VW Offers Its Workers 4-Day Week or Layoffs," *New York Times*, 29 October 1993.

105. Sylvia Nasar, "Where Joblessness Is a Way of Making a Living," *New York Times*, 9 May 1999; Roger Cohen, "Europeans Consider Shortening Workweek to Relieve Joblessness," *New York Times*, 22 November 1993; Mishel, Bernstein, and Schmitt, *State of Working America*, 386; Celestine Bohlen, "Italy's North-South Gap Widens, Posing Problem for Europe, Too," *New York Times*, 15 November 1996.

106. Sabine Lang, "The NGOization of Feminism: Institutionalization and Institution Building within the German Women's Movement," in *Transitions, Environments, Translations: Feminisms in International Politics*, ed. Joan W. Scott, Cora Kaplan, and Debra Keates (New York: Routledge, 1997), 104.

107. Niels Thygesen, Yutaka Kosai, and Robert Z. Lawrence, *Globalization and Trilateral Labor Markets: Evidence and Implications* (New York: Trilateral Commission, 1996), 94–95; Edmund L. Andrews, "The Jobless Are Snared in Europe's Safety Net," *New York Times*, 9 November 1997. Ethnic groups and workers in par-

ticular regions have also suffered. North African immigrants in France have extremely high rates of unemployment. And workers in southern Italy are unemployed at twice the national rate.

108. Edmund L. Andrews, "Making Stock Buyers of Wary Germans," *New York Times*, 17 October 1996; John Tagliabue, "European Giants Set to Close Deal," *New York Times*, 20 April 1999.

109. Edmund L. Andrews, "Germany Proposes Some Tax-Free Stock Sales, Lifting the Market," *New York Times*, 24 December 1999.

110. Andrews, "Germany Proposes Some Tax-Free Stock Sales."

111. John Tagliabue, "Resisting Those Ugly Americans," *New York Times*, 9 January 2000.

112. Rich Miller, "Euro Forces Europe into Industrial Transformation," *USA Today*, 19 July 1999; Suzanne Kapner and Andrew Ross Sorkin, "American Bankers Invade Europe," *New York Times*, 3 February 2001.

113. Gretchen Morgenson, "U.S. Shoppers Shoulder the Weight of the World," *New York Times*, 20 June 1999.

114. Sylvia Nasar, "Economists Shrug as Savings Rate Declines," *New York Times*, 21 December 1998.

115. Saul Hansell, "We Like You. We Care about You. Now Pay Up," *New York Times*, 26 January 1997; Mishel, Bernstein, and Schmitt, *State of Working America*, 275; David Leonhardt, "Belt Tightening Seen as Threat to the Economy," *New York Times*, 15 July 2001.

116. Juliet B. Schor, *The Overspent American: Upscaling, Downshifting, and the New Consumer* (New York: Basic Books, 1998), 72; Maria Fiorini Ramirez, "Americans at Debt's Door," *New York Times*, 14 October 1997.

117. Saul Hansell, "Personal Bankruptcies Surging as Economy Hums," *New York Times*, 25 August 1999.

118. Hansell, "Personal Bankruptcies Surging."

119. *New York Times*, "Debt-Load Growing for College Graduates," 24 October 1997; Ethan Bronner, "College Tuition Rises 4%, Outpacing Inflation," *New York Times*, 8 October 1998; Robert D. Hershey Jr., "Graduating with Credit Problems," *New York Times*, 10 November 1996.

120. Some economists estimate that older generations may transfer $10 trillion to younger generations during the next fifty years. But estimates about the size of intergenerational transfers are the subject of considerable dispute. One study estimated that the older generation may transfer a much greater amount, between $41 trillion and $136 trillion. See David Cay Johnston, "A Larger Legacy May Await Generations X, Y and Z," *New York Times*, 20 October 1999.

121. Klaus Friedrich, "The Real American Savings Rate," *New York Times*, 4 May 1999.

122. Kindleberger, *World Economic Primacy*, 205.

123. Stephanie Strom, "Japan's Investors Become Bullish on Merrill Lynch," *New York Times*, 6 January 2000; Stephanie Strom, "Shopping for Recovery," *New York Times*, 29 May 1998.

2

❧

Dollar Devaluations

At the beginning of the 1970s, Americans faced two economic problems: declining competitiveness and rising inflation. The postwar economic recovery of Western Europe and Japan had enabled businesses there to become more competitive with U.S. firms. As a result, U.S. businesses found it increasingly difficult to sell their goods in foreign and domestic markets. In 1971, for the first time in the twentieth century, the United States posted a trade deficit, meaning that Americans purchased more goods from other countries than they sold to people living in those countries. U.S. policymakers worried about the trade deficit because it signaled that the competitiveness of U.S. firms had declined, that production in the United States was being redistributed to other countries, and that U.S. workers were losing jobs to foreigners.

At the same time, U.S. military spending on the war in Vietnam had pushed up wages and prices, leading to inflation. Policymakers worried about inflation because it is a discriminatory economic process. Some workers can keep up with inflation because they can demand and get higher wages. But other workers cannot easily obtain higher wages, so their real income declines in an inflationary environment. The same is true of businesses. Some firms can raise prices and pass higher costs along to consumers. Oil companies can do this because people have to buy gas to get to work. But other businesses cannot easily do so because consumers will stop buying their goods if the price goes up. If they cannot pass along higher costs, they lose money and face bankruptcy. Because inflation is a discriminatory process, policymakers worked to fight inflation.

On August 15, 1971, President Richard Nixon confronted both problems simultaneously. To improve U.S. competitiveness, he took steps to devalue the dollar in relation to currencies in Western Europe and Japan. And to fight inflation, he introduced wage and price controls, which were designed to limit wage raises and price increases. The "Nixon Shocks," or *shokku* (shocks) as these were called in Japan, marked the beginning of U.S. efforts to solve two serious economic problems.[1] As we will see, the economic "solutions" devised by Nixon and subsequent U.S. presidents had mixed success. Dollar devaluations in 1971 and again in 1985 did not greatly improve competitiveness, redistribute production, or stem job loss in the United States. Moreover, they created some serious, unanticipated economic problems in the United States and around the world.

In the United States, dollar devaluations led to the purchase of U.S. businesses, real estate, and natural resources by foreigners, with important consequences for U.S. workers. Overseas, dollar devaluations contributed to declining incomes for oil-producing countries, which contributed to war in the Persian Gulf. U.S. dollar devaluations also wrecked the international system of fixed exchange rates, which had been established by the United States at Bretton Woods during World War II. The system that replaced it has led to global monetary instability, which created serious problems for countries around the world during the 1990s.

Wage and price controls did not curb inflation in the 1970s. During the 1980s, government economists successfully used high interest rates to curb inflation, but high interest rates contributed to a host of other economic problems. In Latin America, Eastern Europe, and Africa, high U.S. interest rates triggered a massive debt crisis that crippled economies and triggered massive political change (see chapters 4 and 6). In the United States, high interest rates wrecked the savings and loan industry and led to increasing homelessness. The effects of high interest rates are still being felt, in the United States and around the world, today.

To explain these developments in detail, we will look first at currency devaluations and at efforts to improve U.S. competitiveness. In the next chapter, we will examine efforts to curb inflation. Both stories begin in 1971 with Nixon's August 15 speech. Both stories have two parts. The dollar was devalued twice, first in 1971 and again in 1985. And there were two anti-inflationary campaigns, the first in 1971 and the second beginning in 1979. By focusing on successive "solutions" to economic problems, we will discuss the problems with macroeconomic management. U.S. attempts to improve competitiveness and fight inflation have had significant consequences for people across the globe.

Although the story of government efforts to improve U.S. competitiveness and the story about the government's fight against inflation will be told separately, they are joined in important ways. For example, high interest rates in the early 1980s strengthened the value of the dollar, creating problems that led to a second, deeper devaluation of the dollar in 1985. We will note these connections as the stories unfold. Both stories are also about the social costs and unanticipated problems associated with the macroeconomic decisions made by policymakers in the United States, Western Europe, and Japan. In this regard, it is useful to describe how powerful economic institutions can create global changes that affect people's lives. The Group of Five, for instance, played an important role in the 1985 decision to devalue the dollar; the Federal Reserve System was a central actor in the anti-inflationary campaigns of the early 1980s. These institutions are relatively obscure to most people, though their decisions have important impacts on people living in different parts of the world.

1971 DOLLAR DEVALUATION

"At the end of World War II, the economies of the major industrial nations of Europe and Asia were shattered," President Nixon told his television audience on August 15, 1971. "Today, largely with our help, they have regained their vitality. They have become our strong competitors. . . . But now that [they] have become economically strong . . . the time has come for exchange rates to be set straight, and for the major nations to compete as equals."[2]

In this speech, Nixon recognized that the ability of U.S. firms to compete with businesses in Western Europe and Japan had eroded during the postwar period (see chapter 1). To improve U.S. competitiveness and redistribute production from Western Europe and Japan "back" to the United States, Nixon devalued the dollar and altered the relations among currencies in all three regions. By changing exchange rates, Nixon hoped to change consumer behavior in all three regions and persuade U.S. consumers to "Buy American." If consumers could be persuaded to purchase domestic, not foreign goods, U.S. trade deficits would shrink and U.S. jobs would be saved.

Nixon's devaluation of the dollar opened a two-decade campaign to improve the competitiveness of U.S. firms and save jobs by changing global monetary relations. But the monetary policies of Nixon and, later, President Ronald Reagan did not greatly improve the ability of U.S. firms to compete with other foreign businesses, failed to persuade consumers to buy American, and did little to save jobs. Moreover, successive dollar devaluations, the first in 1971 and a second in 1985, created other problems that policymakers did not expect or appreciate.

As Nixon noted, businesses in Western Europe and Japan became strong competitors during the postwar period for a variety of reasons.

They regained their vitality because they adopted policies designed to develop economically and received substantial U.S. economic aid amounting to $14.3 billion, according to Nixon's calculations. They also benefited from favorable exchange rates that were set by the Bretton Woods agreement, an international monetary treaty named after the New Hampshire town where negotiations were held in 1944.

The Bretton Woods agreement made the U.S. dollar the world's monetary standard and fixed the value of other currencies in relation to the dollar. For example, the value of the yen, Japan's currency, was fixed at rate of 360 yen to the dollar between 1949 and 1971. This low rate made it easy for Japanese firms to sell their "cheap" goods in the United States, while making it difficult for Japanese consumers to buy "expensive" American goods in Japan. In effect, exchange rates set at Bretton Woods helped Japanese and Western European businesses sell their goods at home (where U.S. goods were relatively expensive) and abroad, particularly in the United States (where their goods were relatively cheap). The exchange rates fixed after World War II acted somewhat like a golfer's handicap, making the price or "score" of Japanese and Western European firms lower than they would be otherwise.

In golf, the handicapping system is designed to let poor players compete as equals with more skilled players. By allowing poor players to deduct a given number of strokes (their handicap) from their actual score, an unskilled player could compete with a PGA professional playing without a handicap. Likewise, the exchange rates set in the 1940s enabled less productive businesses in Western Europe and Japan to compete as equals with more productive U.S. firms in American markets. The difference between the fixed exchange rate system established at Bretton Woods and the golf handicapping system is that in golf, a player's handicap declines as his or her skills improve. But the Bretton Woods system did not do this because exchange rates were fixed. So, while businesses in Western Europe and Japan improved their "game" during the postwar period, their currency handicap remained the same. By analogy, the result was like giving a handicap to a skilled professional who had just joined the PGA tour. Given an exchange rate "handicap" and their much-improved skills, firms in Western Europe and Japan were able to beat U.S. firms in American markets.[3]

The sale of Volkswagen Beetles and Sony transistor radios to U.S. consumers in the 1960s helped West German and Japanese economies recover and grow. While they grew stronger, producing goods that were of increasing quality, the exchange rates did not change, which kept their products cheap. By the late 1960s, they had begun to compete successfully with American firms in both domestic and U.S. markets, where consumers purchased their goods in increasing volume. And in 1971, the

United States imported more goods than it exported, posting a $2.3 billion trade deficit, the first in decades.[4]

Government officials viewed the trade deficit as a sign that the ability of U.S. firms to compete with businesses in Western Europe and Japan had declined. They blamed the redistribution of production, and the job loss associated with it, on the monetary system, which assigned favorable exchange rates to U.S. competitors. But it was difficult for officials to alter long-standing exchange rates because the values of other currencies were fixed in relation to the dollar. In effect, they could not alter exchange rates unless they were prepared to abandon the dollar's role as the monetary standard for currencies around the world.

By 1971, officials decided to eliminate the dollar as a monetary standard and destroy the Bretton Woods system for two reasons. First, the value of the dollar was weakening. When the dollar became the global monetary standard, the United States agreed to supply dollars to people around the world so they could pay for the imports they needed to rebuild their shattered economies. The United States used the Marshall Plan, foreign aid, and overseas military spending to provide dollars to countries that needed "hard currency" to pay for imports. By providing liquidity and easing cash flow problems, the dollars supplied by the United States helped countries in Western Europe and East Asia grow and prosper. But while U.S. overseas spending helped provide much-needed cash, continued military spending on NATO and the war in Vietnam pumped too many dollars into the world economy. Dollars piled up in central banks around the world. To reduce their stocks of dollars, which they found increasingly difficult to use, some governments decided to return them to the United States and cash them in for gold, at the rate of $35 to the ounce, a rate set by the Bretton Woods agreement.

By the late 1960s and early 1970s, it became clear that the number of dollars in global circulation far outstripped the amount of gold stored at Fort Knox. If too many countries asked to redeem dollars for gold, the U.S. government would soon exhaust its gold reserves and the dollar would lose its value as a standard.

The fact that the world economy needed the United States to pump dollars into the monetary system to maintain liquidity, but that doing so undermined the value of the dollar and its role as a global monetary standard, was known to economists as the Trifflin Dilemma because it was first identified by Robert Trifflin, an economist who published a book on this subject, *Gold and the Dollar Crisis*, in 1960.[5]

To address this problem, Nixon administration officials decided to stop redeeming dollars for gold and force a devaluation of the dollar vis-à-vis other currencies in Western Europe and Japan. By lowering the value of the dollar and raising the value of other hard currencies, U.S. firms could sell more of their (now cheaper) goods overseas and U.S. consumers would be discouraged from buying (now more expensive) products from Western Europe and Japan. By exporting more and importing less, Nixon administration officials reasoned that they could eliminate the trade deficit, restore the competitiveness of U.S. firms, and save jobs in the United States. By abandoning the Bretton Woods system of fixed exchange rates (and the promise to convert dollars for gold at $35 an ounce) and devaluing the dollar—by making "exchange rates to be set straight" as Nixon put it—U.S. businesses could again "compete as equals."

Although Nixon's 1971 policy was designed to devalue the dollar, he was reluctant to describe it that way. He argued that his actions would not result in "the bugaboo of . . . what is called 'devaluation.'" And he told viewers that "if you want to buy a foreign car or take a trip abroad, market conditions may cause your dollar to buy slightly less. But if . . . you buy American-made products, in America, your dollar will be worth just as much tomorrow as it is today." Nixon and Treasury Secretary John Connally argued that this new policy was not a devaluation because Nixon administration officials in previous months had proclaimed they would not devalue the dollar to improve U.S. competitiveness.[6]

But this, and Connally's subsequent press conference statement that the dollar had not been devalued "by presidential action," was misleading. As Paul Volcker, a Treasury Department official who helped draft Nixon's speech, later recalled, a dollar devaluation "was exactly what we had decided was essential."[7] Nixon was reluctant to call it a devaluation because this term was politically charged. If one looks at his own example, though, Nixon describes precisely what a devaluation is all about. The dollar devaluation did not mean that a dollar could buy fewer American goods. But it did mean that Americans could not purchase as many foreign goods, either in the United States or abroad, as they could previously. As a result of changing exchange rates, which weakened the dollar and strengthened first world currencies (franc, yen, deutsche mark), an American tourist in Paris could buy fewer croissants and a shopper in Des Moines could buy fewer Christmas toys labeled "Made in Japan."

As a result of Nixon's 1971 policy, the dollar's value fell and other hard currencies rose during the next few months. By December 1971, the dollar had fallen between 8 and 17 percent, depending on the currency. The yen, for example, rose 16.9 percent, to 308 yen to the dollar.[8] After this initial change, the dollar overall slowly declined to about 25 percent of its August 1971 value by the end of the decade (275 yen to the dollar).

One might have expected Nixon's dollar devaluation and the collapse of the Bretton Woods system (the values of different currencies were no longer fixed and instead began to float in relation to the dollar and to other currencies) to improve U.S. competitiveness, reduce its trade deficit, and save jobs. After all, that was the administration's intent. But it did not. The modest $2.3 billion trade deficit in 1971 grew to $25.5 billion by 1980, a tenfold increase.[9]

U.S. trade deficits increased in part because Americans spent more on imported oil and on high-mileage foreign cars. Rising oil prices, which tripled between 1973 and 1974, also increased the cost of imported oil.[10] In 1973, for instance, the United States spent $23.9 billion for imported oil.[11]

Climbing gas prices, which were the product of the 1973 OPEC oil embargo, persuaded U.S. consumers to buy high-mileage Toyotas and Hondas. During the 1970s, U.S. demand for reliable high-mileage Japanese cars grew dramatically (see chapter 1). In 1970, Japanese automakers sold about four million cars in the United States. By 1975 Nissan had surpassed Volkswagen as the leading foreign car manufacturer, and by the end of the decade, Japanese manufacturers sold nearly twelve million cars.[12]

Rising oil prices contributed to a growing U.S. trade deficit in two ways, first by increasing the cost of imported oil, and second by increasing U.S. demand for foreign autos. Another oil embargo in 1978 tripled prices again. So whatever gains the 1971 dollar devaluation might have achieved were undermined by rising oil prices and increasing consumer demand for imported goods.

While the value of the dollar fell during the 1970s as a result of Nixon's policies, the dollar actually increased in value during the first half of the 1980s, largely as a result of efforts to curb inflation. As we will see in chapter 4, former Nixon aide Paul Volcker, who became head of the Federal Reserve System in 1979, took steps to raise interest rates as a way to slow inflation. As a result, interest rates on government bonds (savings bonds, Treasury bills) rose to very high levels, as high as 20 percent. Attracted by rates that were higher than they could earn on their savings in other countries, investors from around the world bought U.S. bonds. By 1986, the Japanese had purchased $186 billion in U.S. Treasury bonds.[13]

This massive purchase of U.S. bonds, which was stimulated by high interest rates, increased the value of the dollar. The increasing value of the dollar undermined whatever gains had been achieved by Nixon's devaluation. Volcker observed that by the end of 1984, "the yen and the mark, relative to the dollar, had been driven back . . . to their 1973 levels or below, and their car, machinery, and electronics manufacturers were finding the lush American market easy pickings."[14]

U.S. firms found it more difficult to sell their (now more expensive) goods abroad, while (now cheaper) foreign products flooded U.S. markets. As a result, U.S. trade deficits exploded from $25.3 billion in 1980 to $122 billion in 1985.[15] And nearly one-third of this deficit, $50 billion, was with Japan.[16] Princeton economist Robert Gilpin observed, "In the first part of 1986, the United States had achieved the impossible: it had a deficit with almost every one of its trading partners. Not since 1864 had the U.S. trade balance been so negative."[17]

Whereas a $2.3 billion trade deficit had seemed a major problem requiring dramatic solutions in 1971, U.S. officials in 1985 faced a trade deficit sixty times bigger. To deal with declining competitiveness and soaring trade deficits, officials again turned to a dollar devaluation as the cure for economic ills.

1985 DEVALUATION: THE PLAZA ACCORDS

As U.S. trade deficits rose to unprecedented levels in the mid-1980s, government officials once again sought to devalue the dollar to improve the competitiveness of U.S. firms in foreign and domestic markets. But this time, they could not act alone, as they had in 1971. Because exchange rates were no longer fixed, and because world currency markets played a larger role in setting the value of different currencies, the Reagan administration had to secure the cooperation of other governments in Western Europe and Japan to successfully devalue the dollar. In September 1985, the Reagan administration asked the financial representatives of the five leading economic powers, then known as the Group of Five (G-5), to meet at the Plaza Hotel in New York City and hammer out an agreement to devalue the dollar in relation to other hard currencies.

When they convened in the White and Gold Room of the Plaza Hotel on September 22, representatives from the United States, Japan, West Germany, France, and the United Kingdom agreed to devalue the dollar. They issued an innocuous statement saying that "some orderly appreciation of the main non-dollar currencies against the dollar is desirable. They [the G-5] stand ready to cooperate more closely to encourage this when to do so would be helpful."[18]

Reagan administration officials, like Nixon, were reluctant to describe their decision as a "dollar devaluation." Instead they called it an "appreciation" of "non-dollar currencies," which is the same thing. They insisted that "it does not represent a fundamental change in the exchange rate intervention policy," though that is exactly what it was, since the agreement made by G-5 members specified interventionary steps to be taken. As one Reagan official later said, "No country ever likes to say that their currency will depreciate. . . . A government does not make statements that imply weakness."[19]

Although the G-5 ministers said they wanted to see the dollar depreciate, they refused to say how much it should be devalued. They kept this decision secret from the public. In private, G-5 ministers agreed to devalue the dollar substantially. Over the next two years, the dollar would fall to one-half its 1985 value against the yen and the deutsche mark. And it would continue to fall, to only one-third its 1985 value by the early 1990s. So in 1993, for example, the yen traded for 105 to the dollar, down from 250 in 1985.[20]

The decision to devalue the dollar was made by an institution—the G-5—that was largely unknown to the public prior to its 1985 meeting in New York. But the decisions made by this secretive and select group had important global consequences.

SUMMITS AND SHERPAS

Although the 1985 Plaza Accords would have profound consequences for people around the world, the meeting was not widely noticed. On the day it was announced, the *New York Times* gave more attention to an earthquake in Mexico City. Except for financial experts, the economic tremors produced by the Plaza Accords went unregistered by the public. This was because first world economic summits were typically secret and select. As Volcker, one of two U.S. representatives at the meeting, said, "Until that day, [the G-5] had been a secret organization. Nobody outside a very tight official circle knew exactly where and when the five ministers met, what they discussed, and what they agreed. This was the first time a G-5 meeting was announced in advance [it was announced the day before] and a communiqué was issued afterward."[21]

The meeting was so secret that Japanese finance minister Noboru Takeshita "arranged to play golf at a course near Narita airport . . . but then, without playing the back nine," he slipped off to the airport and boarded a Pan Am jet to New York so that the press would not notice his departure from Japan.[22]

The secretive G-5 grew out of a meeting first held in the White House library in April 1973. The "Library Group" consisted of the finance ministers and sometimes central bank governors of the United States, West Germany, France, and the United Kingdom. They added the minister from Japan the following year.[23] In 1975, French president Valery Giscard d'Estaing called a summit meeting of ministers from the United States, France, West Germany, Japan, the United Kingdom, and Italy. They added Canada the following year. At these Group of Seven (G-7) summits, the finance ministers played important roles, crafting the economic agenda for political leaders. Because they worked to prepare presidents and prime ministers for these "summits," insiders referred to them as "sherpas." A British official is said to have coined the term because sherpas are the native porters who help mountaineers scale summits in the Himalayas.[24]

After the Plaza Accords were announced, Canadian and Italian officials complained that they were excluded, and the following year they were added, so that in 1986 the G-5 and G-7 became officially known as the G-7.[25]

Not only has the G-5/7 been secretive, it has been a selective group. Most of the world's 180-plus countries have been excluded from its meetings. Some of them have since formed the Group of 77, which has 120 members and convenes an "alternative economic summit" when the G-7 holds its annual meeting. G-7 members have not opened the door to participation by others because, as West German prime minister Helmut Schmidt explained, "We want a private, informal meeting of those who really matter in the world."[26]

First world countries had other reasons for maintaining secrecy among a select group. If they had announced how much they intended to devalue the dollar after the Plaza meeting, people would have rushed to sell dollars and buy other currencies, which could have caused financial chaos. Because they did not disclose their plans, currency traders reacted "cautiously" to the announcement, according to the *New York Times*.[27]

IMPACT OF THE PLAZA ACCORDS

Although the Plaza Accords initiated a substantial devaluation of the dollar, they created a host of new problems. First and foremost, the dollar devaluation did not reduce the U.S. trade deficit. In 1985, the United States recorded a $122 billion trade deficit. By raising the prices of imported goods, G-5 ministers expected U.S. consumers to buy fewer foreign goods; by lowering the value of the dollar, they expected consumers in other countries to buy more U.S.-made goods. But despite a substantial devaluation, the U.S. trade deficit actually increased to $155 billion in 1986 and $170 billion in 1987. It then decreased slowly, though it remained at pre-1985 levels in 1988 ($137 billion) and 1989 ($129 billion), before falling to $122 billion in 1990.[28] It would rise again in the 1990s. Volcker said of this development, "One of the ironies of [this] story . . . is that, after repeated depreciation of the dollar since 1971 to the point where it is 60 percent lower against the yen and 53 percent lower against the deutsche mark, the American trade and current account deficits are nonetheless much higher than anything imagined in the 1960s."[29]

What went wrong? Why didn't the dollar devaluation accomplish the goals of U.S. policymakers? The reason is that manufacturers and consumers did not respond to macroeconomic changes in the way policymakers expected.

The devaluation of the dollar and the appreciation of the yen should have doubled the price of Japanese imports. But Japanese automakers did not double their prices after the Plaza Accords. As economist Daniel Burstein explained,

> A Nissan automobile that sold in the United States for $9,000 in 1984, and should have sold for $18,000 in 1987 according to changes in yen/dollar exchange rates, actually sold for only $11,000. If it had really sold for $18,000, it might well have been priced out of the market. At $11,000 . . . it was still highly competitive. In fact, Nissan's total U.S. car sales for 1987 fell only 3 percent from the prior year. [30]

Rather than raise prices to conform with post-Plaza exchange rates, Japanese firms kept price increases modest, squeezed costs, and accepted lower profits to retain their share of U.S. markets. U.S. manufacturers, meanwhile, actually increased their prices to keep up with Japanese price increases. "Studies by auto market research firm J. D. Power confirm that while Japanese manufacturers were raising U.S. prices an average of 9–13 percent from 1985 to 1988, General Motors, Ford and Chrysler were raising prices by . . . 12–15 percent."[31]

Essentially, U.S. firms refused to take advantage of changing exchange rates. Rather than keep prices steady, which would have given U.S. automakers a price advantage vis-à-vis foreign car makers, U.S. firms raised their prices so they could make more money per car (rather than expand their production of cars) and swell profits, not increase their market share. U.S. firms emphasized short-term profits rather than increased market share because they wanted to increase stock prices and shareholder dividends (see chapter 1). They did this because the U.S. stock market plays a much greater role in corporate decision making than it does in Japan, where firms do not seek to reward stockholders with dividends and instead concentrate on long-term investment strategies.

Faced with only modest price differentials between imported and domestic goods, differences that often disappeared when quality and brand loyalty were considered, U.S. consumers kept purchasing imported goods from Western Europe and Japan. Given a choice between a Honda Accord and a Chrysler K-car, cars close in price despite the dollar devaluation, American consumers kept buying Hondas.

Because foreign and domestic producers and U.S. consumers did not behave as policymakers expected, U.S. trade deficits increased and the Plaza Accords did not achieve their objectives. As this became apparent in the late 1980s, economists sought to explain the failure of macroeconomic policy. They used the term "hysteresis," which means a "resistance to change," to describe the unwillingness of producers and consumers to act as economic theory predicted and monetary officials expected.[32] But this abstract term, which is drawn from physics, is simply a way of saying that people don't always act as economic theory says they should. Manufacturers and consumers did not do as they were "told," refusing to sell aggressively or "Buy American."

The Plaza Accords resulted in problems in the United States and other countries that G-5 representatives did not anticipate. In the United States, the decline of the dollar reduced the value of U.S. assets, while the appreciation of currencies in Western Europe and Japan provided investors there with the means to purchase U.S. assets at bargain prices. "In 1974, the three largest banks in the world were American while only two of the top ten were Japanese," notes one economist. But by 1988, largely as a result of devaluation, "of the 25 largest banks in the world . . . 17 were Japanese (nine of them were in the top ten), 7 were Western European, and 1 was from the United States."[33]

The growth of Japanese and Western European banks was due, in part, to the dollar devaluation, which increased their assets. Assets rose because Japanese and European banks are larger, on average, than their U.S. counterparts, so they were able to use size to their advantage. Japan has only 158 commercial banks; the United States has 14,000.[34]

As a result of the dollar devaluation, Japanese and Western European investors could buy U.S. banks and businesses, real estate, and natural resources for half price. At post-Plaza prices, foreign businesses rushed to invest in the United States. In the three years after 1985, Japanese firms invested $235 billion in the United States, purchasing government bonds ($30 billion in 1988), U.S. corporations (Sony purchased Columbia Records in 1988), real estate (Rockefeller Center in New York City and Pebble Beach in California), and natural resources (timber from the Pacific Northwest).[35] This development altered postwar investment patterns. Until 1985, investment generally flowed outward from the United States to Western Europe and Japan. After 1985, it began to flow in as well, a process that multilaterized or globalized investment substantially (see chapter 1).

Of course, the sale of U.S. assets to foreign investors may have little or no impact on jobs. Japanese investors did not fire Hollywood actors and filmmakers when they purchased entertainment companies, nor did they fire greens keepers at Pebble Beach or ski instructors at Heavenly Valley. The impact of investment on jobs varies from industry to industry. In one industry, however, the dollar devaluation did contribute to substantial job loss: the Pacific Northwest timber industry.

FALLING DOLLAR, FALLING TREES

The Plaza Accords had a dramatic impact on timber and jobs in the Pacific Northwest, the heavily forested region west of the Cascade Mountains in

Washington, Oregon, and northern California. The dollar devaluation combined with long-standing forestry practices to cut timber, send much of it to Japan, and lay off workers in U.S. mills.

During the postwar period, the federal government, which owns 191 million acres of timber in the United States, adopted policies to make cheap timber available to the logging and housing industry, providing jobs for the most job-intensive industry in America and inexpensive housing for would-be homeowners (see chapter 4).

The Forest Service, which oversees public forests, made cheap timber available to private industry in two ways. First, it sold timber for less than it cost the government to hire forest rangers and build access roads to timber stands. The Forest Service built and maintained 340,000 miles of heavy-duty roads, a network eight times longer than the interstate highway system, able to span the globe thirteen times. Instead of charging buyers for the full cost of its roads and other timber services, the Forest Service (and taxpayers) assumed much of the cost. For example, it amortized the cost of road building over many years, hundreds of years in some cases, so that buyers only had to cover artificially low annual costs. Between 1980 and 1991, the Forest Service lost $5.6 billion from below-cost timber sales.

Second, the Forest Service greatly increased the amount of public timber cut in the postwar period. It increased timber sales from 3.5 billion board feet (1 board foot is 12 by 12 inches square by 1 inch thick) in 1950 to 8.3 billion board feet in 1960, then to 12 billion board feet in the late 1960s, a fourfold increase.[36] The infusion of large public timber supplies into the market kept prices low. Forest managers were encouraged to sell as much timber as possible by laws that based their operating budgets on the volume of timber sales from their districts. In many parts of the country, they did not practice sustained-yield harvesting but cut trees faster than they grew back. Wilderness Society economist Jeffrey Olson estimated that from 1980 to 1985, the Forest Service overcut Northwest woods by 61 percent, and private industry overcut their woods by 126 percent.[37] Over time, this practice led to declining timber supplies and rising prices. The timber sold by the Forest Service in the Pacific Northwest declined by 75 percent, from 8 billion board feet in 1986 to 2.5 billion board feet in 1992.[38] It was in this context that the 1985 dollar devaluation made its appearance in the Pacific Northwest.

The dollar devaluation made Northwest timber available to foreign buyers at bargain-basement prices. For Japanese buyers, it represented a two-for-one sale. (The Japanese were the principal purchasers, though buyers in China, Taiwan, and South Korea also bought heavily.) As the dollar fell, Japanese purchases of Northwest timber increased from 3 billion board feet in 1986 to 4.2 billion board feet in 1988. By 1988, one of every four trees cut in the Northwest was shipped to Japan. In Washington State, 40 percent of the harvest was exported. George Leonard, associate chief of the U.S. Forest Service, admitted that log exports affected the supply of timber in the Northwest. "But if we want to buy Sonys and Toyotas from Japan, we've got to sell them something they want," he said.[39]

Increased U.S. timber exports led to two problems: higher timber prices and fewer jobs. First, the sale of large quantities of timber to overseas buyers reduced domestic supplies and competition between domestic and foreign buyers, who could offer more. This competition led to higher timber prices in the United States. Domestic timber prices increased from about $250 per thousand board feet in the mid-1980s to $350 in 1992 ($474 in 1993), a 30 percent increase.[40] Because a 2,050-square-foot house uses 14,350 board feet of wood, a 30 percent increase in timber prices added $3,000 to the cost of the house. Higher costs made it more difficult for U.S. consumers to purchase a home (see chapter 4).

Second, foreign buyers insisted on buying and shipping whole, raw logs. Japanese buyers did not want U.S. lumber mills to cut the wood before shipping it overseas. Instead they wanted to provide timber to the workers in Japanese mills, where they cut timber into the metric equivalent of two-by-fours. (U.S. mills do not use the metric system and do not cut timber to suit the Japanese construction industry.) Because the United States exported raw logs, not milled wood, employment in U.S. mills declined. Although estimates vary, between three and five jobs are lost for every million board feet of raw timber exported. Oregon representative Peter DeFazio calculated that the export of 4.3 billion board feet in 1988 resulted in the loss of between 13,800 and 23,000 jobs.[41] Between 1986 and 1991, 163 mills were closed in the Northwest.[42] "Decks at Japanese mills are piled high as Mount Fuji with logs from the Northwest, while mills here at home are scrapping for leftovers," said DeFazio. "We're facing the greatest timber supply crisis in our history while Japanese mills are running around the clock."[43]

In addition to export-related unemployment, lumberjacks and mill workers were laid off as the industry automated production, moved some mills to lower-wage countries like Mexico, and ran out of wood to cut in this part of the country. The Forest Service estimated in 1990 that technological change alone would displace 13 percent of the workforce by the end of the century.[44] And Forest Service plans to set aside timber for the protection of spotted owls, a bird threatened by timber cutting in old-growth forests, and salmon, whose streams are threatened by soil erosion from heavily logged forests, will also reduce timber supplies and affect jobs. Although no one at the Plaza meeting considered or anticipated the impact of a dollar devaluation on U.S. natural resources, the one-two punch of Forest Service policy and dollar devaluation resulted in declining timber supplies, higher prices, and growing unemployment in the Pacific Northwest.

GLOBAL CONSEQUENCES OF
DOLLAR DEVALUATIONS, 1970–1990

Successive dollar devaluations did little to reduce U.S. trade deficits. And they had an adverse impact on mill workers and timber prices in the Pacific Northwest. Globally, devaluations had important consequences for oil-producing countries, contributing to inflation in the 1970s, falling oil prices in the 1980s, and war in the Persian Gulf during the 1980s and early 1990s.

Oil

Around the world, dollars are used to buy and sell oil, a legacy of the fact that the United States was the world's first big oil-producing country. Because the world oil trade was and still is conducted in dollars, dollar devaluations have played an important role in the contemporary history of oil, contributing to inflation in the 1970s and to war in the 1980s and 1990s.

The 1971 dollar devaluation lowered the revenues of oil-producing countries because the dollars they were paid with were worth less than before (see chapter 9). Determined to regain lost revenues and to increase the price of oil in real terms, the members of OPEC responded in 1973 with an oil embargo during the Yom Kippur War between Egypt and Israel. This embargo, and a subsequent embargo during the Iranian revolution in 1979, increased oil prices to more than $35 a barrel. During the 1970s, then, the dollar devaluation helped trigger rising oil prices, which spurred inflation in the United States and around the world (see chapter 3).

Although the first dollar devaluation in 1971 contributed to oil-price hikes and increased the power of OPEC, the 1985 devaluation had the opposite effect. The Plaza Accords led to falling oil prices, declining revenues for oil-producing countries, and the outbreak of war between its members.

Between 1980 and 1985, the price of oil declined slowly from $35 to just under $30 a barrel (still a high price compared to the 1970 price of $3 a barrel). Prices fell because countries discovered new oil in the North Sea or expanded production to take advantage of high prices. Increased supplies undermined the price of oil (see chapter 6). But oil prices did not fall below $30 because the Saudi Arabian government was determined to

keep prices high for other OPEC countries. The Saudi Arabian government was willing, for a time, to curb their production so that an oil glut,
and lower prices, did not materialize. But by 1985, increasing production
by OPEC and non-OPEC countries exhausted the patience of the Saudi
Arabians, who had seen their oil revenues decline from $119 billion in
1981 to only $26 billion in 1985.[45] In that year, to Saudi embarrassment, the
United Kingdom produced more oil than Saudi Arabia. After a December
9, 1985, OPEC meeting, only two months after the Plaza meeting, the
Saudi Arabians abandoned their low-production, price-support policy
and increased their production to recapture their market share. Within a
few months the price of oil had collapsed to $10.[46]

U.S. consumers welcomed lower oil prices, but the domestic U.S. oil industry did not because it could not make money at $10 a barrel. Although
the price that producers get for their oil is set at the world level, they each
have different production costs. It costs very little to pump oil from shal-
low wells in Saudi Arabia, more to lift it from deeper wells in West Texas.
At $10 a barrel, U.S. producers could not afford to pump oil from domes-
tic wells. So U.S. firms quit drilling and pumping oil and laid off workers.
The recession in the oil industry depressed the price of real estate in the
Southwest, which contributed to the collapse of savings and loan organizations that had invested heavily in office buildings in the region (see
chapter 3). "Moreover," explained Daniel Yergin, "if prices stayed down,
U.S. oil demand would shoot up [as consumers drove more], domestic
production would plummet, and imports would start flooding in again,
as they had in the 1970s."[47]

To head this off, Vice President George Bush flew to the Middle East to
persuade the Saudi Arabian government to increase oil prices. He later explained:

> I think it is essential that we talk about stability and that we not just have a
> continued free fall [in prices] like a parachutist jumping out without a para-
> chute. . . . I'm absolutely sure . . . that *low* prices would cripple the domestic
> American energy industries, with serious consequences for the nation. [em-
> phasis added][48]

To provide higher prices to U.S. and Saudi Arabian oil producers, the
United States, Saudi Arabia, and some OPEC countries reached a consensus that oil prices should stabilize at $18, a considerable rise from $10.
And their combined efforts eventually established new OPEC quotas,
bringing OPEC and non-OPEC producers into line by 1987.

The decline in oil prices was accompanied in this same period by a devaluation of the dollar. In effect, the price of oil fell twice, first when rising
oil supplies drove down prices, and second when devaluation forced down

the dollar price of oil. In real terms, the price of oil had returned to about what it had been in 1973. As a result, the price of a gallon of gas, in inflation-adjusted 1993 dollars, was $1.12 in 1993 compared to $1.25 in 1973.[49]

But falling oil prices had different consequences for different countries. Oil-importing countries did well, but the oil-producing countries fared badly.

Oil-Importing Countries

The United States saved money because the price of oil was cheaper. This helped reduce the U.S. trade deficit. U.S. consumers were pleased with lower gas prices at the pump. But while the United States saved money, the Japanese and Europeans saved money twice. They benefited from the falling price of oil and from the devaluation of the dollar, which further lowered its cost to them. So while the United States saw its bill for imported oil fall by about 30 percent, Japan saw its bill fall by 50 percent and West Germany by 57 percent in this period.[50] As James Sterngold noted, "The stronger yen also slashed Japan's import bills, since oil is paid for in dollars."[51] Because Japan worked hard to improve its energy efficiency and promote conservation during this period, while the United States did little, Japan actually reduced its dependence on foreign oil (see chapter 13 on global warming).

Japan and West Germany also captured other benefits. In the late 1980s, the U.S. government spent about $50 billion providing military and naval protection to Kuwait and Saudi Arabia, equal to about $100 per barrel of oil imported from the Persian Gulf. Energy economist Amory Lovins notes that "since Germany and Japan depend heavily on Persian Gulf oil (without incurring these tremendous annual military costs) America in effect subsidizes the economies of its two major trading competitors."[52]

Oil-Exporting Countries

For oil-exporting countries, the price decline and dollar devaluation drastically reduced revenues. Despite organizing collectively in OPEC and waging a two-decade campaign to increase oil prices and use the revenue to promote economic development, oil-producing countries in the late 1980s found themselves back where they started in 1973. As a result of falling prices and heavy spending, "The $121 billion in financial reserves amassed by Saudi Arabia [in the early 1980s] have almost vanished," the *New York Times* reported in 1993. "'The Saudis have been drawing down reserves for 10 years,' an American official said. 'They're a mere shadow of their former selves.'"[53]

War in the Persian Gulf

Although oil-producing countries with small populations (Saudi Arabia, Kuwait, Libya) were still relatively prosperous, the heavily populated oil-producing countries like Nigeria, Iraq, and Iran saw their economic fortunes decline (see chapter 9). Iraq, the second largest oil-producing country in OPEC, saw its oil revenues decline from $26 billion in 1980 to $12 billion in 1988, at a time when it was spending heavily to wage war with neighboring Iran. Iraqi dictator Saddam Hussein had earlier invaded Iran to capture its oil fields. If he had succeeded, he would have controlled enough of world oil production to demand higher prices in OPEC and recapture the revenues Iraq lost to price cuts and then dollar devaluation. But Hussein's decade-long campaign failed to defeat Iran's army and capture its oil. So having failed to capture Iran's oil, Hussein invaded Kuwait to seize its oil fields in 1990. The United States and its allies then assembled a multinational army to drive Iraqi forces from Kuwait in 1991.

Although the 1985 dollar devaluation contributed to lower world oil prices, which benefited consumers in Western Europe, Japan, and the United States, it also contributed to war in the Middle East, leading to U.S. military intervention. In this context, the Plaza Accords had consequences and repercussions that policymakers did not intend, anticipate, or even imagine.

GLOBAL CONSEQUENCES, 1990–2000

The reverberations from dollar devaluations continued to be felt in the 1990s. They altered economic relations between the United States and Japan, contributed to global monetary instability and economic crises in countries around the world, and persuaded other countries to abandon their own currencies and adopt the U.S. dollar as their own, official domestic currency.

United States and Japan

Although the 1985 Plaza Accords considerably reduced the value of the dollar in relation to the yen (from ¥250 = $1 to ¥125 = $1), they did little to reduce the U.S. trade deficit and the job loss associated with it during the rest of the 1980s and beginning of the 1990s. But the value of the dollar fell again in the mid-1990s. It fell to ¥105 = $1 in 1993 and then to ¥80 = $1 in 1995, its lowest rate ever.

As the dollar devalued, the price of Japanese goods in U.S. markets rose sharply. Japanese manufacturers, particularly automakers, began losing ground to U.S. producers. Here's why. It cost Japanese automakers 1.43

million yen, on average, to build a car in Japan, plus $2,600 to ship and market the car in the United States. When the exchange rate was ¥110 = $1, the cost to Japanese manufacturers, in dollars, was $13,000. But when the exchange rate fell to ¥80 = $1, the cost to Japanese car makers rose to $17,875, a substantial price increase.[54] As the dollar fell in 1995, the *New York Times* reported that "Japanese corporate executives seem shellshocked by market trends over which they have no control."[55] And many small manufacturing firms in Japan began to go out of business or move abroad, where exchange rates and labor costs were more favorable.[56] This is what officials had hoped the 1985 Plaza Accords would accomplish. It just took a deeper devaluation and a longer time (ten years) than they anticipated.

But just when exchange rates fell to the point where they did stem imports, reduce U.S. trade deficits, and redistribute manufacturing jobs, this time from Japan to the United States, policymakers in the United States and Japan reversed course. In mid-1995, monetary officials in the United States and Japan worked together to increase the value of the dollar, forcing it back up to ¥114 = $1 in 1996. It then stayed at about this level for the rest of the 1990s and into 2001.

Of course, the rising dollar benefited Japanese, not U.S., manufacturers. One Japanese study found that for every dollar rise in the exchange rate, Japanese automakers reaped about $360 million in economic benefits.[57] By 1996, the *New York Times* reported that "Japanese manufacturers have [again] found themselves in the catbird seat with a choice of grabbing bigger profits, cutting prices to build market share, or both."[58]

Why on earth would U.S. and Japanese officials collaborate in 1995 to reverse the devaluation of the dollar, just when it had begun to alter the relation between U.S. and Japanese manufacturers? They did so for two reasons.

First, Clinton administration officials worried that a devalued dollar would hurt Japanese businesses, which were mired in a recession that had begun in 1990 (see chapter 1). If Japanese auto sales faltered in the United States, the recession in Japan might deepen. U.S. officials reasoned that this would slow the world economy and dim U.S. economic prospects. Because the U.S. economy was growing during this period, largely as a result of gains made from the reorganization of American business (see chapter 1), Clinton administration officials were in a generous mood. "I don't think there's any question about the health and competitiveness of [U.S.] industries compared with industries [in Japan and] all over the world," Treasury Secretary Robert Rubin argued.[59]

Second, U.S. officials did not want to antagonize the Japanese, who had made substantial purchases of U.S. government bonds during the decade since the 1985 Plaza Accords. Japanese negotiators warned that if U.S. officials did not help strengthen the dollar, Japanese investors might sell off

U.S. bonds.[60] If Japanese investors sold U.S. bonds, the U.S. government would have to raise interest rates to attract new buyers. As we will see in chapter 4, higher interest rates can trigger a recession and force people out of work. This was something the Clinton administration was unwilling to do as the 1996 presidential campaign approached.[61]

Because the Japanese economy remained stuck in a recession, and because the U.S. government still needed foreign investors to purchase U.S. bonds, exchange rates remained at the 1996 level into 2001, and U.S. trade deficits grew to new heights. By 1998, the U.S. trade deficit rose to more than $350 billion, a staggering sum compared with the $2.3 billion deficit that triggered the first devaluation in 1971, and the $122 billion trade deficit that forced a second devaluation in 1985. Although they were first identified as a problem by Nixon in 1971, exchange rates remain an ongoing problem for the United States and its allies today.

Gender and Exchange Rates

Most imported products are manufactured goods (autos and oil) made by men in factories, not services performed by women in offices, hospitals, or restaurants. So changing exchange rates generally affect men in manufacturing industries more than they do women in service industries. This being the case, the monetary policies adopted by government officials to alter exchange rates were really about the redistribution of jobs held by men in Western Europe, the United States, and Japan. Put simply: monetary policy is gendered; changing exchange rates primarily affect men in manufacturing.

The loss of mill jobs in the Pacific Northwest provides a good example of this because the workers hired to cut, transport, and mill timber are predominantly male. Of course, the loss of male jobs also affected the women who lived with men employed in these industries. When men lost manufacturing jobs during the 1970s, 1980s, and 1990s, many women in working- and middle-class households looked for work in growing service industries. But in the Pacific Northwest, this was difficult to do because there were few service industries located in mill and logging towns. Under these conditions, lost income from male employment was not easily replaced by female partners, living standards fell, divorce rates climbed, and mill towns emptied as people searched elsewhere for jobs.

Monetary Instability and Economic Crises

When Nixon first devalued the dollar in 1971, he also destroyed the system of fixed exchange rates established at Bretton Woods in 1944. After 1971, the values of currencies were not fixed in relation to each other, but "floated" up and down depending on the health of their respective economies. They floated up and down fairly slowly, largely because governments did much of the buying and selling of currencies on global markets. But that began to change in the 1980s and 1990s, as investors began pouring money into currency markets, and trading currencies at an increasingly frenetic pace. They were able to invest heavily in currency markets because many governments were forced to make their currencies "convertible" as part of debt-crisis management plans, and because they agreed to open their economies to foreign investment and trade (see chapter 6). By the 1990s, decisions by private investors (whether to buy or sell a particular currency) could have a rapid and dramatic impact on exchange rates, triggering a serious economic crisis. During the 1990s, when currency traders sold off Mexican pesos, Thai bahts, South Korean wons, Indonesian rupees, and Russian rubles, they forced rapid and deep devaluations of these currencies. Foreign investors who owned government bonds saw their value decline and sold them off, leaving governments without the money they needed. To persuade investors to buy bonds, government officials raised interest rates to very high levels. But this sent their economies into deep recessions, leading to widespread bankruptcy and job loss. "This is off the radar screens in terms of severity," Allen Sinai, a global economist explained. "It is the single most negative event since the Great Depression in the United States."[62]

Because the monetary instability associated with the floating system of exchange rates can cause serious economic problems, some effort has been made to develop a new Bretton Woods system of fixed rates. The Bretton Woods Commission, led by Paul Volcker, former chairman of the Federal Reserve (see chapter 3), argues that "there is good evidence that exchange rates fluctuate excessively" and do not accurately reflect a currency's underlying economic health.[63] The commission proposed reforms that would address some of the problems created by floating rates. But other economists opposed the project, arguing that "the world isn't capable of maintaining a true fixed exchange-rate system."[64] So far, efforts to reform global monetary relations have made little headway.

Countries Adopt the Dollar

In the absence of any global monetary reform, a growing number of countries have adopted the U.S. dollar as their own official currency in an effort to stabilize their exchange rates and curb inflation. In 2000, Ecuador abandoned the sucre and adopted the dollar as its official currency; El Salvador and Guatemala "dollarized" their economies in 2001.[65] In Argentina, Brazil, and Lithuania, governments have linked or "pegged" their currency to the dollar, keeping both their own currency and the dollar in circulation.[66] And in other countries like Russia, the dollar plays a strong though unofficial role as a medium of exchange, largely because billions of U.S. dollars circulate in the economy.

Why have countries dollarized their economies? First, they have done so because they "have a tremendous need for a stable currency, and we [the United States] are providing them with a benefit they cannot get any other way," explained William Poole, an economist at Brown University.[67]

Second, they dollarize to curb inflation, which has reached astronomical rates in some countries. Because the supply of dollars is limited, and foreign governments cannot print more dollars on their own, a switch to dollars makes it difficult to raise prices because people do not have enough dollars to pay higher prices. And third, countries adopt the dollar or link their currencies to it as a way to legalize the widespread use of dollars to conduct business or bank savings.

U.S. officials have welcomed the dollarization of foreign economies. In 2000, Treasury Secretary Lawrence Summers applauded El Salvador's decision to dollarize: "[T]his step should help contribute to financial stability and economic growth in El Salvador and further its integration into the global economy."[68]

But there are several problems associated with dollarization. First, when a country adopts the dollar as its own, or links its own currency to the dollar, it surrenders its ability to make independent monetary policy and assigns this authority to another country: the United States. Moreover, when U.S. officials make decisions about monetary policy, they do so without consulting other countries or considering the impact of their decisions on other economies. In effect, governments that dollarize surrender monetary policy to the U.S. Federal Reserve Board (see chapter 3).[69] "The Fed uses its control of interest rates to stimulate or cool the American economy, but does not directly consider the needs of other countries that use the dollar," observers have noted.[70] So officials in dollarized countries give up their ability to stimulate the economy, create jobs, reduce unemployment, or encourage investment or savings.

Second, when countries dollarize, they create discriminatory, two-tier economies. People with ready access to dollars obtain an important advantage over people without access to dollars. In many cases, access is closely related to occupation, gender, and geography. In Russia, for example, male taxicab drivers in big cities can obtain dollars from businessmen and tourists; female prostitutes in Moscow and other big cities do the same. "Those with access to hard currency or 'valyuta' flaunt that access in their choice of dress, shampoo or entertainment," one observer wrote, leading to what he called "dollar apartheid."[71]

People who have jobs or live in areas where dollars are hard to come by are disadvantaged, in part because people with dollars bid up the prices of local goods and services. These small-scale or selective inflations create hardships for people with few dollars. In Ecuador, for example, rural Indian populations protested violently against the dollarization in 2000 because they had little access to dollars and were preyed on by currency traders and counterfeiters.[72] Many of them are illiterate, making it difficult for them to decipher U.S. coins, which do not use numbers to indicate their values (quarter dollar not 25 cents, dime not 10 cents).[73] As a result, dollarization tends to discriminate between groups based on gender, occupation, ethnicity, and geography.

Third, when countries dollarize, they tie their economic fortunes to U.S. exchange rates. So when rates change, as they did in 1971, 1985, and 1995 when the dollar was devalued, countries can experience serious problems. Panama provides a good example of this because it has long used the U.S. dollar as its official currency.

Panama

After 1985, Panamanians watched helplessly as the value of the dollar fell dramatically. This made it more difficult for them to import manufactured goods or oil from other countries. For rich people, this meant that "they consumed more [domestic] rum and less [imported] whiskey. 'I used to eat caviar,' one wealthy Panamanian put it, 'Now I eat ham.'"[74] For the poor, it meant that the cost of tortillas, cooking fuels, and transportation increased, resulting in a rapid decline in their standard of living.

The devaluation of the dollar came at a time of rising indebtedness and deteriorating relations between Panama and the United States. The U.S. government wanted to oust or arrest Panama's dictator, General Manuel Noriega, on drug trafficking charges, and the Bush administration applied economic sanctions on the country to force him to surrender power. The combination of intentional sanctions and unintentional devaluation crippled the economy, leading to a 17 percent decline in gross domestic product and 25 percent unemployment by 1989.[75] When Noriega refused to surrender power, and his soldiers attacked U.S. military personnel, the United States invaded the country on December 20, 1989. Noriega was captured and deported for trial in Miami. Although civilian democrats assumed power after Noriega was deposed, the economy remained in difficult circumstances as a result of debt, devaluation, embargo, and invasion.

The 1971 dollar devaluation marked the beginning of a twenty-five-year effort to improve U.S. competitiveness. But it also contributed to global monetary instability and led to serious problems for domestic workers in Northwest forests and Texas oil fields. It created problems for oil-producing countries and contributed to war in the Middle East. And it contributed to monetary crises in other countries around the world.

The year 1971 also marked the beginning of a long campaign against inflation. In some respects, the fight against inflation was more successful than the fight against trade deficits and job loss in the United States. Efforts to curb inflation during the early 1980s successfully reduced inflation rates in the United States. But this was not without cost. The battle against inflation crippled the domestic savings and loan industry, which contributed to rising homelessness, and triggered a global debt crisis for countries around the world. The battle against inflation in the United States will be examined in chapter 3, and the far-reaching consequences of the debt crisis will be examined in chapter 4.

NOTES

1. Paul A. Volcker and Toyoo Gyohten, *Changing Fortunes: The World's Money and the Threat to American Leadership* (New York: Times Books, 1992).

2. *New York Times*, "Transcript of President Nixon's Address on Moves to Deal with Economic Problems," 16 August 1971.

3. Shigeto Tsuru, *The Mainsprings of Japanese Growth: A Turning Point* (Paris: Atlantic Institute for International Economics, 1989), 18.

4. Berch Berberoglu, *The Legacy of Empire: Economic Decline and Class Polarization in the United States* (New York: Praeger, 1992), 56.

5. Volcker and Gyohten, *Changing Fortunes*, 38–39.

6. Volcker and Gyohten, *Changing Fortunes*, 79–80.

7. Volcker and Gyohten, *Changing Fortunes*, 81.

8. Volcker and Gyohten, *Changing Fortunes*, 346.

9. Berberoglu, *Legacy of Empire*, 56.

10. Bill Orr, *The Global Economy in the '90s: A User's Guide* (New York: New York University Press, 1992), 261.

11. Matthew L. Wald, "After 20 Years, America's Foot Is Still on the Gas," *New York Times*, 17 October 1993.

12. Andrew Pollack, "A Lower Gear for Japan's Auto Makers," *New York Times*, 30 August 1992.

13. Robert Gilpin, *The Political Economy of International Relations* (Princeton, N.J.: Princeton University Press, 1987), 331.

14. Volcker and Gyohten, *Changing Fortunes*, 229.

15. Berberoglu, *Legacy of Empire*, 56; Gilpin, *Political Economy*, 157.

16. Gilpin, *Political Economy*, 194.

17. Gilpin, *Political Economy*, 194.

18. Yoichi Funabashi, *Managing the Dollar: From the Plaza to the Louvre* (Washington, D.C.: Institute for International Economics, 1989), 263.

19. Funabashi, *Managing the Dollar*, 231.

20. Daniel Burstein, *Yen! Japan's New Financial Empire and Its Threat to America* (New York: Simon & Schuster, 1988), 142; Orr, *Global Economy*, 167.

21. Volcker and Gyohten, *Changing Fortunes*, 256.

22. Volcker and Gyohten, *Changing Fortunes*, 252.

23. Robert D. Putnam and Nicholas Bayne, *Hanging Together: The Seven-Power Summits* (Cambridge, Mass.: Harvard University Press, 1984), 18; Volcker and Gyohten, *Changing Fortunes*, 329–30.

24. Putnam and Bayne, *Hanging Together*, 45–46, 48, 237; Peter I. Hajnal, *The Seven-Power Summit: Documents from the Summits of Industrialized Countries, 1975–1989* (Millwood, N.Y.: Kraus International, 1989), xxiii, xxiv.

25. Volcker and Gyohten, *Changing Fortunes*, 329–30.

26. Putnam and Bayne, *Hanging Together*, 17.

27. Peter T. Kilborn, "U.S. and 4 Allies Plan Move to Cut Value of Dollar," *New York Times*, 23 September 1985.

28. Orr, *Global Economy*, 91.

29. Volcker and Gyohten, *Changing Fortunes*, 294.

30. Burstein, *Yen!* 147.

31. Burstein, *Yen!* 148.

32. Volcker and Gyohten, *Changing Fortunes*, 270; Dilip K. Das, *The Yen Appreciation and the International Economy* (New York: New York University Press, 1993), 25–28.

33. Berberoglu, *Legacy of Empire*, 42–43.

34. Das, *The Yen Appreciation*, 77.

35. James Sterngold, "Intractable Trade Issues with Japan," *New York Times*, 4 December 1991; James Sterngold, "Japan Shifting Investment Flow Back toward Home," *New York Times*, 22 March 1992.

36. *The Economist*, "The Forest Service: Time for a Little Perestroika," 10 March 1988.

37. Jeffrey T. Olson, *National Forests: Policies for the Future*, vol. 4, *Pacific Northwest Lumber and Wood Products: An Industry in Transition* (Washington, D.C.: Wilderness Society, 1988), 10.

38. Jeff Pelline, "Timber Shortage Chops Industry," *San Francisco Chronicle*, 13 July 1992.

39. Timothy Egan, "With Fate of the Forests at Stake, Power Saws and Arguments Echo," *New York Times*, 20 March 1989.

40. Timothy Egan, "Export Boom Dividing Pacific Timber Country," *New York Times*, 23 April 1988.

41. Egan, "Export Boom."

42. Pelline, "Timber Shortage."

43. Egan, "Export Boom."

44. Ted Gup, "Owl vs. Man," *Time*, 25 June 1990.

45. Daniel Yergin, *The Prize: The Epic Quest for Oil, Money and Power* (New York: Simon & Schuster, 1991), 747.

46. Yergin, *The Prize*, 750.

47. Yergin, *The Prize*, 755.

48. Yergin, *The Prize*, 756–57.

49. Orr, *Global Economy*, 261; Wald, "After 20 Years."

50. Orr, *Global Economy*, 302–3; Das, *Yen Appreciation*, 18.

51. James Sterngold, "Leaders Come and Go, but the Japanese Boom Seems to Last Forever," *New York Times*, 6 October 1991.

52. Amory B. Lovins and Joseph J. Romm, "Fueling a Competitive Economy," *Foreign Affairs* (Winter 1992–93), 49.

53. Stephen Engelberg, Jeff Gerth, and Tim Weiner, "Saudi Stability Hit by Heavy Spending over the Last Decade," *New York Times*, 22 August 1993.

54. Keith Bradsher, "Falling Yen Puts Car Makers in Japan in the Driver's Seat," *New York Times*, 15 July 1996.

55. Andrew Pollack, "Shellshocked by Yen, Companies in Japan Still Find Ways to Profit," *New York Times*, 18 April 1995.

56. Andrew Pollack, "Japan Inc.'s Dying Bit Players," *New York Times*, 27 May 1995.

57. Bradsher, "Falling Yen."

58. Bradsher, "Falling Yen."

59. Bradsher, "Falling Yen."

60. John Judis, "Dollar Foolish," *New Republic*, 9 December 1996, 23–24.

61. Judis, "Dollar Foolish," 23.

62. David E. Sanger, "The World Looks at Bali and Sees Krakatoa," *New York Times*, 18 January 1998.

63. Peter Passell, "A Blast from the Exchange-Rate Past," *New York Times*, 21 July 1994.

64. Passell, "A Blast."

65. Larry Rohter, "Ecuador's Use of Dollars Brings Dollar's Problems," *New York Times*, 5 February 2001.

66. Thomas L. Friedman, "Never Mind Yen. Greenbacks Are the New Gold Standard," *New York Times*, 3 July 1994.

67. Friedman, "Never Mind Yen."

68. Joseph Kahn, "U.S. and I.M.F. Welcome Salvador's Adoption of Dollar," *New York Times*, 25 November 2000.

69. Rohter, "Ecuador's Use of Dollars."

70. Kahn, "U.S. and I.M.F."

71. Steven Erlanger, "'Dollar Apartheid' Makes a Few Russians Rich but Resented," *New York Times*, 23 August 1992.

72. Rohter, "Ecuador's Use of Dollars."

73. Rohter, "Ecuador's Use of Dollars."

74. Steve C. Ropp, "Military Retrenchment and Decay in Panama," *Current History* (January 1990), 39.

75. Steve C. Ropp, "Panama: The United States Invasion and Its Aftermath," *Current History* (March 1991), 116.

3

⚛

Fighting Inflation

When President Nixon devalued the dollar to improve U.S. competitiveness, he also introduced wage and price controls to fight inflation. "The time has come for decisive action," he said in his August 15, 1971, speech, "action that will break the vicious circle of spiraling prices and costs."[1] His orders to "freeze . . . all prices and wages throughout the United States for a period of 90 days" opened the government's attack on inflation.

During the following decade, the government would wage two major campaigns to slow inflation. The first, in 1971, proved to be a failure. The second, which began in 1979, succeeded in curbing inflation. But the cost of victory was high. The policies used to fight inflation prompted a debt crisis in the third world and contributed to the collapse of the savings and loan (S&L) industry in the United States, and this led to rising homelessness in America.

In his speech, Nixon blamed inflation on the war in Vietnam. "One of the cruelest legacies of the artificial prosperity produced by the [Vietnam] war is inflation," he argued. "For example, in the four war years between 1965 and 1969, your wage increases were completely eaten up by price increases. Your paychecks were higher but you were not better off."[2]

Economists agree that government spending on the war in Vietnam contributed to rising inflation. But they argue that other factors also contributed to it. In the postwar period, most countries in Western Europe, North America, and Japan experienced modest inflation. This inflation was caused by government policies designed to keep unemployment low and to prevent the recurrence of prewar depression. During the 1930s, businesses had responded to recession by laying off workers to cut costs. But high levels of unemployment reduced the demand for goods. Without consumers to buy their goods, businesses could not increase production, rehire workers, and begin the steps to economic recovery.

After the war, governments in Western Europe, North America, and Japan developed programs designed to maintain demand and prevent widespread unemployment when normal business-cycle recessions occurred. They did this by pumping money into the economy, either through defense, social service, or public works programs. These policies, generally described as "Keynesian" after the British economist John Maynard Keynes who developed them, helped avert depression. But by pumping money into the economy, they also produced modest rates of inflation. When money was plentiful and demand high, businesses could raise prices. And because unemployment rates were low, and labor was relatively scarce, workers could demand and get higher wages. These developments, and the fact that the United States was pumping dollars into Western Europe and Japan to promote economic recovery (see chapter 1), produced modest rates of inflation in all three regions during the 1950s and 1960s.[3]

When the United States began waging the Vietnam War in earnest in 1965, U.S. military spending in the United States and overseas soared. But U.S. officials were unwilling to raise taxes to pay for the war. If they had, taxes would have taken away some of the money the government was putting into the hands of businesses and into the pockets of workers, which would have lowered their demand for goods. But because taxes stayed low, demand remained high. And when demand stayed high, businesses could raise prices and workers could ask for higher wages. As a result, inflation rose sharply.

If prices and wages rose in tandem, for everyone, inflation would not be regarded as a terrible social problem. But inflation is a discriminatory economic process, hurting some people more than others. Some businesses, for example, were better able to raise their prices than others, usually because what they produced was more of a necessity than other products. Oil, for example, was something that homeowners in wintry New England or drivers in suburban Los Angeles could not do without. It was more of a necessity than lawn chairs or vintage wine. By the same token, some workers were better able to demand and get higher wages, usually because they were organized in unions or performed services regarded as essential to others. So workers who belonged to the United Auto Workers Union or worked for the local fire department or collected garbage were better able to bargain for pay raises than restaurant waiters or office workers who were not unionized or employed by the government.

Other groups were also disadvantaged by inflation. People living on fixed incomes or pensions—some twenty million Americans in 1971, according to Nixon—found it difficult to increase their incomes to keep pace with inflation. And people who derived their income from savings accounts and government bonds found that inflation eroded the value of their assets because the interest they received was fixed at fairly low levels (often below the rate of inflation). So, for example, if the rate of inflation was 6 percent annually, a savings account offering 4 percent was losing value and a ten-year savings bond that provided a 6 percent return was not earning a dime.

Because inflation is discriminatory, affecting businesses, workers, pensioners, and investors in different ways, government officials regarded it as a social problem. Although people adversely affected by inflation despair of its consequences, even those who kept up with inflation complained about it. As economist Anthony Compagna notes, "Someone's income increased by $1,000 (which he or she regards as due to merit, conveniently forgetting that inflation boosts other people's income as well) and rising prices take away $500 of the $1,000, the person is still better off but feels cheated anyway [because] $1,000 at the old prices would have meant a [more] significant increase in living standards."[4]

For these reasons, Nixon introduced wage and price controls to curb inflation, then at about 4 percent annually. When inflation rises to 7 percent annually, as it did during the rest of the decade, consumer prices and monthly wages double in just ten years.[5] This is what occurred in the 1970s.

Nixon's wage and price controls, which remained in effect until April 1973, briefly slowed but did not curb inflation.[6] As journalist William Grieder observed, "The inflation rate subsided for a time, but still remained about 3 percent. By 1973, prices were escalating rapidly again and the consumer price index rose by a new postwar record, 8.89 percent. The following year, 1974, OPEC pushed up oil and the price level rose 12.2 percent."[7]

Several developments frustrated the Nixon administration's efforts to slow inflation. Many economists believe that the wage-and-price-control program was not effectively managed, allowing exemptions to some businesses and workers but not others. And when the controls ended, everyone scrambled to recover lost gains.[8] Soviet crop failures in 1973 and 1974 increased the demand for grain and sent food prices soaring (see chapter 13). At the same time, dramatically increased oil prices followed successive oil crises—the first in 1973 following the Yom Kippur War, and the second during the 1979 revolution that overthrew the shah of Iran and resulted in the capture of hostages at the U.S. Embassy in Tehran (see chapter 10). The simultaneous rise of food and oil prices pushed inflation to record heights. After each oil crisis, inflation in the United States hit double-digit figures: 12 percent in 1974, 13.3 percent in 1979, and 12.4 percent in 1980.[9] At these rates, prices doubled every five or six years.

The burst of inflation at the end of the 1970s prompted government officials to launch a second campaign against inflation. But instead of using wage and price controls administered by the federal government, officials used high interest rates and the Federal Reserve, a semipublic agency, to curb inflation.

1979: THE SECOND BATTLE AGAINST INFLATION

At the beginning of 1979, inflation was running at a rapid 11 percent annual rate. "In a year's time, a dollar would buy only 89 cents' worth of goods. A $6,000 car would soon cost $660 more. And every wage earner would need a pay raise of more than 10 percent simply to stay even," noted Greider.[10] By the summer of 1979, new OPEC price increases began to kick in, pushing the inflation rate to 14 percent. Rising inflation and lengthening lines at gas stations drove down President Jimmy Carter's popularity. By July, "barely a fourth of the voters approved of his performance as President."[11]

Faced with rising inflation and declining popularity, Carter took two steps. First, he made a stern speech criticizing American materialism:

> In a nation that was proud of hard work, strong families, close-knit communities and our faith in God, too many of us now tend to worship self-indulgence and consumption. Human identity is no longer defined by what one does, but by what one owns. But . . . owning things and consuming things does not satisfy our longing for meaning. We have learned that piling up material goods cannot fill the emptiness of our lives which have no confidence or purpose.[12]

Overnight, this speech boosted Carter's popularity by 10 percent, and "75 percent of voters agreed with the President's warning of spiritual crisis."[13] His increased popularity proved to be only temporary. After Iranian students seized hostages at the U.S. Embassy in Tehran on November 4, 1979, his popularity again declined.

In addition to his speech, Carter took another step. On July 25, he appointed Paul Volcker, who had helped shape the Nixon administration's 1971 dollar devaluation and introduce wage and price controls (see chapter 2), to head the Federal Reserve System.[14] Volcker's subsequent decision to raise interest rates to curb inflation would have a long and lasting impact on U.S. economic fortunes. Although his high interest rate policies succeeded in bringing down inflation, they created other problems for the United States and other countries around the world. One important problem was an economic recession during an election year. This and the hostage crisis in Iran led to Carter's electoral defeat by Ronald Reagan one year later.

THE FEDERAL RESERVE SYSTEM AND HIGH INTEREST RATES

The Federal Reserve System, established in 1913, acts as the central bank for the United States, controlling the supply of money and credit to private banks and financial institutions, supervising the industry, and managing the sale of U.S. bonds, which are used (along with taxes and fees) to raise money for the government so that it can pay its bills.[15] Its governors are appointed by the president, subject to Senate confirmation, to serve fourteen-year terms. As a result, the Federal Reserve System has considerable autonomy to shape economic policy.

In general, the "Fed" can use its control over money and credit to affect U.S. economic fortunes. If it increases the supply of government money and credit going to private banks, investors, and businesses, the stimulated economy usually grows. If the Fed decreases the money supply, making money and credit harder to get, then the price or interest rates that banks, investors, and businesses have to pay for money rises. The higher the price of money, and the higher the interest rate, the harder it is to borrow money, invest, or build new factories. As a result, the economy usually slows and unemployment increases.

After he was appointed to the Fed, Volcker adopted an anti-inflationary strategy. By tightening the supply of money and credit, he hoped to force up interest rates, slow economic growth, and curb inflation. Although he knew this would trigger an economic recession and increase unemployment, Volcker thought it necessary to act. "After years of inflation," he told an audience in the autumn of 1979, "the long run has caught up with us."[16]

So on October 6, 1979, Volcker announced that he would fight inflation by restricting the supply of money and credit and raising interest rates. "Appropriate restraint of the supply of money and credit is an essential part of any program to achieve the needed reduction in inflationary momentum and in inflationary expectations," he announced. "Such restraint . . . will help to restore a stable base for financial, foreign exchange and commodity prices."[17]

During the next six months, interest rates nearly doubled, rising from about 11 percent when Volcker became chairman to 20 percent in the summer of 1980.[18] But when the Fed eased off, inflation resumed, so Volcker pushed interest rates back up. And during the next two years, until the summer of 1982, interest rates rocketed up and down as the Fed used interest rates to wrestle with the tag team of inflation and recession.[19] In the end, the high interest rate policy pinned inflation, though recession remained standing. As Greider noted,

The Gross National Product contracted in real terms by more than $82 billion from its peak and, since 1979, the country had accumulated as much as $600

billion in lost economic output. The excess supply of goods, the declining incomes, the surplus labor—all had worked to force down wages and prices.
Price inflation fell dramatically: from above 13 percent [in 1979] to less than
4 percent [in 1983].[20]

Volcker's high interest rate policies, which triggered the deepest recession in the postwar period, had returned the U.S. economy to the kind of
modest inflation that had first triggered Nixon's wage and price controls
in 1971.[21] Recall that Nixon took action to curb inflation when it was running at about 4 percent. After 1982, inflation remained at this level, running about 4 percent during the 1980s and 1990s.[22]

In addition to a deep recession, the Fed's high interest rate policies also
affected the fortunes of different social groups in the United States. Although everyone complained about its effect, inflation had been good for
some groups—middle-income homeowners who had seen the value of
their homes rise sharply—but bad for others. Wealthy investors, for instance, had seen the value of their assets, particularly bonds, decline
sharply in the 1970s. As New York University economist Edward N. Wolf
reported, "Inflation acted like a progressive tax, leading to greater equality in the distribution of wealth."[23]

But high interest rates and falling inflation changed that. High interest
rates rewarded the wealthy, primarily because the top 10 percent of the
population "owned 72 percent of corporate and federal bonds . . . plus 86
percent of state and local bonds."[24] As interest rates rose to record highs,
and inflation fell to modest lows, their assets increased. "According to the
U.S. Census, only families in the top 20 percent of the economic ladder enjoyed real increases in their after-tax household incomes from 1980 to
1983. The others, the bottom 80 percent, actually lost."[25]

Volcker anticipated this development. When farm representatives
asked him to lower interest rates, Volcker responded, "Look, your constituents are unhappy, mine [banks and bond holders] aren't."[26]

Farmers were unhappy with high interest rates because many of
them had borrowed heavily (at government request) during the 1970s
to expand production and reap the high prices associated with repeated
crop failures in the Soviet Union (see chapter 13). But during the early
1980s, Soviet harvests recovered and world grain prices fell, just when
U.S. interest rates skyrocketed. The combination of falling prices and
rising interest rates drove 400,000 farmers out of business during the
early 1980s.

The ruin of small farmers across the Midwest had a huge impact on
other small businesses and rural communities because farmers are really
big consumers. The average farm family consumes more in a year than
a steel worker household consumes in a lifetime. A farm household

might borrow $100,000 each year to buy the fertilizer, seed, crop insurance, and equipment they need to plant and harvest their crops. A steel worker family might borrow this sum to buy a house only once every thirty years. So the bankruptcy of large numbers of small farmers had a devastating impact on other small businesses that provided goods and services to farmers. Moreover, the entry of men and women from struggling or bankrupt farms into local labor markets, as farmers sought "off-farm" income to keep their land, usually depressed wages for the jobs that remained. Small wonder, then, that farmers were angered by Volcker's high interest rate policies. Many felt betrayed because these policies were pursued by a Republican administration, which they had supported during the 1980 election (Midwest farm states traditionally vote Republican).

By squeezing the supply of money and credit and raising interest rates, the Federal Reserve triggered a deep recession and curbed inflation. But while high interest rates curbed inflation, they also contributed to an overseas debt crisis, rising U.S. budget deficits, and declining U.S. competitiveness.

DEBT, DEFICITS, AND DEVALUATION

When the Fed raised interest rates to record highs in the early 1980s, foreign and domestic investors rushed to buy U.S. bonds or "securities." They did so because they viewed them as safe—nothing is safer than U.S. government-backed securities—and profitable: a 15 to 20 percent annual return was higher than more risky investments in stock markets or real estate. High U.S. interest rates, which were substantially higher than what other governments offered in this period, acted like a magnet, attracting monies from around the world. The magnetism created by high U.S. interest rates had important consequences for different countries.

In Latin America, high U.S. interest rates resulted in increased debt and "capital flight." During the 1970s, businesses and governments borrowed money from the United States and from banks in Western Europe and spent it on economic development projects (see chapter 4). The interest rate they paid on borrowed money was tied to U.S. interest rates. So when the Fed pushed up U.S. interest rates, borrowers in Latin America saw their interest payments soar, which made it more difficult for them to repay their debts.

High U.S. interest rates also attracted Latin American investors, who spent their money on U.S. bonds rather than on development projects in their own countries. In 1978, before U.S. interest rates rose, Latin American investors sent about $7 billion overseas. But in 1980, Latin Americans invested nearly $25 billion overseas, most of it in the United States.[27] "Capital flight," as it is called by economists, was a problem for Latin American countries because it reduced domestic investment, which resulted in unemployment, and deprived governments of the currency they needed to run their countries and repay debts. In August 1982, the Mexican government ran out of money to manage its affairs or repay its $80 billion debt to foreign countries, and the Federal Reserve had to take emergency measures to prevent it from defaulting on its loans. If Mexico had declared bankruptcy, major U.S. banks would also have been forced into bankruptcy and a global financial crisis would have ensued.[28] In subsequent years, the Fed and the U.S. government had to address a series of financial crises in Latin American countries, Eastern Europe, and Africa, known collectively as the debt crisis, which was partly a product of the Fed's high interest rate policies.

High U.S. interest rates also acted like a magnet for other first world investors, drawing huge sums of money from Western Europe and Japan. Although these countries did not have foreign debts, like Latin American countries, U.S. economists thought that the flight or migration of capital from Western Europe and Japan would deprive them of money to invest in public works or new factories in their countries, causing increased unemployment and reducing their ability to compete with the United States. But despite massive purchases of U.S. securities, these problems did not materialize in Western Europe and Japan because the U.S. government gave back to them through military spending what the Federal Reserve took away from them in capital flight.

When President Reagan took office, he promised to increase military spending and cut taxes. As foreign capital flooded into U.S. securities markets as a result of high U.S. interest rates, the administration found that it could deliver on both its promises. With high interest rates, the Reagan administration could sell bonds and raise the money it needed to increase military spending. By using the sale of U.S. bonds to borrow money from foreigners, the government increased military spending 50 percent, from $201 billion a year in 1980 to $311 billion in 1987.[29] Moreover, it could do this without raising taxes to pay for it. In fact, the Reagan administration cut taxes dramatically during much of this period.

Put another way, in 1985, the U.S. government spent about $79 billion more on defense than it had in 1980. And it received $71.4 billion from foreign investors. Thus, the increases in military spending were almost entirely paid for by foreigners, which meant that the government did not have to use domestic taxes to raise this money.

This policy—increased military spending and lower taxes—had several important consequences in Western Europe, Japan, and the United States.

The Military Rebate

Western Europe and Japan have been U.S. military allies since the end of World War II. To protect them from invasion by communist countries, the U.S. government had stationed troops and spent money on defense in these countries throughout the postwar period. Economists estimate that between 60 and 70 percent of all U.S. military spending was devoted to the North Atlantic Treaty Organization (NATO), which defended Western Europe.[30] The United States spent a smaller, though still large, amount defending U.S. allies in East Asia, Japan among them. As U.S. military spending increased under the Reagan administration, its spending in Western Europe and Japan also increased. By purchasing equipment and supplies from its allies, by paying the salaries of about 351,000 U.S. soldiers in Europe, and by providing military aid to its allies, the U.S. government injected huge sums of money into the economies of its allies.[31]

In 1985, for example, foreigners (mostly from Western Europe and Japan but also from Latin America) purchased $71.4 billion in U.S. securities.[32] That year, the United States spent $278.9 billion on the military.[33] If the United States spent 60 percent of its military budget for the defense of its first world allies (a low figure since some estimates of U.S. spending on NATO are higher and this figure does not include U.S. spending on Japan), then about $167.34 billion was spent on U.S. allies. This means that the United States took in less capital from its allies than it gave back in military spending. Total U.S. giving to U.S. allies in that year amounted to $95.94 billion, a kind of massive military rebate. So while the Federal Reserve's high interest rate policy pulled money out of European and Japanese economies, the Reagan administration's defense spending policies put much of it back. Moreover, the high U.S. interest rates, which U.S. economists expected to hurt other U.S. competitors, did not result in recession or high unemployment in either Europe or Japan.

High U.S. interest rates did not greatly reduce the availability of capital in Western Europe and Japan for another reason. Workers in these countries saved more of their money than Americans. They were more thrifty because high tariffs often made imported goods expensive, because their governments and banks did not make consumer credit as easily available as they did in the United States, and because they were more reluctant to go into debt than Americans. Because they put a higher percentage of their income in their savings accounts, their banks had more money available to invest. As a result, Japan substantially increased its domestic investments—building new roads and factories and creating more jobs—while also increasing its purchases of U.S. government securities in this period (see chapter 2).[34]

As a result of U.S. defense spending and their own thriftiness, Western Europe and Japan were able to benefit three times from high U.S. interest rates. First, they profited from interest rates that were higher than they could obtain at home. Second, they benefited from increased U.S. military spending in their countries. And third, the flood of foreign currency into the United States increased the value of the dollar, making it easier for them to sell their wares in the United States. During the early 1980s, the stronger dollar made it more difficult for U.S. firms to sell their goods abroad, while making it easier for Western European and Japanese businesses to sell their products in the United States. Propped up by high U.S. interest rates, the stronger dollar undermined U.S. competitiveness and led, in 1985, to a second devaluation of the dollar through the Plaza Accords.

These developments were the product of two sets of policies. It was the combination of the Fed's monetary policy, which used high interest rates to fight inflation, and the Reagan administration's fiscal policy, which borrowed money from abroad to increase military spending while cutting taxes, that contributed to these different global developments: the debt crisis in Latin America, Africa, and Eastern Europe (see chapter 4) and the redistribution of production from the United States to Western Europe and Japan (see chapter 1).

For the U.S. economy, the combination of the Fed's high interest rate policy and the Reagan administration's policy of increased military spending but lower taxes had important consequences. By increasing its spending and cutting taxes, the Reagan administration created large and growing budget deficits that contributed to a rapidly growing national debt. Because the government borrowed money to cover annual budget deficits at high rates of interest, interest payments grew, which also contributed to the size of total debt.

High interest rates also contributed to the collapse of the domestic S&L industry. Widespread bankruptcies in this industry reduced investment in the housing industry. This led to a housing shortage and to rising home prices and rents. And this led, by decade's end, to rising homelessness in America.

HOUSING AND HOMELESSNESS

During the thirty years before 1979, the housing industry built millions of inexpensive homes and apartments, making it possible for two-thirds of all Americans to purchase and own their homes. But in the ten years after 1979, the S&Ls that provided money to the construction industry and to home buyers collapsed. Home building slowed, prices and rents rose, and homelessness increased. The Federal Reserve's 1979 decision to raise interest rates marked a turning point and played an important role in reversing the housing industry's fortunes.

High U.S. interest rates not only attracted money from Latin America and investors in Western Europe and Japan, they also drew money out of domestic savings accounts. The flight of capital from the passbook savings accounts of domestic S&Ls created problems that led to the collapse of the industry, which had long been a mainstay of the housing industry. Although capital flight from the domestic S&L industry had been a minor problem since 1965, it became a major problem after 1979.

During the postwar period, S&Ls provided much of the money used by private construction companies and independent contractors to build homes and apartments. S&Ls differed from commercial banks in several important respects. Unlike commercial banks, they did not offer checking accounts or provide services to merchants or loans to businesses. Instead they offered passbook savings accounts to local depositors and attracted customers by paying interest rates that were slightly higher than those offered by banks. (The federal government set these rates and made sure they were higher than those offered by commercial banks.) The S&Ls then took the money deposited in savings accounts and lent the money to contractors and home buyers at a slightly higher rate so they could build and buy homes and so S&Ls could profit from the loans. The income they received from construction loans and mortgage payments enabled the 5,500 S&Ls in the United States to pay their depositors interest on their savings accounts and make a small profit, which they used to pay salaries, rent, and dividends to shareholders.

Between 1950 and 1970, "31 million housing units were built, including 20 million single-family homes."[35] The large supply of inexpensive housing and the availability of cheap, low-interest, thirty-year home loans made it possible for most Americans to purchase homes. Although only 43.6 percent of Americans owned homes in 1940, 64.4 percent owned homes in 1980, a 50 percent increase.[36] By collecting the savings of small depositors and lending it out to builders and buyers, the S&Ls played an important role in postwar prosperity.

The industry's first real problems began in the mid-1960s. To fight the war in Vietnam, President Lyndon Johnson needed to increase military spending. But he was reluctant to increase taxes to pay for the war because he worried that tax increases would make the war more unpopular. To raise the money, Johnson persuaded the Fed to raise interest rates on government securities, much as Volcker did fifteen years later. Interest rose to a rate that was slightly higher than the rate S&Ls offered depositors on passbook savings accounts. In 1966, for example, the government's three-month Treasury bills (T-bills) paid 5.28 percent interest, while S&Ls provided only 4.75 percent interest on savings accounts (and commercial banks offered only 4 percent).[37]

As a result, some depositors began withdrawing their money from S&L savings accounts and investing it in government securities that offered a higher rate of return. When investors withdrew money from a financial institution, they reduced its assets and weakened its ability to make loans. The technical term for this process is "disintermediation," but it might also be called a "slow run on the bank" or "capital flight," which undermines the ability of financial institutions to operate as intermediaries between investors and borrowers.

Initially, investors drawn by higher U.S. interest rates withdrew only modest amounts of money from S&Ls, only $2.5 billion in 1966, a small sum compared to the more than $500 billion held by S&Ls. But disintermediation continued, growing to $4 billion in 1969. The federal government responded to this slow flight of capital by raising the interest rates S&Ls could offer on savings accounts to 5 percent in 1970 and 5.25 percent in 1973. They also made it more difficult to purchase T-bills by setting a $10,000 minimum on purchases, which was more than most small savers could afford.[38] But because the government's interest rates also increased, remaining one or two percentage points higher than S&L rates for much of the 1970s, the flow of money out of S&Ls continued at a moderate pace.

But in 1979, the Federal Reserve raised interest rates, and the return on three-month T-bills reached 12.07 percent in 1979, 15.66 percent in 1980, and 16.30 percent in 1981.[39] As a result, money flooded out of S&Ls offering depositors only one-half or one-third as much. In 1981, investors withdrew $21.5 billion from S&Ls, five times as much as they had in 1969. To stop this massive capital flight and to prevent the wholesale disintermediation of the S&L industry, government officials took two steps that would have fateful consequences.

As its first step, Congress in 1980 passed and President Carter signed the Depository Institutions Deregulation and Monetary Control Act. This bill allowed S&Ls to increase their interest rates on savings accounts (the federal government had previously limited interest rates) so they could win back runaway investors, and it increased the government's insurance on investors' deposits from $40,000 to $100,000.

Although higher interest rates prompted some investors to redeposit their money in S&L accounts, they created another problem. Recall that payments on home mortgages provided the income for S&Ls. If they took in money from borrowers at 8 percent and paid depositors 5 percent on their savings accounts, the S&Ls earned 3 percent. But when they raised interest rates on savings accounts, to say 10 percent, this increased their expenditures. But they could not easily raise their income from mortgage payments because they had made home loans at fixed rates for long periods of time. In the 1980s, S&Ls' income came from people who had borrowed money at 8 percent in the 1960s. Because they could not raise the mortgage payments of long-term borrowers to increase their income, the S&Ls began paying depositors (10 percent) more than they earned from borrowers (8 percent). As a result, they began to lose money, about $4.6 billion in 1981. And bankruptcies began to mount: 17 S&Ls failed in 1980, 65 in 1981, and 201 in 1982. The assets of insolvent S&Ls grew from one-tenth of a billion dollars in 1980 to $49 billion in 1982, a 500 percent increase.[40]

Although disintermediation had been slowed, government policy had contributed to increasing bankruptcy. To address this problem, the Garn-St. Germain Depository Institutions Act was passed in 1982. This bill allowed S&Ls to offer checking accounts, issue credit cards, loan money to consumers for autos and personal purchases, make commercial loans to businesses, and invest in stocks and bonds. By allowing S&Ls to offer these services and become more like commercial banks, government officials expected S&Ls to increase their income. For example, the interest rates on credit cards or business or auto loans are much higher than interest rates on home loans. So if the S&Ls could make 15 percent from their credit card customers, government officials thought, they could pay depositors 10 percent and still make money. For a time it worked. S&Ls loaned money in new ways, at higher rates of interest, and paid depositors higher rates on their savings accounts. For a brief time they attracted investors and turned a profit. But two problems soon emerged.

First, increased commercial lending led to the widespread construction of office buildings, golf courses, and resort developments, particularly in the Southwest, where high oil prices in the early 1980s encouraged the expansion of the domestic oil industry and created a booming market for commercial real estate in cities like Dallas. But the massive construction of office towers and shopping malls created a glut of commercial properties, and the fall of oil prices after 1985 led to the collapse of the domestic oil industry, which crippled the real estate market (see chapter 2). The value of commercial real estate in the Southwest fell by nearly one-half between 1984 and 1989.[41] As the value of real estate fell, builders and developers found it difficult to repay their loans and many went bankrupt. When the S&Ls could not recover their loans from bankrupt borrowers, they too went bankrupt. As bankruptcy threatened, depositors began withdrawing their money, which led to renewed disintermediation. Profits plummeted, and the S&L industry began to collapse wholesale.

It did not help matters that the Reagan administration cut the budget for bank examiners during this period, which actually reduced the number of field agents and cut the number of federal examinations by 50 percent between 1980 and 1984.[42] Nor did it help that regulators ignored a

1983 government report warning that "the deregulation of the past few years . . . has substantially reduced the ability of regulatory agencies to constrain the risk-taking of insured institutions. . . . In light of the competitive pressures the industry will face in the next few years, this deregulation could result in substantial losses."[43]

When S&Ls went bankrupt, the Federal Reserve and U.S. government agencies seized control and paid off depositors, who were insured up to $100,000 (as a result of changes in the 1980 law). Investors who owned shares of bank stock were not covered by government insurance, and many lost their investments. The cost of repaying depositors in failed S&Ls—the government's bailout, as it was called—was high. In 1990, Treasury Secretary Nicholas Brady testified to Congress that one thousand S&Ls, or 40 percent of the industry, would have to be seized and depositors repaid. He estimated that this would cost the government between $89 billion and $130 billion. Taxpayers would eventually cover this cost, amounting to $1,300 for each American household.[44] Other cost estimates were higher. Some economists calculated that the bailout cost taxpayers between $159 billion and $203 billion. If one included the interest payments on this debt, the cost climbed to between $325 billion and $500 billion, or about $5,000 for each American household.[45]

The government tried to recover some of these costs by selling off the assets of seized S&Ls. By 1992 it had sold off assets worth $144 billion. But the sale of 2,300 square miles of real estate, an area twice the size of Rhode Island, was difficult in a sluggish market.[46]

Second, the S&Ls' changed lending practices reduced the amount of money going to the housing industry. In the early 1970s, S&Ls loaned 60 percent of their assets to home builders and buyers. But as they shifted their emphasis to consumer and commercial loans, at higher rates of interest, they made only 40 percent of their money available to home buyers in 1984 and only 30 percent in 1988.[47] As money for the housing industry dried up, and the money that was available cost more (because of higher interest rates), fewer homes were built and fewer people could borrow money to purchase homes.

In 1972, the housing industry built 2.4 million new homes. But in 1984, when the population was larger and the demand for housing had grown, the industry built only 1.7 million homes.[48] And during the 1990s, the industry built 1.1 million a year on average, only one-half as many as were built each year in the 1970s.[49] This was a critical development because the population was bigger and the housing supply smaller. This pushed up housing prices and rents across the country.

Inflation had pushed up the cost of housing during the 1970s. In the 1980s, inflation abated, but a shrinking supply of houses and a growing demand for houses continued to push prices up. As the price of housing and the cost of money to purchase a home rose in the early 1980s, fewer people could afford to buy a home and the percentage of homeowners began to decline for the first time since 1940.

Many people who might have bought a home in previous decades kept on renting apartments. This development, and the decline in the construction of apartment units, increased the demand for rental units. The Joint Center for Housing Studies at Harvard University reported in 1989 that the number of poor renters had grown, but rental housing stock had declined, and this helped drive up rents. As a result, rents began to rise sharply after 1980, increasing from about $350 a month in the Northeast in 1980 to $420 a month in 1986, and from $380 a month in the West in 1980 to $480 a month by 1986.[50]

While rents rose, federal housing assistance to the poor declined. The Reagan administration cut housing assistance from $27 billion in 1980 to less than $8 billion in 1987, and the number of federally subsidized housing units declined from 200,000 to 15,000.[51]

As a result, the demand for rental housing outstripped the supply and rents rose. "In 1978 there were 370,000 more low-cost units (renting for $250 a month) than there were low-income renter households, but by 1985, there were 3.7 million fewer low-cost units than there were low-income renter households."[52] Government cuts in housing assistance and stagnant wages made it difficult for poor people to pay higher rents or compete for the available housing. In 1997, 5.4 million families paid more than one-half of their income for housing, a dangerously high level. "To make matters worse, the number of affordable housing units is shrinking just when it needs to expand," a federal government study warned.[53]

Poor people who could not afford rising rents were forced out of the housing market. Before people hit the streets, they try to crowd into apartments with others or go to live with relatives if they can (this is a process familiar to college students, who cram into apartments or share houses to save on rent). But when these alternatives are exhausted, poor people, many of them working in minimum wage jobs, become homeless.

By 1990, there were between six hundred thousand and three million homeless people in the United States.[54] Of course, they were not all forced to wander the streets of American cities by high interest rates and the collapse of the S&L industry. A small homeless population had long existed in the United States, and its number increased as a result of personal choice or misfortune, economic recession, or government policies, such as the de-institutionalization of mentally disabled patients from state hospitals in the 1970s. But the growing percentage of homeless families, about 40 percent of the homeless population in 1993, indicated that economic developments during the 1980s played an important role in increased homelessness during the 1980s and 1990s.[55]

Gender and the Housing Crisis

High interest rates triggered the collapse of the S&L industry and crippled the residential construction industry, which had long relied on S&Ls to finance new construction. Construction is a very gendered industry. Of the 4.4 million workers employed in construction, the most labor-intensive industry in America, the overwhelming majority (more than 90 percent) are men.[56] Moreover, many of the workers in related industries—timber, furniture, landscape, hardware, plumbing, electrical, air conditioning, heating, painting, paving, flooring, and roofing—are also predominantly male. So the decline of the residential construction industry since 1971 has hammered occupations long identified with male workers. Although the S&L industry provided service jobs (tellers) for women, the number of women employed by S&Ls was small by comparison.

The rising price of homes and rents had rather different gender consequences. During the 1980s, the homeless population was predominantly male. This was due in part to the fact that the "old" homeless population, which consisted of derelict, alcoholic, and transient streetpeople, was overwhelmingly male. During the 1980s, many of the "new" homeless were poor men who had lost their jobs in manufacturing and also construction industries. So in the 1980s, homelessness was closely associated with men. But this began to change in the 1990s, as the number of homeless women and children increased. By 2000, the number of women and children in the homeless population had grown considerably, to perhaps 40 percent of the total. In New York City, for example, of the 25,000 people who sought city-provided shelter each night in 2000, 18,000 of them were children and a parent, most of them women.[57] Women and parents with children were more likely to seek, and obtain, shelter than single men. But the homeless population today is not identified only with men, as it once was, but with men, women, and children.

Although the Federal Reserve's high interest rate policy successfully curbed inflation, the victory was costly. The collapse of the S&L industry and the rise of homelessness have become major and continuing problems in the United States. Overseas, rising interest rates triggered a massive debt crisis, which is still a serious problem. It is to this development that we now turn.

NOTES

1. *New York Times*, "Transcript of President's Address on Moves to Deal with Economic Problems," 16 August 1971.
2. *New York Times*, "Transcript of President's Address."

3. Michael R. Smith, *Power, Norms and Inflation: A Skeptical Treatment* (New York: Aldine de Gruyter, 1992).

4. Anthony S. Campagna, *The Economic Consequences of the Vietnam War* (New York: Praeger, 1991), 122.

5. Berch Berberoglu, *The Legacy of Empire: Economic Decline and Class Polarization in the United States* (New York: Praeger, 1992), 61.

6. Campagna, *The Economic Consequences of the Vietnam War*, 89.

7. William Greider, *Secrets of the Temple: How the Federal Reserve Runs the Country* (New York: Touchstone, 1987), 91.

8. Campagna, *The Economic Consequences of the Vietnam War*, 114.

9. Paul A. Volcker and Toyoo Gyohten, *Changing Fortunes: The World's Money and the Threat to American Leadership* (New York: Times Books, 1992), 115; Berberoglu, *The Legacy of Empire*, 61.

10. Greider, *Secrets of the Temple*, 14.

11. Greider, *Secrets of the Temple*, 14.

12. Greider, *Secrets of the Temple*, 14.

13. Greider, *Secrets of the Temple*, 15.

14. Greider, *Secrets of the Temple*, 46–47.

15. *The World Almanac and Book of Facts 1990* (New York: Pharos Books, 1990), 83; Greider, *Secrets of the Temple*, 32–33.

16. Greider, *Secrets of the Temple*, 104.

17. *New York Times*, "Test of Fed's Announcement on Measures to Curb Inflation," 8 October 1979; Steven Rattner, "Anti-Inflation Plan by Federal Reserve Increases Key Rate," *New York Times*, 7 October 1979.

18. Greider, *Secrets of the Temple*, 148–49.

19. Greider, *Secrets of the Temple*, 219.

20. Greider, *Secrets of the Temple*, 507.

21. Bill Orr, *The Global Economy in the '90s: A User's Guide* (New York: New York University Press, 1992), 257.

22. Orr, *The Global Economy*, 258.

23. Greider, *Secrets of the Temple*, 44.

24. Greider, *Secrets of the Temple*, 372.

25. Greider, *Secrets of the Temple*, 577.

26. Greider, *Secrets of the Temple*, 676.

27. Manuel Pastor Jr., *Capital Flight and the Latin American Debt Crisis* (Washington, D.C.: Economic Policy Institute, 1989), 9.

28. Greider, *Secrets of the Temple*, 517.

29. Orr, *The Global Economy*, 287.

30. Ruth Sivard, *World Military and Social Expenditures, 1987–88* (Washington, D.C.: World Priorities, 1987), 37.

31. *New York Times*, "U.S. Official Affirms a 40% Cut in Troops Based in Europe by '96," 29 March 1992.

32. Norman J. Glickman and Douglas P. Woodward, *The New Competitors: How Foreign Investors Are Changing the U.S. Economy* (New York: Basic, 1989), 116.

33. Orr, *The Global Economy*, 287.

34. David E. Sanger, "Japan Keeps Up the Big Spending to Maintain Its Industrial Might," *New York Times*, 11 April 1990.

35. David Kotz, "S&L Hell: Loan Wolves Howl All the Way to the Bank," *In These Times*, 8 August 1989.

36. Kotz, "S&L Hell."

37. Lawrence J. White, *The S&L Debacle: Public Policy Lessons for Bank and Thrift Regulation* (New York: Oxford University Press, 1991), 63.

38. White, *The S&L Debacle*, 62, 64.

39. White, *The S&L Debacle*, 68.

40. Kotz, "S&L Hell."

41. White, *The S&L Debacle*, 111.

42. White, *The S&L Debacle*, 88–89.

43. White, *The S&L Debacle*, 92.

44. David E. Rosenbaum, "How Capital Ignored Alarms on Savings," *New York Times*, 6 June 1990.

45. Rosenbaum, "How Capital Ignored Alarms."

46. Leslie Wayne, "The Great American Land Sale," *New York Times*, 30 November 1992.

47. Kotz, "S&L Hell."

48. Greider, *Secrets of the Temple*, 654.

49. Richard W. Stevenson, "The Gospel according to Greenspan: Rising Home Prices," *New York Times*, 3 November 1999.

50. Ann Mariano, "Fewer Can Buy Homes, Study Finds: Poor Seen Trapped in Rent Cost Squeeze," *Washington Post*, 24 June 1989.

51. Kevin Phillips, *The Politics of Rich and Poor: Wealth and the American Electorate in the Reagan Aftermath* (New York: Random House, 1990), Appendix 1.1; Richard Sweeney, *Out of Place: Homelessness in America* (New York: HarperCollins, 1993), 89.

52. E. J. Dionne Jr., "Poor Paying More for Their Shelter," *New York Times*, 18 April 1989.

53. Michael Janofsky, "Home Prices Are out of Reach for Many," *New York Times*, 12 June 2000.

54. Michael Levitas, "Homelessness in America," *New York Times Magazine*, 10 June 1990.

55. William Clairborne, "Big Increase in Homeless Families," *San Francisco Chronicle*, 22 December 1993.

56. Lawrence Mishel and David M. Frankel, *The State of Working America, 1990–91* (Armonk, N.Y.: Sharpe, 1991), 104.

57. Nina Bernstein, "Shelter Population Reaches Highest Level since 1980s," *New York Times*, 8 February 2001.

4

⚛

Debt Crisis and Globalization

On March 27, 1981, Polish government officials in London told representatives of five hundred Western banks that Poland could not repay the $27 billion it had borrowed from them. In July, the Romanian government followed suit, suspending payments on its more modest $7 billion debt to Western banks.[1] The financial problems created by these defaults were dwarfed a year later, in August 1982, when Mexico's finance secretary, Jesús Silva Herzog, announced that Mexico could no longer make payments on its $90 billion foreign debt. During the next year, more than forty other countries, most of them in Latin America, ran out of money and announced they could no longer repay the interest or principal on huge debts owed to private banks and government lending agencies in first world countries. Collectively, countries in Latin America, Africa, and Eastern Europe owed $810 billion in 1983, a twelvefold increase from the $64 billion they owed in 1970.[2] "Never in history have so many nations owed so much money with so little promise of repayment," *Time* magazine observed.[3]

The sudden inability of so many countries to repay their debts created a "debt crisis" that threatened first and third world countries alike. If countries like Poland, Mexico, and Brazil could not repay loans made by banks in Western Europe and North America, then major banks could fail, creating widespread bankruptcy, financial chaos, and possibly, global economic depression. Moreover, if borrowing countries in Latin America, Africa, and Eastern Europe defaulted on their loans and declared bankruptcy, they could no longer obtain the money they needed to pay for essential food and oil imports, or develop the industry they needed to provide jobs for growing populations.

Although the debt crisis, which became acute in the early 1980s, threatened rich and poor countries alike, measures taken to address the crisis had different consequences for northern creditors and southern debtors. The threat of bankruptcy for lenders in the North has receded. In 1994, the *New York Times* even announced that the debt crisis was officially over. But while the crisis may have ended for lenders, it continues for debtors, who found themselves even deeper in debt in 2000, despite having made every effort to repay debts accumulated in the 1970s.

But how did a collective crisis produce such different outcomes? As we will see, debts in Latin America, Africa, and Eastern Europe increased rapidly during the 1970s, both because rich countries wanted to lend and because poorer countries wanted to borrow large sums of money. Although the transfer of money from northern lenders to southern borrowers was beneficial to both in the 1970s, it proved troublesome in the early 1980s as a result of two developments. First, rising interest rates, which were designed to fight inflation in the United States, increased the amount that borrowers were expected to pay northern lenders. Second, falling commodity prices for the goods southern countries exported to the North decreased the incomes of countries in the South, making it more difficult for them to repay northern lenders. Increasing costs and falling incomes made it difficult for borrowers to repay their debts, and a debt crisis ensued. To solve this crisis, northern creditors and the global monetary institutions that represented them (the International Monetary Fund [IMF] and the World Bank) demanded that southern borrowers adopt strenuous economic measures to repay their debts. These global institutions took the opportunity presented by the debt crisis to remake the economies of debtor countries along neoliberal market lines. So the debt crisis indirectly became a force for globalization. Although the creditors were able to avert a financial crisis, the debt crisis caused enormous economic hardship for borrower countries and left them, for decades, even deeper in debt. Debt doubled from $639 billion in 1980 to $1,341 billion in 1990.[4] In Latin America, the region with the largest share of debt, "total indebtedness . . . now equals about $1,000 for every man, woman and child" on the continent—this in a region where $1,000 is more than most families earn in a year.[5]

Although countries around the world experienced a debt crisis, Latin America will be the focus of the discussion here because the countries with the largest outstanding foreign debts (Mexico and Brazil, with $90 billion each) are in Latin America and because the continent owes more than half of the total outstanding debt.[6] By contrast, for example, Eastern European countries collectively owed $92.8 billion in 1981, equal to Mexico's debt,[7] and African countries together owed $82 billion in 1985, less than either Mexico or Brazil.[8]

GETTING INTO DEBT

In the 1970s, the amount of money loaned to countries in Latin America, Africa, and Eastern Europe increased dramatically. Between 1970 and 1973, banks in Western Europe and the United States lent $23.4 billion to Latin America, more money than had been loaned in the previous thirty years.[9] During the next decade, Latin America multiplied its debts more than twelve times. Debt expanded rapidly in the 1970s because lenders had large supplies of money that they were eager to lend and because countries around the world had great demand for borrowed money. "Indebtedness is a two-sided relationship," New York investment banker Richard Weinert observed. "It depends not only on a willing borrower, but equally on a willing lender. Indebtedness results as much from the need of lenders to lend as from the need of borrowers to borrow."[10] But conditions that in the 1970s encouraged rich countries to lend, and poorer countries to borrow, changed dramatically in the 1980s.

THE LENDERS

After World War II, government agencies and institutions like the IMF and the World Bank made the majority of the loans to poor countries. They did not lend large amounts (about $20 billion to Latin America between 1950 and 1970), they attached strict conditions to the loans, and they loaned money primarily to promote financial stability or to finance large-scale development projects like dams and ports. During the 1970s, private banks in Western Europe and North America began lending increasing amounts of money, increasing their share of total lending from about one-third to more than one-half of all loans by the end of the decade.[11] Private banks lent large sums of money to Latin American countries because they saw it as a way to invest profitably the growing pool of money available to them in "Eurodollar" or European currency markets.

During the 1970s, governments and private investors from around the world deposited U.S. dollars and other hard currencies they had earned in trade with the United States in Western European banks and in U.S. banks with subsidiaries in Europe. Some of the first dollar deposits were made by the Soviet Union. They were joined by investors in Latin America, Japan, and other countries around the world who deposited dollars in these accounts because they regarded them as safe and because they were not subject to the same kind of government regulations that applied to currencies deposited in the accounts of domestic banks.[12]

The money available in this Eurodollar banking pool grew from about $10 billion in 1960 to $110 billion in 1970.[13] Then, in the 1970s, money from another source began to deepen and expand this monetary pool. After the 1973 OPEC oil embargo sent oil prices soaring, OPEC countries received huge amounts of dollars from industrialized countries in payment for their oil, as much as $100 billion a year. "Since $100 billion a year is hard to spend," one writer observed, "even on Cadillacs, private 747s, and sophisticated missiles," the OPEC countries deposited much of their money in Western European and U.S. banks, and this money found its way into the Eurodollar market.[14] OPEC countries did this because they wanted to earn interest on their newfound wealth and because they regarded Western European banks as safe havens for their money. With the influx of dollars from oil-producing countries, often called petrodollars because they were dollars used to pay for OPEC oil, the pool of money in the Eurodollar market grew to $1,525 billion by the 1980s.[15] (Precise estimates vary enormously because government regulatory agencies have a difficult time monitoring or tracking this money. Still, the rate of increase during the 1970s is the same regardless of the figures used.[16])

As the money available to Western banks grew, bank officials searched for profitable ways to invest or loan it. Large U.S. banks became particularly active in Latin America, where banks had numerous subsidiaries and a fairly long history of involvement in local economies. "The nine largest U.S. banks, whose total capital is $27 billion, have lent over $30 billion (or more than their net worth) to private and government borrowers in just three countries: Mexico, Brazil and Argentina," the *Wall Street Journal* wrote in 1984.[17] The banks loaned money from Eurodollar pools, from U.S. depositors in their branch banks, and from smaller banks that joined loan syndicates.

Public and private lenders in the North lent money to countries in the South for a variety of reasons. Banks made loans so that poor countries could purchase goods made in Western Europe and North America. In the 1970s, for example, "42 percent of [Britain's] construction equipment, 33 percent of new aircraft and 32 percent of British textile machinery went to third world markets. In the United States, by 1980, the third world market accounted for . . . 20 percent of U.S. industrial product and about one-quarter of gross farm income."[18]

The U.S. government's Export-Import Bank, for example, loaned money to Latin American governments so they could purchase U.S. airplanes. As Boeing Aircraft president Malcolm Stamper explained, "The Ex-Im Bank . . . was created to help promote exports . . . to help foreign firms and their nations to buy big-ticket goods that would be of social and economic benefit. Airplanes certainly meet this description. . . . Airplane exports are also very good business for this country's own economy, by the way."[19]

Private lenders also discovered that they could make more money loaning money to foreign borrowers than to domestic borrowers.

> While the ten largest U.S. banks had a phenomenal expansion of international earnings [from Latin American loans] in 1970 to 1976, profitability in the domestic market ran generally flat. By the mid-1970s, most of the large banks had 50 percent or more of their earnings from abroad. In the case of Citicorp . . . by 1970 over 80 percent of their earnings came from their international operations.[20]

U.S. bankers in the 1970s did not worry greatly about the risks associated with foreign loans for several reasons. First, most of their money was loaned to Latin American dictatorships, which maintained close and friendly ties to the United States and seemed unlikely to renege on their debts (see chapter 6).[21] Second, they observed that the prices of many southern commodities, particularly oil, were rising in the 1970s, which helped their economies grow. This suggested that as their incomes grew, borrower countries would be able to repay old debts and shoulder new ones without difficulty. And third, because governments had the authority to raise money by taxing their citizens, they could still repay loans should economic problems develop. Explaining why his bank was bullish on foreign loans, Citicorp chairman Walter Wriston told the *New York Times* in 1982, "A country does not go bankrupt."[22]

Not everyone was so optimistic. *Euromoney* observed in 1975 that

> a purely technical analysis of the current financial position [of many borrowing countries] would suggest that defaults are inevitable; yet many experts feel this is not likely to happen [because] the World Bank, the IMF and the governments of major industrialized nations . . . would step in rather than watch any default seriously disrupt the entire Euromarket apparatus.[23]

Despite their enthusiasm for foreign loans, northern banks worried about the risks associated with mounting debt. So they hedged their bets, insisting in the late 1970s that borrowers agree to readjust interest rates on new and old loans every six months and bring interest rates into line with current market rates.[24] And by 1983, nearly 70 percent of all loans in Latin America were subject to floating interest rates, which would rise or fall depending on the interest rates set in the United States.[25] Although interest rates were then stable, which meant that borrowers did not worry greatly about accepting this new condition, the bankers' insistence that floating interest rates be adopted by borrowers would have important consequences for both lenders and borrowers in the early 1980s.

THE BORROWERS

Not only were bankers willing to lend, governments and corporations in Latin America were eager to borrow money in the 1970s. Public and private borrowers had substantial and diverse needs for northern loans. Much of the money they borrowed was simply used to repay lenders. "Between 1976 and 1981," Sue Branford and Bernardo Kucinski wrote,

> Latin America borrowed an enormous $272.9 billion. But over 60 percent of this, $170.5 billion, was immediately paid back to the banks as debt repayments or interest. Another $22.9 billion remained with [northern] banks as reserves [against potential losses], which were a kind of additional guarantee for the debt itself. And an estimated $56.6 billion was quickly sent abroad as capital flight. Only $22.9 billion effectively entered the continent to be used (or not) in productive investment.[26]

Of the $88 billion Mexicans borrowed between 1977 and 1979, only $14.3 billion was actually available for use in the country.[27]

Although estimates of the amount of borrowed money actually available for use in any given country vary considerably, the money that remained was put to different uses by public and private borrowers.

In their effort to promote economic growth, governments borrowed money to pay for essential imported goods like oil, food, and machinery. Rising oil prices in the 1970s forced countries without oil to pay more for imported oil. U.S. economist William Cline estimated that oil price increases cost southern countries an extra $260 billion in the years between 1974 and 1982, a figure comparable to the $299 billion acquired by these same countries during this period.[28]

Some Latin American countries, like Mexico, had large oil supplies of their own. But while Mexico did not pay more for imported oil, it borrowed heavily to develop its oil fields and become a major producer, expecting that increasing oil prices would enable it to pay off mounting debts. As we will see, this expectation did not materialize, and falling oil prices after 1980 helped trigger Mexico's debt crisis.[29]

The cost of imported food also rose in the 1970s. Rising oil prices increased the cost of growing food because farmers rely heavily on gasoline-powered tractors and petroleum-based fertilizers and pesticides. Moreover, poor harvests in the Soviet Union during the mid-1970s increased the demand and therefore the price of food on world markets (see chapter 13). "For low-income countries, the increased cost in these years . . . of food imports from [first world] countries far exceeded the increased cost of oil imports," argued Shahid Burki.[30]

As we have seen, with money provided by northern lenders, southern governments also purchased tractors and textile machines to expand commodity production in fields and factories and built roads, ports, and airports—and the aircraft to use them—to facilitate the transport of commodities, business managers, and bankers. Many of these activities provided jobs to northern manufacturers of imported goods and employment for domestic users of these products in the South. By building huge mining, hydroelectric, irrigation, and industrial projects, governments could put people to work and increase their income from project revenues and worker taxes.

In addition to paying for essential imports, governments used borrowed money to build up hard currency reserves and stabilize their currencies, to subsidize or lower the cost of fuel, food, and transportation so that domestic consumers would not be adversely affected by rising oil and food prices, and sometimes to balance their budgets.[31] As one Latin American finance minister recalled, "I remember how the bankers tried to corner me at conferences to offer me loans. If you are trying to balance your budget, it's terribly tempting to borrow money instead of raising taxes."[32]

Not all the money was used for essential or legitimate government purposes. Some of it was used to increase military expenditures, wasted on boondoggle development projects, or siphoned off for personal gain. Military spending by Latin American countries doubled during the 1970s, despite the fact that they faced no external threats. Military spending in Africa increased by one-third.[33] Many development projects proved to be boondoggles. A huge development project providing electricity from the Inga Dam on the Zaire River to a copper-cobalt mining complex in Shaba province cost nearly $1 billion, but when it was finished, the electricity it delivered was no longer needed at the mines.[34] And in some countries, government corruption was widespread. In Zaire, a country described by some writers as an "absolutist kleptocracy," President Mobutu Sese Seko stashed away about $5 billion in personal Swiss bank accounts, a sum equal to his country's total foreign debt.[35] In Brazil, President Fernando Collor de Mello was impeached for corruption in 1992.

Governments were not the only borrowers. Private borrowers acquired a substantial portion of Latin American debt. In Latin America, "private debt rose from $15 billion in 1972 to $58 billion in 1981," accounting for about 20 percent of the total ($272.9 billion in 1981).[36] During the 1970s, domestic owners of Latin American farms and factories, often "the principal national monopolistic groups of the country," borrowed heavily to finance the expansion of their businesses.[37] In Mexico, these groups acquired one-quarter of the country's total debt.

Alongside private domestic borrowers, subsidiaries of businesses in Western Europe and North America also borrowed money, and when they did, they increased the debt of southern countries. So, for example,

General Motors, Ford, Union Carbide, Pepsico, and Volkswagen were all important borrowers in Mexico, adding $750 million of debt to Mexico's total.[38]

Like northern lenders, southern borrowers were confident they could repay mounting debts. Inflation in northern countries meant that real interest rates were fairly low and stable in the 1970s, commodity prices for the raw materials and goods they produced were rising, and their economies were growing. But these favorable conditions, which encouraged both lenders and borrowers in the 1970s, did not last. When conditions changed—when interest rates rose and commodity prices fell—in the 1980s, they triggered a crisis that proved earlier assumptions wrong.

THE CRISIS: RISING INTEREST RATES, FALLING COMMODITY PRICES

When Paul Volcker, head of the Federal Reserve, raised U.S. interest rates in 1979 to fight inflation in the United States, he did not intend to create a global debt crisis. But rising U.S. interest rates and the rising London Interbank Offered Rate (LIBOR), which set interest rates for Eurodollar lending, greatly increased the cost of southern loans, most of them now tied to floating rates set by the United States or LIBOR.[39]

Rising interest rates had two important consequences. First, they increased interest payments on accumulated debt. "Mexico's interest bill tripled from $2.3 billion in 1979 to $6.1 billion in 1982. . . . For the region as a whole, interest payments more than doubled, from $14.4 billion in 1979 to $36.1 billion in 1982."[40] High interest rates made it harder for borrowers to pay back their debts. U.S. economist William Cline estimated that high interest rates in the 1980s cost indebted countries $41 billion more than they would have paid had interest rates remained at the average level between 1961 and 1980.[41] Other economists have estimated that Latin American countries paid out more than $100 billion in "excessive" interest between 1976 and 1985.[42]

A second problem was that high U.S. interest rates acted like a magnet, attracting money from around the world. U.S. officials understood that capital flight from other countries would reduce investment abroad and undermine the competitiveness of other countries. As we have seen, it did not greatly weaken Western Europe and Japan because they had higher savings rates, which meant they had more capital available to them and because the U.S. government returned some of this capital to them in the form of U.S. military spending. Unfortunately, countries in Latin America, Africa, and Eastern Europe did not have these advantages, because they had low savings rates and a huge demand for capital (which is why they had been borrowing money from abroad). And, except for Panama, where the United States stationed a large military force, the United States returned little of the money it acquired from Latin American investors in the form of military spending. As Volcker observed, "In many [indebted countries], their excessive debt burdens can be traced in large part to a flight of capital by their own citizens discouraged from investing at home."[43] He might have added that U.S. policies, which were under his control, also encouraged them to invest their capital in the United States.

High interest rates attracted $150 billion in capital from Latin America between 1973 and 1987, the bulk of it after 1979, when as much as $25 billion annually "flew" to the United States to purchase Treasury bonds.[44] Massive capital flight created several problems for Latin American countries: it deprived them of money they might have used to invest in their own countries, pay for imports, or repay debt, and it eroded their country's tax base as investors withdrew taxable savings from Latin American banks and placed them in tax-free deposits in U.S. banks.[45] During the height of Mexico's debt crisis, "a Mexico City newspaper published the names of 537 Mexicans each with over a million dollars on deposit with foreign banks."[46] Thus, capital flight deprived indebted countries of money at a time when they needed it most.

Just as interest rates increased, commodity prices began to fall. During the 1970s, the price of commodities typically exported by third world countries—metals, raw materials, and foodstuffs—generally rose. They could then use the hard currencies they earned by selling these goods to northern countries to repay their loans, which had to be repaid in hard currencies. Lenders insisted on repayment in dollars or other hard currencies (deutsche marks, pounds, yen), not in pesos or astrals, because they worried that indebted governments would simply print more money and use inflation to repay loans in worthless, depreciated currency.

Generally speaking, the prices Latin American countries could get for their commodities fell slowly between 1950 and the mid-1970s, when the OPEC embargo and weather-related food shortages began to increase commodity prices, particularly of oil and food. Commodities then began to fall dramatically in the 1980s.[47] Between 1980 and 1982, world commodity prices fell by more than one-third, "to their lowest level in 30 years, a disastrous development for countries that expected commodity exports to pay their way," noted sociologist John Walton.[48] "The beef that Argentina [exported] fell from $2.25 a kilogram . . . in 1980 to $1.60 by the end of 1981. Sugar from Brazil and the Caribbean fell from 79 cents a kilo to 27 cents by 1982. And copper, a big-ticket item for the likes of Chile and Zaire, fell from $2.61 a kilo to $1.66," one writer observed.[49]

Falling prices reduced the ability of borrower countries to repay debts, which were being pushed up by higher interest rates. Prices continued to fall during the rest of the 1980s. A World Bank index of raw material prices, which started at 168.2 in 1980, fell to 100 by 1990, and 86.1 in 1992, the lowest prices in real terms since 1948.[50]

The price of oil also fell, slowly after 1980 and then sharply after 1985. Mexico, which borrowed heavily to become a major oil producer because it believed oil prices would continue to climb, found itself with mounting debt and declining revenues.[51] "Given the deterioration in the terms of trade, Latin Americans sell more and get less," observed Mexico's finance minister, Jesús Silva Herzog.[52]

Why did commodity prices fall so dramatically in the 1980s, crippling the ability of borrowers to repay their debts? They did so because high U.S. interest rates triggered a global recession that reduced demand for their goods. They also fell because northern countries had begun to develop new supplies or to substitute materials for southern commodities (we will examine these developments in greater detail in the next chapter). In the case of oil, the discovery of new oil fields in the North Sea increased the supply and helped lower global prices, while energy conservation measures reduced demand. Commodity prices also fell because southern countries collectively produced more of these goods in the 1980s. Remember that in the 1970s borrowers used northern money to expand their production of oil (Mexico), coffee (Colombia), frozen orange juice (Brazil), beef (Argentina), copper (Chile, Zaire), and tin (Bolivia). With money and hard work, they succeeded in producing more of these goods. But as production expanded, supplies increased and prices fell. The irony is that the harder they worked and the more they did what they set out to do, the less they earned and the more deeply they fell into debt.

CRISIS MANAGEMENT: THE IMF TAKES OVER

Mounting debt and a growing inability to repay loans threatened to bankrupt the major U.S. and Western European banks. If that occurred, financial chaos and a global economic crisis would have ensued. To avert such a catastrophe, lenders acted quickly to manage the crisis. The IMF and the World Bank quickly took the lead, assuming responsibility for managing the debt crisis and ensuring that borrower countries repay all of their debts, both public and private. As Princeton economist Robert Gilpin observed, "Interest payments on the debt would not be decreased across the board nor world commodity prices received by debtors be increased. The burden of solving the problem would continue to rest squarely on the debtors."[53]

For much of the postwar period, the World Bank and IMF were fairly obscure institutions. The World Bank made modest loans to promote economic development in poor countries, the IMF helped manage the occasional crisis that emerged when poor countries ran out of the hard currencies they needed to purchase foreign goods. But the debt crisis thrust these Bretton Woods institutions into the forefront of efforts to manage the debt crisis. The IMF, in particular, was asked to assume a new, expanded, global role. Not only was it asked to manage crisis in dozens of countries simultaneously, it was also expected to reshape the government policies and economic structures of those countries. The IMF required governments to abandon decades-old development policies and adopt new, neoliberal market policies in their place. These new policies and practices contributed to the globalization of these economies.

If the IMF had not taken the lead, if individual banks had tried to collect debts or seize assets on their own, chaos would have ensued and debtors might have been able to play lenders against each other. Instead, by forming what Gilpin calls a "creditor's cartel," which was led by the IMF and World Bank, northern lenders could practice a "divide and conquer strategy" and "impose their will on the debtors."[54] They were able to do this because private lenders could speak with one voice, through the IMF, in negotiations with foreign borrowers. The lenders also possessed two important advantages: they alone could lend borrowers the money they needed to make ends meet and they alone possessed accurate information on the debts and economic conditions of borrowing countries (most debtor governments lacked key financial information on private debt in their own country).[55] The existence of powerful global institutions, unity of purpose, and control of economic data enabled the lenders to bargain with debtors from a position of strength.

While IMF officials managed the debt crisis in dozens of countries, private lenders moved to protect themselves from the consequences of the crisis by reducing credit and shifting the burden of financing new loans to public agencies and making taxpayers assume some losses.

During previous Latin American debt crises, lenders simply stopped lending to borrowers, sometimes for decades. During the 1980s, private banks greatly reduced their lending, though they did not cut off credit entirely. Capital flows to Latin America fell by one-third between 1980 and 1984 as private lenders began to cut and run.[56] The problem was that borrowers desperately needed new loans, at least in the short term, so they could get their finances in order and take steps that would eventually enable them to repay debts. Public lending agencies urged private bankers to continue lending money. U.S. treasury secretary James Baker, whose 1985 Baker Plan attempted to advance a comprehensive settlement of the debt crisis, argued that "increased lending by the private banks in support of comprehensive economic adjustment programs" was essential in order to make it possible for borrowers to repay their debts. As Baker told bankers, "I would like to see the banking community make a pledge to provide these amounts" ($20 billion over the next three years) on a "voluntary basis."[57]

Because private banks did not respond to Baker's invitation, the Baker Plan failed.[58] So the U.S. government and international lending agencies had to pick up the slack, which meant that taxpayers in Western Europe and the United States had to shoulder increasing responsibilities for debt crisis management.

Private lenders also protected themselves by declaring losses on foreign loans, which enabled them to reduce their taxes. But while they claimed losses, they could still demand full repayment from borrowers, so they could declare losses, receive tax breaks, and recover their original investment.[59] Although the tax laws that allow banks to take "provisions," or make "loan-loss reserves" as they are called, differ from country to country, the savings to banks can be substantial. One economist estimated that between 1987 and 1990, "over $20 billion of U.S. bank debt on the third world was charged off and provisioned under federal mandate. Since the corporate tax rate on U.S. banks is 34 percent, this sum would give rise to tax credits of at least $6.8 billion."[60] British banks received about $7 billion, German banks $10 billion, and French banks $10.9 billion as a result of similar laws.[61] Altogether, private banks probably received between $44 billion and $50.8 billion in tax credits in this period, all at taxpayer expense.[62]

Taxpayers not only assumed responsibility for revenues lost in this fashion, they also had to foot the bill when their governments agreed to provide debt relief to some borrowers, as the U.S. government did when it discharged $7 billion of Egypt's debt for agreeing to participate as a U.S. ally in the 1990–1991 Persian Gulf War.[63]

Although the stockholders of some banks experienced losses when it became apparent that their banks had lent heavily to debtors and the value of their bank stocks declined, private banks emerged from a potentially devastating crisis relatively unscathed. No major Western bank failed as a result of the debt crisis. But while lenders averted serious problems, borrowers did not.

DEBTORS FALL APART

When borrowers in Latin America, Africa, and Eastern Europe ran out of money to repay their debts, they faced serious problems. Without foreign currency, they could not pay for imported fuel or food, and owners of domestic capital began to send it abroad. Without imported or domestic capital, agricultural and industrial businesses would grind to a halt and lay off workers, and the economy would collapse. To avert these economic disasters, borrower governments, under IMF direction, took steps to get the hard currency they needed to purchase imported goods and repay lenders.

As a condition for receiving a continued influx of money, borrower governments were asked to assume responsibility for repaying private debts that they did not themselves incur. In Venezuela and Argentina, nearly 60 percent of the total debt had been acquired by private businesses, domestic and foreign.[64] Although private borrowing in Latin America as a whole accounted for 20 percent of the outstanding debt, about $58 billion, governments and taxpayers were asked to repay this debt as if it were their own. According to Harvard economist Jeffrey Sachs, "In country after country, governments took over the private debt on favorable terms for the private sector firms, or subsidized the private debt service payments, in order to bail out the private firms. This 'socialization' of the private debt resulted in a significant increase in the fiscal burden of the nation's foreign debt."[65]

This was unfair because it imposed on the people of these countries debts not of their making. Not only did governments bail out private sector firms in their own country, many of them subsidiaries of northern corporations, they effectively bailed out private northern banks because these banks would not otherwise have been able to recover private debts in foreign countries. Once they knew potential losses were averted and private debt responsibilities assumed by southern governments, private lenders agreed to continue making loans during the debt crisis, though, as we have seen, they reduced their share of new loans.

With the money they needed to avert the immediate crisis, southern governments were then forced to adopt painful economic policies, which the IMF insisted would allow them to repay debts in the long term. Although the specific IMF policies varied, most governments adopted similar "structural adjustment programs" (SAPs), or "austerity programs," as they were called, trying to create trade surpluses and government budget surpluses to raise the money they needed to repay their debts.

Indebted governments were first required to increase their trade surpluses. If they could export more goods than they imported, they could acquire a larger amount of hard currency, which they could then use to repay old debts and reduce their need to borrow money to pay for imported goods. To create trade surpluses, they tried simultaneously to increase exports and reduce imports. To increase exports, governments urged agricultural and industrial businesses to expand production and export more of their goods. They assisted in this process by devaluing their currencies. When the United States devalued the dollar, government officials hoped that this would make U.S. exports cheaper abroad and make Japanese

goods more expensive in America (see chapter 2). They expected the devaluation to increase U.S. exports and discourage U.S. consumers from
buying imported Japanese goods, thereby reducing the U.S. trade deficit.
Latin American governments hoped their own currency devaluations
would have the same effect, helping them create a trade surplus that
would provide them with much-needed currency.

Latin American countries did export more goods, though falling prices
for those commodities, global recession, which reduced demand, and
northern restrictions or tariffs on many southern goods meant they had a
difficult time keeping exports at 1980 levels. From 1980 to 1985, Latin
American countries increased the volume of goods they exported by 23
percent, but the value of these exports remained about the same.[66] Latin
American countries exported between $90 billion and $100 billion worth
of goods between 1980 and 1984. Exports then fell to about $78 billion
from 1984 to 1986, mostly as a result of falling oil prices, before recovering to the $100 billion level by 1988.[67]

With exports holding steady (despite increased efforts), the only way
Latin American governments managed to create trade surpluses in the
1980s was by cutting back on imported goods. Whereas Latin American
countries imported between $90 billion and $100 billion worth of goods in
1980, they imported only $60 billion by 1982, staying at this level through-
out much of the mid-1980s.[68] By slashing imports, they created a trade
surplus that gave them between $30 billion and $40 billion, which they
used to repay lenders. As Mexican finance minister Silva Herzog observed, "The much heralded improvement in Latin America's current accounts therefore is attributable mostly to import reduction, rather than to
export increase."[69]

Because indebted governments were responsible for repaying public
and private debts, they also had to find ways of raising money to repay
lenders. The IMF instructed them to raise money by selling off state assets
to foreign or domestic buyers and by creating budget surpluses.

During the 1960s and 1970s, many southern governments created state-
run business to promote economic development. Governments could borrow money and derive revenue from their operations. These businesses—
government-owned phone, airline, bank, oil, cement company, or state
coffee board—often enjoyed monopoly status, either because they offered
services that private businesses could not profitably provide or because
monopoly eliminated "wasteful" domestic competition and allowed
these firms to compete with large transnational corporations (TNCs). In
debt crisis negotiations, the IMF insisted that indebted governments sell
off or "privatize" state-owned businesses, both to increase "competition"
and to raise money to pay off debts. They also insisted that governments
change their laws so that TNCs could purchase these assets when they

were offered for sale. Many southern governments had long restricted foreign investment because they worried that key sectors of the economy would fall under the control of foreign owners.

At IMF insistence, Latin American governments began selling off state-owned businesses. By 1990, for example, the Mexican government sold off 875 of the 1,155 enterprises that it had owned all or part of in 1982.[70] And this pattern was repeated around the continent, with governments selling off airlines, port facilities, phone companies, and chemical plants. Currency devaluations played an important role in this process.

As in the United States, where the 1985 dollar devaluation made U.S. assets available for sale to Japanese investors at one-half their previous price, currency devaluations in Latin America enabled foreign investors to purchase important economic assets at bargain-basement prices. Privatization and currency devaluations worked more to the advantage of foreign investors, though Latin American investors who had placed their money in the dollar accounts of Western banks during the great capital flights of the early 1980s could also acquire state assets at advantageous prices. So, for example, Mexico sold Teléfonos de México, the government's telephone company, for $1.76 billion to a French, American, and Mexican communications consortium.[71] Because the Mexican government had devalued the peso, foreign investors got a real bargain.

Although governments could raise money to repay debts by selling public assets, this was a one-time way to raise money. To raise the money they needed, governments had to create continuing budget surpluses. They did this by increasing taxes and cutting public spending. The burden of tax increases and spending cuts typically fell on poor and middle-income taxpayers.

During the 1970s, many third world governments used borrowed money to keep oil and food prices low so that transportation, cooking fuel (kerosene), and basic foodstuffs would remain affordable for poor and working people at a time when world oil and grain prices were climbing. But to create budget surpluses, they were forced in the 1970s to eliminate these subsidies, which accounted for a considerable proportion of government spending, and to increase taxes. Generally speaking, taxes on corporations and the rich were reduced (as they were in the United States in this same period), while excise and sales taxes, which fell most heavily on the poor, and income taxes on middle-income groups increased. "A 1986 study of 94 [IMF]-supported adjustment programs implemented between 1980 and 1984 found [that] 63 percent . . . contained wage and salary restraints; 61 percent included transfer payment [for Social Security and unemployment programs] and subsidy [for food and fuel] restraints; . . . and 46 percent included personal income tax measures," writes economist Howard Lehman.[72]

The SAPs administered by the IMF provided important benefits for the North. First, they ensured that lenders would be repaid by southern borrowers, which averted a financial crisis in the North. Second, they increased the flow of goods from South to North. Because the increased supplies helped lower the price of these goods, producers and consumers in the North paid less, reaping a substantial benefit. Third, the sale of industries in the South, at cheap, devalued prices, meant that northern investors could snatch up some real bargains and increase their control of economies in the South. SAPs contributed to the globalization of the South, a development that provided important benefits for the North.

THE CONSEQUENCES

The steps taken by lenders and borrowers had important economic, social, and political consequences for indebted countries. It undermined economic development, setting them back decades, accelerated environmental destruction, and adversely affected women and female children. The silver lining in this otherwise black cloud was that debt crisis also contributed to the fall of dictatorships and the rise of democracy in many countries (a development we will examine in chapter 6). Let us review some of the important social consequences of the debt crisis.

Deeper in Debt

In economic terms, indebted countries managed to repay their debts, but found themselves deeper in debt. Moreover, their strenuous efforts to repay debt exhausted their economies, prompting some economists to describe the 1980s as a "lost decade." As Volcker said, "Even a decade later, the wounds in Latin America itself have not fully healed. For some of those countries (and for those similarly affected in Africa), the 1980s was a lost decade in terms of growth and price stability."[73]

How could borrower countries repay their debts, yet end up deeper in debt? Between 1982 and 1990, lenders sent $927 billion to southern borrowers. In the same period, borrowers repaid lenders $1,345 billion in principal and interest. As a result, indebted countries paid $418 billion more than they received. British economist Susan George argues that this sum is six times greater in real terms than the amount of money the U.S. transferred to postwar Europe through the Marshall Plan.[74] Yet despite these massive repayments, borrowers found themselves "61 percent more in debt than they were in 1982."[75]

Mexico, for example, paid lenders $100 billion in debt service between 1982 and 1988, $10 billion more than it owed when the crisis struck in 1982. But while it repaid vast sums to first world lenders, it owed even more: $112 billion in 1988. How could this happen? It is similar to what happens when people buy a house. Home buyers understand that if they borrow $100,000 at 10 percent interest for a thirty-year period, they will actually pay $300,000 in all, two-thirds of it as interest and one-third as the principal (the amount of the original loan). The bank, of course, insists that the borrower repay the interest first. After ten years, the borrower has repaid $100,000, but still owes $200,000, which is larger than the original loan. In the same way, Mexico had made substantial payments, but still had a lot left to repay.

Latin American debt increased from about $280 billion in 1982 to $435 billion in 1993, and total third world debt climbed from $639 billion in 1980 to $1,341 billion in 1990. Most borrowers will continue to repay debt into the foreseeable future.

In Argentina, debt grew from $40 billion in 1982, when the debt crisis began, to $132 billion in 2001. At IMF request, the government introduced repeated austerity programs. But its debts grew anyway, despite its two-decade effort to repay them. The most recent austerity program, announced in 2001, required the government to cut salaries and pensions for government workers. Teachers were not paid for months, schools could no longer afford to boil water to make powdered milk for malnourished children, and public health officials no longer vaccinated dogs for rabies, leading to a widespread outbreak of the disease.[76] "Argentina is a country without credit," President Fernando de la Rua admitted.[77]

Although indebted governments successfully repaid lenders in the 1980s, they drained their economies. Instead of growing, most Latin American economies actually shrank by about 10 percent while their populations continued to grow.[78] In Mexico, the real incomes of average workers fell 40 percent between 1981 and 1988, while the incomes of government employees fell even more, nearly 50 percent.[79] In most Latin American countries, wages fell, while unemployment rose, prices and taxes increased, and hunger grew. In 1986, twenty million more people in Latin America were living below the poverty line than in 1981, 150 million people in all.[80] Not surprisingly, declining incomes and rising unemployment persuaded many Latin Americans to emigrate to the United States in search of jobs. According to Sachs, "As for the debtor countries, many have fallen into the deepest economic crisis in their histories. . . . Many countries' living standards have fallen to levels of the 1950s and 1960s. A decade of development has been wiped out throughout the debtor world."[81]

Gender and Debt Crisis

The debt crisis adversely affected men and women across the South. But, for a variety of reasons, SAPs were particularly hard on women and female children.[82]

First, the IMF encouraged governments to expand the production of export crops so they could earn hard currency to repay debt. But the expansion of land devoted to export crops often reduced the amount of land devoted to subsistence agriculture and in-common uses: forests for firewood, land for gardens, water for domestic consumption (see chapter 13). Women in many countries grow food for their families, forage for firewood to cook their meals, and draw water to bathe their children and wash their clothes. The conversion of agricultural land and forests from subsistence production to export agriculture, and the use of water from rivers to grow water-intensive crops, has made it harder for women to provide these goods and resources for their families.[83] Women and female children have had to walk farther, forage longer, and pay more for resources they need. So SAPs have increased female work burdens substantially.

Second, governments eliminated subsidies for food, fuel, and transport, forcing people to pay more for these goods. Higher costs meant that many families must do with less of each. But when families cut back, men cut back less and women more. Women cut back more because in most patriarchial families, men command a greater proportion of household income and resources than women. Women typically work longer hours (twelve to fourteen hours a day compared to eight to twelve hours for men), and devote a greater share of their earnings to the household. Economists have found that women in Mexico contributed 100 percent of their earnings to the family budget, but men contributed only 75 percent of theirs. As the World Bank reported, "It is not uncommon for children's nutrition to deteriorate while wrist watches, radios and bicycles are acquired by the adult male household members."[84] Throughout the South, the adverse impact of rising prices, a product of SAPs, was disproportionately felt by women.

Third, governments cut back on public services, particularly education and health care. Again, these cuts adversely affected men, but hurt women more because in patriarchial households, families more often send male children to school or send males to seek medical treatment, neglecting the needs of women and girls. In hard times, women and girls do without. The result is that fewer girls attend school and illiteracy among women has increased. As public health care services have declined, government-sponsored campaigns against AIDS or female genital mutilation (in Africa) have languished, and infant mortality rates, particularly for girls, have increased.[85] The IMF-directed decline of public services has been particularly detrimental for women and female children.

Environmental Destruction and Debt

Governments also increased the rate of deforestation so they could export hardwood timber or beef raised on cleared rainforests. Brazil and Mexico, the two largest debtors, are also major deforesters. Brazil is ranked number one and Mexico number six in the world. In Mexico, much of this deforestation has occurred in Chiapas, the southern state where Zapatista peasants revolted in 1994. Both have increased deforestation rates dramatically in the past two decades: Brazil up 245 percent, Mexico up 15 percent.[86]

Debt and Protest

These social and environmental problems frequently led to social conflict, what some scholars have called "IMF riots," when people protested government SAPs. University of California sociologist John Walton recorded fifty major "protest events" in thirteen countries between 1976 and 1986. He found that when governments cut subsidies for food and basic necessities, increased fares on public transportation, or eliminated government jobs, riots sometimes resulted. In September 1985, for example, "hundreds of Panamanian workers invaded their legislature chanting: 'I won't pay the debt! Let the ones who stole the money pay!'"[87]

Debt and Democracy

Although the debt crisis had disastrous economic and social consequences for indebted countries, it had some positive political consequences. The debt crisis and SAPs imposed by the IMF discredited the dictators who had borrowed and ruled most Latin American countries. When debt crises struck, civilian democrats demanded and received political power in return for their support for arduous debt crisis management programs. As we will see (chapter 6), debt crises contributed to the democratization of much of Latin America in the 1980s. So while the debt crisis was an economic disaster, it was also a political opportunity.

Impact on the North

Although indebted countries experienced great difficulties as a result of debt crises, people in northern countries also experienced debt-related problems. As we have seen, Latin American borrowers increased trade surpluses, which provided them with much-needed cash, by reducing their imports. Because many of these imports were goods made or grown in northern countries, import reductions contributed to unemployment in Western Europe and the United States.

Between 1980 and 1986, U.S. exports to Latin America fell by $10 billion. One economist estimated that this resulted in the loss of 930,000 jobs in the United States.[88] U.S. trade representative William Brock calculated that 240,000 U.S. jobs were lost as a result of the Mexican debt crisis alone.[89] Senator Bill Bradley observed that Latin American debtors had made a "Herculean effort" to service their debts. But he noted, "The price the United States has paid for Latin America's ability to meet its new debt schedules has been the collapse of Latin American markets for U.S. products . . . and the loss of more than one million [U.S.] jobs."[90] So while private lenders managed to cover their assets, workers and taxpayers have had to foot some of the bill.

Debt Relief?

Although lenders averted a global economic crisis, borrowers continue to wrestle with the consequences of the debt crisis. From the lenders' perspective, borrowers still owe them a great deal. But from the borrowers' perspective, they have already repaid their debts. Some economists have suggested that remaining debts could be forgiven or reduced without great harm to lenders. They also note that continued indebtedness undermines the ability of indebted countries to purchase imports, which is essential for the health of economies in Western Europe and the United States. Former World Bank president Robert McNamara argued, "The evidence that growth and progress in the developing countries now has a measurable impact on the economy of the United States reflects the importance of the developing countries to the United States as export markets and as customers of U.S. commercial banks."[91]

The continued insistence on full repayment of debt, the objective of bankers, conflicts in the long run with the sale of northern goods in southern markets, which is the objective of farmers and manufacturers in the North. The problem in coming years will be how to resolve the conflicting objectives and needs of different groups, North and South.

One proposal, advanced by the IMF in the late 1990s, would be to provide debt relief to some of the poorest countries. The money would come in part from the sale of gold reserves held by the IMF.[92] For extremely poor countries like Uganda, which "spends $3 per inhabitant on health annually, and about $17 a person on debt repayment," debt relief would be extremely welcome.[93] But German, Japanese, and other officials in the G-7 have objected to the plan, arguing that the IMF should not sell off even a small part of its $40 billion in gold.[94] Countries in Africa, the poorest of the debtor countries, would receive most of the relief outlined in recent IMF plans. But they would receive only a partial reduction of their debt, and then only if their governments adopted new SAPs.

The debt crisis was not anticipated either by lenders in the North or borrowers in the South. But when it occurred, institutions in the North seized the opportunity to reshape the South along neoliberal lines, a process that contributed to contemporary globalization.

NOTES

1. Iliana Zloch-Christy, *Debt Problems of Eastern Europe* (Cambridge: Cambridge University Press, 1987), 29, 34; Christopher A. Kojm, *The Problem of International Debt: The Reference Shelf*, vol. 56, no. 1 (New York: Wilson, 1984), 8.

2. John Walton, "Debt, Protest and the State in Latin America," in *Power and Popular Protest: Latin American Social Movements*, ed. Susan Eckstein (Berkeley: University of California Press, 1989), 301.

3. Robert Gilpin, *The Political Economy of International Relations* (Princeton, N.J.: Princeton University Press, 1987), 317.

4. Steven Greenhouse, "Third World Markets Gain Favor," *New York Times*, 17 December 1993.

5. Paul Kennedy, *Preparing for the 21st Century* (New York: Random House, 1987), 204.

6. Kojm, *The Problem of International Debt*, 10.

7. Zloch-Christy, *Debt Problems*, xiii.

8. Trevor W. Parfitt and Stephen P. Riley, *The African Debt Crisis* (London: Routledge, 1989), 16–17.

9. Sue Branford and Bernardo Kucinski, *The Debt Squads: The US, the Banks and Latin America* (London: Zed Books, 1988), 47.

10. Michael Moffitt, *The World's Money: International Banking from Bretton Woods to the Brink of Insolvency* (New York: Simon & Schuster, 1983), 98.

11. Branford and Kucinski, *The Debt Squads*, 47; Jacobo Schatan, *World Debt: Who Is to Pay?* (London: Zed Books, 1987), 9.

12. Branford and Kucinski, *The Debt Squads*, 58.

13. Branford and Kucinski, *The Debt Squads*, 58. See Stephany Griffith-Jones and Osvaldo Sunkel, *Debt and Development Crises in Latin America: The End of an Illusion* (Oxford: Clarendon Press, 1986), 72.

14. Kojm, *The Problem of International Debt*, 36.

15. Branford and Kucinski, *The Debt Squads*, 58.

16. Robert Cherry, *The Imperiled Economy*, vol. 1, *Macroeconomics from a Left Perspective* (New York: Union for Radical Political Economics, 1987), 202; Gilpin, *The Political Economy*, 315.

17. Kojm, *The Problem of International Debt*, 18.

18. Richard W. Lombardi, *Debt Trap: Rethinking the Logic of Development* (New York: Praeger, 1985), 91.

19. Lombardi, *Debt Trap*, 90.

20. Robert Devlin, *Debt and Crisis in Latin America: The Supply Side of the Story* (Princeton, N.J.: Princeton University Press, 1989), 36, 38.

21. Lombardi, *Debt Trap*, 76–77.

22. Lombardi, *Debt Trap*, 74.

23. Cherry, *The Imperiled Economy*, 201.

24. Branford and Kucinski, *Debt Squads*, 59; Schatan, *World Debt*, 7.

25. Branford and Kucinski, *Debt Squads*, 56.

26. Branford and Kucinski, *Debt Squads*, xiv; Cherry, *The Imperiled Economy*, 200.

27. Branford and Kucinski, *Debt Squads*, 78; Griffith-Jones and Sunkel, *Debt and Development Crises*, 107.

28. Branford and Kucinski, *Debt Squads*, 64; Vincent Ferraro, "Global Debt and Third World Development," in *World Security: Trends and Challenges at Century's End*, ed. Michael T. Klare and Daniel C. Thomas (New York: St. Martin's Press, 1991), 329.

29. Nora Lustig, *Mexico: The Remaking of an Economy* (Washington, D.C.: Brookings Institution, 1992), 20.

30. Shahid Javed Burki, "The Prospects for the Developing World: A Review of Recent Forecasts," *Finance and Development* 18, no. 1 (March 1981): 21.

31. Walton, "Debt, Protest and the State," 305.

32. Walton, "Debt, Protest and the State," 305.

33. Branford and Kucinski, *Debt Squads*, 75.

34. Schatan, *World Debt*, 83.

35. Lombardi, *Debt Trap*, 86.

36. Parfitt and Riley, *The African Debt Crisis*, 79.

37. Arturo R. Guillen, "Crisis, the Burden of Foreign Debt and Structural Dependence," in *Latin American Perspectives* 60, vol. 16, no. 1 (Winter 1989): 38.

38. Guillen, "Crisis."

39. Guillen, "Crisis," 40.

40. Jeffrey D. Sachs, *Developing Country Debt and the World Economy* (Chicago: University of Chicago Press, 1989), 302; Devlin, *Debt and Crisis*, 50; Jackie Roddick, *Dance of the Millions: Latin America and the Debt Crisis* (London: Latin America Bureau, 1988), 35.

41. Branford and Kucinski, *Debt Squads*, 95; Kojm, *The Problem of International Debt*, 15; Schatan, *World Debt*, 10.

42. Ferraro, "Global Debt and Third World Development," 329.

43. Schatan, *World Debt*, 110; Branford and Kucinski, *Debt Squads*, 96.

44. Frank E. Morris, "Disinflation and the Third World Debt Crisis," in *World Debt Crisis*, ed. Michael P. Claudon (New York: Ballinger, 1986), 83.

45. Manuel Pastor Jr., *Capital Flight and the Latin American Debt Crisis* (Washington, D.C.: Economic Policy Institute, 1989), 8–9.

46. Manuel Pastor Jr., *Capital Flight*, 11–12.

47. Roddick, *Dance of the Millions*, 65.

48. Walton, "Debt, Protest and the State," 306; Griffith-Jones and Sunkel, *Debt and Development Crises*, 13; Howard P. Lehman, *Indebted Development: Strategic Bargaining and Economic Adjustment in the Third World* (New York: St. Martin's Press, 1993), 15; Parfitt and Riley, *The African Debt Crisis*, 2–3.

49. Darrell Delamaide, *Debt Shock: The Full Story of the World Credit Crisis* (New York: Doubleday, 1984), 28; Schatan, *World Debt*, 41.

50. Lombardi, *Debt Trap*, 10; Filipe Ortiz de Zevallos, "Mañana Has Arrived: Latin America Recovers from the Lost Decade," *The World Paper* (October 1993), 5.

51. Kojm, *The Problem of International Debt*, 16–17; Robert A. Pastor, *Latin America's Debt Crisis: Adjusting to the Past or Planning for the Future?* (Boulder, Colo.: Lynne Rienner, 1987), 13; Lustig, *Mexico*, 24–25, 39.

52. Robert A. Pastor, *Latin America's Debt Crisis*, 35.

53. Gilpin, *The Political Economy of International Relations*, 326.

54. Gilpin, *The Political Economy of International Relations*, 319–20.

55. Roddick, *Dance of the Millions*, 10.

56. Branford and Kucinski, *Debt Squads*, 8; Gilpin, *The Political Economy of International Relations*, 327.

57. Robert A. Pastor, *Latin America's Debt Crisis*, 151–52.

58. Branford and Kucinski, *Debt Squads*, 120–21.

59. Susan George, *The Debt Boomerang: How Third World Debt Harms Us All* (Boulder, Colo.: Westview, 1992), 65–66.

60. George, *The Debt Boomerang*, 74.

61. George, *The Debt Boomerang*, 79–81.

62. George, *The Debt Boomerang*, 82.

63. George, *The Debt Boomerang*, 156.

64. Roddick, *Dance of the Millions*, 71, 110–11.

65. Sachs, *Developing Country Debt*, 13.

66. Branford and Kucinski, *Debt Squads*, 5.

67. Robert A. Pastor, *Latin America's Debt Crisis*, 35–36; Clyde H. Farnsworth, "Latin America Records Some Economic Gains," *New York Times*, 11 September 1989.

68. Farnsworth, "Latin America Records Some Economic Gains."

69. Robert A. Pastor, *Latin America's Debt Crisis*, 36.

70. Lustig, *Mexico*, 105.

71. Lustig, *Mexico*, 106.

72. Lehman, *Indebted Development*, 44–45.

73. Paul A. Volcker and Toyoo Gyohten, *Changing Fortunes: The World's Money and the Threat to American Leadership* (New York: Times Books, 1992), 187.

74. George, *The Debt Boomerang*, xv.

75. Clifford Krauss, "Argentina's Provinces Struggle to Stay Afloat," *New York Times*, 18 November 2001.

76. Clifford Krauss, "Argentine Leader Announces Debt Revision and Subsidies," *New York Times*, 2 November 2001.

77. Krauss, "Argentine Leader Announces Debt Revision."

78. Griffith-Jones and Sunkel, *Debt and Development Crises*, 6.

79. Lustig, *Mexico*, 69.

80. Branford and Kucinski, *Debt Squads*, 24.

81. Sachs, *Developing Country Debt*; Mariarosa Dalla Costa and Giovanna F. Dalla Costa, eds., *Paying the Price: Women and the Politics of International Economic Strategy* (London: Zed, 1993).

82. C. George Caffentzis, "The Fundamental Implications of the Debt Crisis for Social Reproduction in Africa," in *Paying the Price: Women and the Politics of International Economic Strategy*, ed. Mariarosa Dalla Costa and Giovanna F. Dalla Costa (London: Zed, 1993), 28.

83. Lester Brown, *State of the World 1993* (New York: Norton, 1993), 64.

84. Caffentzis, "The Fundamental Implications of the Debt Crisis," 34; Andre Michel, "African Women, Development and the North-South Relationship," in *Paying the Price: Women and the Politics of International Economic Strategy*, ed. Mariarosa Dalla Costa and Giovanna F. Dalla Costa (London: Zed, 1993), 65–66.

85. Ferraro, "Global Debt and Third World Development," 333.

86. George, *The Debt Boomerang*, 10–11.

87. Walton, "Debt, Protest and the State," 200, 316.

88. Ferraro, "Global Debt and Third World Development," 335.

89. Lombardi, *Debt Trap*, 91; Delamaide, *Debt Shock*, 12.

90. Robert A. Pastor, *Latin America's Debt Crisis*, 70.

91. Paul Lewis, "Debt-Relief Cost for the Poorest Nations," *New York Times*, 10 June 1996.

92. Paul Lewis, "World Bank Moves to Cut Poorest Nations' Debts," *New York Times*, 16 March 1996.

93. Lewis, "World Bank Moves to Cut Poorest Nations' Debts."

94. *New York Times*, "Rich Nations Pledge to Double Countries Getting Debt Relief," 17 September 2000.

5

✺

An Age of Migrations

According to demographers Stephen Castles and Mark Miller, we live in "The Age of Migration."[1] The phrase suggests that cross-border migration is a new development and that there is one kind of global migration.[2] But cross-border migration is not new. The current period is not the first but the third "age" of global migration since Europeans began migrating overseas, and it is not characterized by one kind of migration but by many different kinds, each driven by different economic, environmental, political, and cultural developments. It would be more appropriate to describe the current period as "an age of migrations."

Today, about 200 million people have migrated across borders and live in countries outside their homelands.[3] As a percentage of world population, the number of migrants—3 percent—is about the same today as it was in 1950.[4] Most of these people—150 to 175 million—have migrated primarily for economic and sometimes environmental reasons. Another 25 million have moved for political reasons, as a result of partition, government migration policies, or ethnic conflict and war. This number has fluctuated. The United Nations estimated that there were 17 million political refugees in 1991, 27 million in 1995, but only 11 million in 2007.[5] About half of economic and political migrants are women, a recent development that demographers have described as the "feminization of migration."[6] Women migrate not only for economic and political reasons, but also for sociocultural reasons based on their identities as women. The migration of nurses, brides, orphaned girls, and trafficked women are gender-specific migrations based on attributes assigned to women and girls by different cultures.

To appreciate the causes and consequences of economic, political, and gender-specific migrations, it is important first to describe historical migration patterns and identify the distinctive features of the current age.

PATTERNS OF MIGRATION: 1500–2000

Before 1500, people around the world lived in insular geographic and political worlds. Except for nomads, gypsies, caravan traders, and travelers like Marco Polo, few people migrated between the continents and there was very little, if any, contact between people living in Africa, Asia, Europe, the Americas, and Australia. That changed after 1492, when Europeans explored, conquered, and colonized other parts of the world, a development that started the first "age" of global migration.

During the first age of migration, from 1500 to 1850, people generally migrated from the "East" (Europe and Africa) to the "West" (the Americas). The Europeans who conquered and colonized the Americas wanted workers to mine precious metals; raise cattle; and grow sugar, tobacco, and cotton that could be shipped to Europe. But it was extremely difficult for colonial rulers to find workers anywhere who were willing to make difficult journeys and work under arduous conditions for little pay in the mines and fields of the Americas. To obtain the labor they required, the Spanish government organized the *encomienda* system to force indigenous peoples to work in the mines and fields of their empire in the Americas. The *encomienda* system forced indigenous communities to devote some of their time to producing export commodities for their European overseers.[7] They received some meager compensation for the work they were required to perform. Between 1500 and 1850, European states also enslaved and transported fourteen million Africans to the Caribbean and the Americas, where they produced sugar, tobacco, and cotton. Of these, five million died during the journey and the first year of captivity in America.[8] European states also used conviction and indenture to deliver European workers to the Americas, though they later used the offer of cheap land (taken from the Indians) to persuade poor and dissident Europeans to migrate west. The East-to-West migration that characterized the period between 1500 and 1850 might be described as "The Age of Coercion." The British and American abolition of the slave trade in 1807 and 1808 signaled the eclipse of this period, though it took a civil war in the United States and revolutionary wars across South America to bring it to an end.[9]

A new pattern of global migration took shape in the mid-nineteenth century. Starting in 1850, Europeans migrated in large numbers to North America, where they streamed into rural areas to farm or congregated in cities to work in industry. They also migrated to European colonies in

South America, Africa, South and Southeast Asia, and Australia, where they worked as bureaucrats, entrepreneurs, soldiers, settlers, and missionaries. This migration from the European "North" to the American "West" and global "South" might be described as "The Age of European Migration."

The United States was an important destination for European migrants, though small numbers of people from China and Japan also migrated to the United States in this period, which triggered legislation restricting Asian immigration. Fifty million Europeans migrated to the United States between 1850 and 1920, when the U.S. government curtailed immigration for Europeans. At the high point, in 1910, 1.5 million Europeans migrated to the United States and 14.7 percent of the population was foreign born.[10] Although immigration in the United States reached 1910 levels in 2000, only 12.3 percent of the population was foreign born.[11]

The flight from Europe was interrupted by World War I, which closed the sea lanes to international migration. After the war, anti-immigration and anti-communist sentiment in the United States led to severe restrictions on immigration. Later, the Great Depression persuaded countries around the world to restrict immigration as a way to prevent migrants from competing with domestic workers for scarce jobs. Then, when World War II broke out, countries sealed their borders to prevent the passage of soldiers, migrants, and refugees, and the age of European migration, which had begun in 1850, came to an end.

The pattern of global migration after 1945 might be called "The Great Reversal." It began with the "return" of eight million ethnic Germans from Eastern Europe and the Soviet Union to occupied Germany, and the "return" of 2.6 million ethnic Japanese living across southeast Asia to Japan.[12] Many of these German and Japanese refugees had lived for generations outside their respective "homelands" and they were either persuaded or forced to return "home" by their host countries after the collapse of the German and Japanese empires.

After the war, European empires were forced or persuaded by the superpowers and independence movements in their colonies to "decolonize" and depart from colonial territories. Many of the Europeans who lived in the colonies "returned" home, though like German and Japanese migrants, many of them had lived in the colonies for generations and had never visited their European "homelands."

Decolonization also facilitated the migration of non-European people in the colonies to Europe, and they often followed in the footsteps of retreating Europeans. While people from the South migrated "North" to Europe and Japan, people also migrated "North" to the United States.

During World War II, the U.S. government encouraged the migration of Mexican workers to the United States through the Bracero Program, which directed male workers into agricultural jobs. This program paved the way for the legal and illegal migration of Mexican and other South American workers to the United States in the years after the war, much as colonialism paved the way for non-European migrants to Europe.[13] During the postwar period, the number of migrants to the United States slowly increased and reached 1910 levels in 1990. By 2005, the foreign-born population in the United States reached 32 million.

Although the South-to-North pattern of migration during the postwar period is clear, there are important, secondary migrations that complicate this picture. First, while contemporary migrants typically head "North" to Western Europe and the United States, they do not go to Japan, a country that in other respects is part of the developed "North." Except for a small number of migrants from Korea, the Japanese government restricts migration and only 1 percent of the population in Japan is foreign born.[14]

Second, starting in the 1970s, people from the South also began migrating to the oil-rich Gulf States: Saudi Arabia, Kuwait, and the United Arab Emirates (UAE).[15] In economic terms, these countries might be considered part of the "North," even though they are located in the Middle East. The Gulf States encouraged the migration of Egyptian, Palestinian, and Jordanian men to work in their oil and construction industries. During the 1990 Gulf War, many of these immigrants were forced to leave because they sympathized with Iraq. To replace them, the Gulf states recruited male Muslim workers from countries in South Asia—Pakistan, India, and Indonesia—to work in industry, and recruited female workers from Sri Lanka, Indonesia, Malaysia, and the Philippines to work in the service industry and as domestic workers: nurses, maids, and nannies. Immigrants now comprise 30 percent of the population in Saudi Arabia, 60 percent in Kuwait, and 80 percent in the UAE.[16]

Third, many people from poor countries in the South migrate to other, not-so-poor countries in the South. Haitians migrate to the Dominican Republic, where they now make up 11 percent of the population; poor people from Zimbabwe and Mozambique move to South Africa; people from Tajikistan move to Russia; people from Nicaragua move to Costa Rica; and people from Nepal migrate to India.[17] The World Bank estimates that there are 74 million "South-to-South" migrants.[18] Of course, some of these migrants will move on to destinations in the North, though many will remain in the South.

Why have so many people from the South migrated to the North in the postwar period? Most do so for economic reasons. They do so because there is a demand for unskilled labor in the North and because there is a large supply of unskilled workers in the South who are ready and willing to migrate North. To appreciate the economic causes of contemporary migration, it is important first to describe the developments that increased the demand for migrant labor in the North, and then discuss developments that have created a large supply of labor in the South. We will then look at the consequences of migration in the North and South.

ECONOMIC CAUSES: DEMAND IN THE NORTH

Most demographers argue that wage disparities are an important reason why people migrate from the South to the North.[19] Workers with few skills receive poor wages in the South, where unemployment is high. They migrate North to find higher-paying jobs. Studies have shown that wage disparities between North and South are large and that they have grown during the postwar period.[20] One study found that manufacturing workers in the United States were paid $9.87 an hour in 1980, while workers in the Philippines were paid 53 cents per hour, workers in Thailand received 31 cents, and workers in China 25 cents.[21] By 1995, manufacturing wages in the United States had increased to $17.20 an hour. Wages in some of these other countries rose too, but not as fast—in the Philippines to 71 cents per hour, Thailand to 46 cents—while wages in China stayed the same. As a result, the wage gap grew wider.[22]

Take Mexico, which has been the primary source of migrants to the United States in the postwar period. In 1995, wages for Mexican workers in manufacturing were only 26 percent of manufacturing wages in the United States. But by 2000, wages in Mexico had fallen to only 12 percent of U.S. wages.[23]

Of course, wage disparities alone cannot explain why people from some countries in the South are more likely to move North than people in other countries in the South, where wages are even lower. Proximity matters. People in Mexico live close to the United States; people in Bolivia live much farther away, which makes it more expensive and difficult to migrate North. History matters. Countries in the North established colonial or neo-colonial relations with particular countries in the South. During the Spanish-American War, the United States seized colonies in Puerto Rico, Cuba, and the Philippines. Today, migrants from Puerto Rico can travel freely to the mainland because they are U.S. citizens; migrants from Cuba can seek political asylum if they can set foot on U.S. soil (they are returned to Cuba if they are intercepted at sea); and it is easier for migrants from the Philippines to enter the United States and work as nurses (see below) than it is for migrants from most other countries, in part because the United States had military bases in the Philippines and there are close political connections between the two countries. Long-standing political ties between countries in the North and South have created pathways that migrants use to journey to the North.

Demographers have also observed that social networks play an important role in migration. First-generation migrants help friends and relatives migrate by providing them with information about employment, giving them money to pay for their travel, offering advice on how to obtain visas, and so on. These social networks facilitate the migration of particular groups from particular countries and help migrants cluster in particular neighborhoods, where they set up grocery stores, restaurants, travel agencies, and legal services, which are characteristic features of dense social networks.

But while wage disparities are an important part of the explanation, it is important to explain why wage disparities have increased, why the demand for unskilled workers in the North has grown, why supplies of labor in the South have grown, and what role noneconomic factors—the environment, language, and gender—play in these developments.

First, why are wages in the North higher than they are in the South? And why have wage disparities grown?

After World War II, the U.S. Marshall Plan, military aid, and private investment fueled economic reconstruction and growth in Western Europe, Japan, and the United States (see chapter 1). Governments in the North invested heavily in education, which increased worker skills, and businesses invested in tools to increase worker productivity. New skills and technologies helped workers produce more goods, and increased productivity allowed businesses to pay workers higher wages. In the United States, worker incomes doubled between 1945 and 1970; in Western Europe and Japan, they increased even faster. Economic growth and mutually reinforcing investment and trade increased the wealth and wages of workers in the North and distanced them from countries in the South, where these developments did not occur.

Initially, economic growth in the North increased the demand for low-skill male workers in agriculture, construction, and manufacturing. Then in the 1970s, the growth of the service sector increased the demand for male and, increasingly, female cooks, cashiers, security guards, janitors, truck drivers, nursing aides, laborers, and domestic service workers: child care providers and maids.[24] Wages in these industries were low, so governments in the North used "minimum wage" laws to keep them from falling below subsistence levels; one minimum-wage job in the United States was not sufficient to support a family living in the United States, although it was sufficient to support a migrant worker who maintained a household in another country.

Economic growth in the North also increased the demand for some skilled labor from the South. The U.S. computer industry has lobbied Congress to expand its high-end visas and provide green cards to skilled engineers from South and East Asia. Google founder Sergey Brin, an immigrant from the Soviet Union, has said that "to stay competitive in a 'knowledge-based economy,'" Google needs to hire more immigrants as software engineers, mathematicians, and computer scientists.[25] University administrators have urged Congress not to restrict the admission of foreign students to U.S. colleges, both because they pay $14.5 billion annually in tuition and because students may stay in the United States and increase its intellectual capital.[26]

The Hollywood film industry has exhibited a strong demand for skilled actors and directors from countries around the world. In recent years, Hollywood studios have welcomed migrants from Australia (Mel Gibson, Heath Ledger, Eric Bana, Hugh Jackman, Russell Crowe, Nicole Kidman, and Cate Blanchett), China (Jet Li, Jackie Chan, Chow Yun Fat), and South Africa (Charlize Theron). The biggest barrier to migrant actors from Spanish-speaking countries (Salma Hayek, Penélope Cruz, Javier Bardem) is their English, which they must master to be considered for most Hollywood films. There have been exceptions. Austrian immigrant Arnold Schwarzenegger, an actor with a hardened physique but flabby acting skills and heavily accented English, managed to make blockbusters by playing an inarticulate killer robot, which played to his strengths. Of course, while Hollywood needs only a small number of migrant actors, the demand is real and the rewards are huge for migrants with the requisite skills.

Postwar economic growth in the North raised wages and increased the demand for (mostly) low-skill migrants in agriculture, construction, manufacturing and service industries, and the domestic sector, and for some skilled migrants in high-tech industries. At the same time, economic developments in the South contributed to persistently low wages *and* a growing supply of workers who were willing to migrate North in search of employment and higher wages.

ECONOMIC CAUSES: SUPPLY IN THE SOUTH

Wages in the South have been low for historical, political, and economic reasons. Historically, colonization by European empires impoverished people in the South and discouraged the kind of investment in education, infrastructure, and industry that might have promoted economic growth, improved productivity, and raised wages. Although decolonization created independent states across the South, dictators controlled most governments during much of the postwar period (see chapter 6). Capitalist and communist dictatorships abolished unions, suppressed wages, and jailed political dissidents, which helped keep wages low. But even after they democratized, it was difficult for governments in the South to promote economic development and increase wages. Most countries in the South have relied heavily on the production of agricultural goods and raw materials to provide jobs and generate income. But the prices they received for these commodities have fallen for most of the postwar period. Commodity prices have fallen because countries in the South produced more of these goods, which increased the supply of oil, coffee, bananas, sugar, copper, gold, and cotton, and because the demand for these goods in the North has fallen. Demand in the North fell because producers and consumers there have introduced new technologies or adopted new preferences that replaced or reduced their demand for goods from the South. So, for example, the use of high-fructose corn sweetener in foods and beverages in the North reduced the demand for cane sugar grown in the South; the advent of wireless phone technologies and widespread recycling reduced the demand for mined copper.

Falling commodity prices reduced the wages that farmers and miners in the South received for producing commodities. Falling wages have increased the disparity between wages paid in the North and South.

Governments in the North were aware of these developments and took several steps to promote economic development in the South. They introduced "Green Revolution" agricultural technologies, provided public and private loans to finance economic growth, and used private investment to create jobs and produce goods in the South for foreign and domestic markets. But while these measures contributed to some economic development in the South, they also contributed, ironically, to the displacement of unskilled workers and to a growing supply of workers in the South who could migrate north.

Green Revolution

Beginning in the 1960s, the U.S. government and private foundations financed the transfer of "Green Revolution" technologies—hybrid seeds, chemical fertilizers and pesticides, and farm machinery—to increase the production of food and fiber. These technologies greatly increased productivity. India increased grain production from 131 million metric tons (mmt) to 190 mmt between 1980 and 1989; Indonesia from 33 mmt to 49 mmt in the same period.[27] But these technologies also allowed farmers to produce more food and fiber with fewer workers. Farmers who used Green Revolution technologies also expanded and took over land from subsistence farmers, which displaced other workers.[28] Some of the displaced farmers went to work on Green Revolution farms and plantations as wage workers; others migrated to marginal land in rain forests and desert-fringe areas to practice subsistence farming, often with disastrous environmental consequences; many migrated to cities in the South to find jobs in the formal or informal sector; and some of these migrants headed North to finds jobs in agriculture and industry.[29] As a result, the Green Revolution helped increase the supply of low-skill worker-migrants in the South.

Debt

During the 1970s, the United States provided public and private loans to dictatorships in the South (mainly in Latin America and Eastern Europe) to develop new infrastructure and expand their production of agricultural commodities, natural resources, and manufactured goods for foreign and domestic markets. But when U.S. officials raised interest rates in 1979, they triggered a debt crisis that forced most dictators from power. The civilian governments that replaced them were nonetheless required to repay the debts incurred by their predecessors by increasing the production of export commodities (they used labor-displacing technologies to increase productivity) and selling off or "privatizing" many state-owned businesses, which resulted in widespread layoffs. This led to the displacement of workers in agriculture and industry and increased the supply of unskilled and skilled workers. Many of them joined migratory streams to the North. Displaced workers in Latin America headed to the United States; displaced workers from former communist dictatorships in Eastern Europe and the Soviet Union migrated to Western Europe.

Foreign Direct Investment (FDI)

During the 1980s and 1990s, corporations based in the North invested heavily in the North (see chapter 1). Corporations also made substantial investments in the South to produce goods for foreign and domestic markets.[30] About 25 percent of all foreign investment is directed at developing countries in the South, though China has received the lion's share (see chapter 7).[31] So, for example, Coca-Cola built bottling plants to produce beverages for local markets, while Ford built auto-parts factories that exported components to auto-assembly plants in the United States. Proponents of free trade agreements have argued that foreign investment would create jobs in the South and help prevent workers from migrating North. But sociologist Matt Sanderson has shown that the impact of FDI on migration depends on how it is invested. If FDI is used to finance the expansion of manufacturing in the South, it does provide jobs and helps *curb* migration. But if it is invested in agriculture, forestry, or mining, where it is used to deploy technologies that increase productivity, FDI has typically displaced workers and increased the supply of unskilled workers, which has *contributed* to migration.[32] So instead of curbing migration from the South, FDI has often encouraged migration.

OTHER CAUSES

Generally, migration is triggered by economic developments in the North and South. But environmental degradation, language acquisition, and gender identities also contribute to migration in some settings, playing roles that demographers often ignore.

Environmental Degradation

In Senegal, local fishermen and European fishing fleets, which have paid the government millions of dollars for the right to fish for bottom-dwelling species and tuna, have depleted the fishery. Fish stocks have declined 75 percent during the last 15 years, a serious form of environmental degradation that has impoverished the 200,000 people in Senegal who depend on the fishery.[33] In 2007, 31,000 took their wooden pirogues and attempted the hazardous voyage to the Canary Islands in hope of migrating on to Europe. More than 6,000 of the voyagers died or disappeared at sea.[34]

In other countries, the expansion of shrimp farms, which export the shellfish to the North, has contributed to the decline of local fisheries in the Philippines, China, Thailand, and Indonesia, while the expansion of salmon farming has ruined local fisheries in Chile.[35] The degradation of fisheries, and of rain forests and marginal lands by logging companies and subsistence farmers, has displaced people across the South and forced many to migrate North.

Language Acquisition

In general, people in the South try to migrate to language-affiliated countries in the North, where they can speak their native language. But the rise of English as the global standard has put a premium on English-language skills. As the economic and social value of English has increased, many people have moved so they can learn to speak English. In Europe, 90,000 French migrants have found work in England and 20,000 live in Ireland.[36] Laurent Girad-Claudon, who moved to Dublin, said he and many of his friends did so "to gain experience and improve their English," so they could eventually find better jobs back in France.[37]

The acquisition of English is also seen as an important skill by South Korean families. Many women have migrated with their children to English-speaking countries like New Zealand and Australia, leaving their husbands behind, so their children can attend elementary schools where they learn to speak English. More than 40,000 South Korean schoolchildren live outside South Korea where they study English.[38] South Korean parents migrate because they think that Korean schools fail to teach English and creative-thinking skills. They send their children abroad to improve their English-language test scores so they can eventually study in the United States, where 103,000 South Korean students now attend college. As Kim Soo-in, who took her two sons to New Zealand, explained, "It was never a question of whether to do it, but when. We knew we had to do it at some point."[39]

English-language acquisition is so important for families like the Kims that parents are prepared to live separately for years. South Koreans call them "wild geese" families because separated parents fly back and forth like birds. Because it is South Korean mothers who typically take their children and leave their husbands behind, these "wild geese" migrations have an important gender dimension. Oh Oookwhan, a professor at Ehwa Women's University in South Korea, argues that women migrate so they can raise the children; so their husbands can keep working in Korea; and, in many cases, so the women can "get as far away as possible from their mothers-in-law," who are seen as very demanding of daughters-in-law (see migrant brides, below).[40]

GENDER-BASED MIGRATIONS

In the 1950s and 1960s, most migrants from the South were men. They were drawn North to work in agriculture and construction, jobs that demanded skills commonly associated with men. But the expansion of service jobs in the formal and informal sector increased the demand for workers with skills and attributes culturally assigned to women. Today, half of all cross-border migrants are women. As the South Korean example illustrates, many contemporary migrations are very gender-identified. The fact that some migration streams are closely identified with women suggests that gender plays an important causal role, as it does in the Philippines, where migration has been described as "arguably the best-documented case of gender-induced and sculpted migration."[41]

In 1974, Ferdinand Marcos, dictator of the Philippines, issued a decree encouraging Filipinos to seek work overseas and established a government bureaucracy, the Overseas Workers Welfare Association, to facilitate migration.[42] The number of migrant workers from the Philippines grew from 19,221 in 1976 to 523,000 in 1989, and 700,000 in 1995, more than any other country in Asia.[43] Most of the migrants were women. Filipina women migrated to the Middle East and East Asia to work as domestics (the Greek word for maid is now "Filipineza"); to Japan to work as "entertainers" in the sex industry (see below); and to work as nurses in the United States, where they have established a major presence in the field.[44]

In New York City, 30 percent of the 173,000 Filipinos in the city work as nurses. Most of the rest are their spouses, children, or aging parents. "'If you meet a Filipino man, he'll probably say, "my wife is a nurse,"' explained Pio Pannon, a nurse manager at Montefiore Medical Center in the Bronx and president-elect of the Philippine Nurses Association of New York."[45] American hospitals have "aggressively recruited nurses from the Philippines . . . and prize Filipina nurses for their English-language skills and their education in public and professional schools that are modeled on their American counterparts. . . . They also value the nurses for their work ethic, their loyalty to employers and a tenderness that seems to stem from a culture where people insist on caring for their own aging or sick relative."[46]

The demand for Filipina nurses in the United States is shaped not only by colonial relations, economic necessity, and English-language preferences but also, importantly, by the particular skills associated with Filipina women.

Hospitals are not the only U.S. institutions to encourage the migration of foreign women. Recently, women's colleges have begun recruiting women student-migrants from the Middle East. Bryn Mawr, Barnard, Mount Holyoke, Wellesley, and Smith colleges, known collectively as "the Sisters," have tried to increase the number of female applicants from the Middle East because they want to boost their enrollments (the number of women-only colleges in the United States has declined from 300 in 1960 to only 60 in 2008) and because they think that single-sex education in the United States might appeal to families in many Muslim countries, where education and employment are sex segregated.[47] The difficulty, of course, is that this gender-based recruitment strategy, if successful, would bring women from gender-conservative societies to "liberal strongholds where students fiercely debate political action, gender identity, and issues like 'heteronormativity.'. . . Pesangi Perera Weerasingag [a student from Dubai] said that when she arrived at Mount Holyoke, she was shocked by the presence of so many lesbians among the students. But she adjusted, she says, and now loves the environment, with the widespread willingness to discuss race and class ('so refreshing') and her classmates' engagement in politics."[48]

There are other kinds of women-identified or women-only migrations: of brides, orphaned girls, and trafficked women. But these migrations occur not only for economic reasons but also for social and cultural reasons, which have to do with characteristics ascribed to women.

As we have seen, women migrate to the North to find jobs as nurses in the formal sector and as maids and child care providers in the informal sector. In the North, the growing demand for workers with skills that are culturally attributed to women has contributed to the "feminization of migration." There are also gender-specific migrations made by women and girls. Women from the South migrate to marry men in the North, female orphans in the South migrate North with adoptive parents, and women from the South are trafficked across borders to work in the sex industries of the North. These migrations cannot be understood only in economic terms because they are shaped, in important ways, by demographic, cultural, and social developments.

Brides

A growing number of women migrate across borders to marry: "A majority of international marriage migrants are women, and most of these women move from poorer countries to wealthier ones, from the less developed global 'South' to the more industrialized 'North.'"[49] Between 1960 and 1997, "marriage migration to the United States almost tripled," and in 1997, the 202,000 legal marriage migrants to the United States amounted to 25 percent of all migration.[50]

Several developments contributed to marriage-migration flows. U.S. military policy after World War II played an important role. After 1945, U.S. servicemen based in Asia and Europe met and married women, who returned with them to the United States after their overseas tours of duty ended. By the early 1980s, 200,000 Asian-born wives of American servicemen had migrated to the United States.[51] Marriage migration created a problem for the military during the 1950s and 1960s because interracial marriages were banned in Southern states, where many U.S. military bases were located. As one solution, the Pentagon assigned servicemen with Asian wives (but not European brides) to places like Fort Riley, in Kansas, where tolerant local communities accepted "mixed marriages."

During the 1980s, mail-order bride services (with names like "Cherry Blossom" and "China Doll") emerged to introduce U.S. men seeking brides to women living in Asia. In the 1990s, mail-order services introduced men to women living in Eastern Europe and the former Soviet Union.[52] The "mail-order" industry, now increasingly conducted over the Internet, produces between four and six thousand marriage migrants to the United States each year.[53]

Why do U.S. men seek mail-order brides? Some scholars argue that men seek foreign women who have attributes they regard as "exotic," who are more "compliant" than women raised in the United States, and more "dependent" on the husband for their emotional, economic, and social needs.[54]

Although a large number of the migrant marriages in the United States consist of interracial couples, many women marry and migrate with men from the *same* ethnic group. Hispanic women marry Hispanic men in the United States, Filipina women marry Filipino men, and Vietnamese women marry Viet Kieu (Vietnamese living outside Vietnam) men.

The marriage-migration of Vietnamese women to the United States has been driven largely by a demographic imbalance that has caused an unusual "double marriage squeeze." Demographer Daniel Goodkind has said that "a high male mortality rate during the Vietnam war, combined with the migration of a larger number of men than women [as boat people and refugees after the war—see above] . . . has produced a low ratio of men to women in Vietnam [which has created a marriage squeeze for women there], as well as an unusually high ratio of men to women in the Vietnamese diaspora [which has created a marriage squeeze for men in the United States]."[55]

The demographic and political developments associated with the war in Vietnam have created pressure for Vietnamese women to marry Viet Kieu men. Typically, when women from the South marry men in the North, they "marry up," or improve their economic standing—what sociologists call "hypergamy."[56] But for many Vietnamese women who marry Viet Kieu men, they marry both "up" and "down." That is, many of the Vietnamese women are well educated and have a "high" socioeconomic status in their homeland. But many of the Viet Kieu men in the United States are less educated, relatively unskilled, and have, by U.S. standards, the kind of relatively low-paying jobs associated with unskilled, first-generation immigrants. So when Vietnamese women marry Viet Kieu men, they are marrying "up" by moving to the North, but they are also marrying "down" by joining someone with a lower socioeconomic status. In one study, Hung Cam Thai found that "70 percent of the brides . . . were college educated and that 80 percent of the men were low-wage earners."[57]

The different social backgrounds and cultural expectations of people in cross-border migrant marriages are a problem for couples in the United States. It has also been a problem in Japan—where the number of marriages between Japanese men and women from the Philippines, South Korea, China, and Thailand increased from 5,000 in 1970 to 27,000 in 1993—and in South Korea, where male farmers in rural areas seek brides of ethnic Korean descent in China.[58] Many of these cross-border Korean marriages end in divorce. South Korean men accuse their brides of "using" them to migrate for economic reasons and then leaving them as "runaway brides" to go to the city; women accuse men of "using" them as farmhands and domestic servants, and of demanding that the woman care for the man's elderly parents, who are seen as overbearing and demanding.[59]

To prevent "mercenary" marriage migrations, some countries have passed laws against "bogus" and "sham" marriages, which are made with an eye on economic gain, not love.[60]

Orphaned Girls

Every year, American households adopt 100,000 boys and girls who are born in the United States.[61] For different reasons, the demand for young children in the United States exceeds the domestic supply of children available for adoption, though many children—disabled, teenage, and African American children—still go unadopted. Moreover, adoption in the United States is complicated by policies that for many years restricted interracial adoption (1972–1994) and then permitted it (1994 to present).[62] In any event, many U.S. families travel overseas and adopt 20,000 orphaned children—almost all of them girls—and return with them to the United States.[63] As a result, "Americans adopt more foreign-born children than all countries in the world combined, and movie stars like Angelina Jolie, Madonna, and Mia Farrow are prominent among them."[64]

In recent years, cross-border adoption by U.S. parents has been characterized by three important developments. First, the number of children adopted annually from other countries grew from 7,000 in 1990 to 22,884 in 2004, and then declined to 19,292 in 2007.[65] It declined because many countries imposed new restrictions on adoptions (governments did not want to be seen as being unable to provide homes for their orphans) and because receiving countries wanted tighter regulation to "standardize process, procedures and safeguards to reduce corruption in the largely unregulated adoption marketplace."[66] The ratification in 2007 of the Hague Convention on Intercountry Adoption slowed or eliminated cross-border adoptions in some countries and reduced the total number of adoptions, which has been difficult for U.S. parents who have been waiting, sometimes for years, for adoptions to go through.[67]

Second, there has been a shift in the origin of adopted children. For most of the postwar period, half of all the children adopted by U.S. parents came from South Korea: 3,552 of the 7,948 children adopted in 1989.[68] But after 1989, the number of adopted children from South Korea declined and the number of children adopted from Russia, China, Romania (briefly), and Guatemala increased. In 1998, 4,491 orphans were adopted from Russia, 4,206 from China, and 911 from Guatemala.[69] In 2007, these numbers had shifted again: "China sent 5,453 children to American families in 2007, down from 7,906 in 2005. Russia's total dropped to 2,207, down from 3,706 in 2006 . . . [but] adoptions increased from Guatemala (to 4,728 from 4,136 in 2006), Ethiopia (to 1,255 from 732) and Vietnam (up to 626 from 163)."[70]

These changes reflect new policies and practices in each country and competing views of adoption. Some policymakers argue that cross-border adoptions are "bad" for the country's international reputation, while others maintain that it is "good" for orphaned children and for the public and private agencies that collect adoption fees, which can range between $10,000 and $30,000 per child.[71]

Third, cross-border adoption is a gender-specific migration. Boys are rarely, if ever, put up for adoption. They are readily adopted by relatives or non-family members in the countries where they are born. This practice reflects the prevailing male-oriented cultural values in societies that prize boys and, typically, devalue girls. (In the United States, by contrast, orphans of both genders are readily adopted, though black orphans of either sex are not as readily adopted, either because of racism or because child-welfare officials discourage the practice.) The cultural devaluation of girls is most evident in China, which is now the leading source of adopted children for U.S. parents.

In the late 1970s, the Chinese dictatorship adopted a one-child-per-family policy to curb the country's rapid population growth (see chapter 7) and enforced this policy by monitoring pregnant women; forcing women who had given birth to undergo sterilization or insisting that they insert metal IUDs, which could be monitored by X-rays to make sure they remained in place; and performing abortions on women who had become pregnant with a second child.[72] Between 1983 and 1991, "more than 30 million women were forcibly sterilized."[73]

For various cultural reasons, Chinese families preferred boys and took steps to prevent the birth of infant girls. They used ultrasound machines, which were widely introduced after 1979, to identify girls so they could be aborted, and they frequently drowned or abandoned infant girls.[74] One Chinese women's organization found that "in one village alone, forty baby girls had been drowned in 1980 and '81."[75] As an alternative, many parents took young girls to distant train or bus stations and abandoned them there.[76]

These political and cultural practices had two important consequences. First, they altered sex ratios dramatically, so there were more boys born every year than girls (113 boys for every 100 girls in 1990).[77] Second, a large number of abandoned girls were placed in Chinese orphanages. Some scholars estimated that "about 150,000 are abandoned/orphaned every year."[78] The government has admitted to having only 100,000 children in orphanages, though the United Nations has counted 50,000 in one province alone.[79]

Chinese orphanages are poor, local institutions that are financed by local, municipal authorities, not by the central government. These institutions were unprepared to handled the influx of orphaned girls generated by the central government's one-child policies, and could not afford to provide adequate medical care for the infants and children in their custody. By the early 1990s, mortality rates in Chinese orphanages ranged from 20 to 80 percent annually.[80] Officials at one orphanage near Shijiazhuang "disclosed that they had a death rate of 90 percent among their predominantly female foundlings."[81] The mortality rates at Chinese orphanages in the early 1990s were comparable to those in Nazi concentration camps during the Holocaust. The difference, of course, was that mortality rates in German camps were the product of design; in China, they were the product of neglect. Mortality rates in China were high primarily because orphanages could not afford to pay for the medicines needed by their inmates.

In the early 1990s, the central government decided to let Americans adopt orphaned girls as a way to address this problem. The central government charged U.S. parents $10,000 to adopt an orphaned girl. The central government kept $7,000 of the fee and gave $3,000 to the orphanage, and the orphanage then used the money to pay for the medical treatment of girls who remained in their care. As a result, the adoption fees paid by U.S. parents helped subsidize local orphanages and reduce mortality rate among their inmates. Although proponents have depicted this as a "win-win" situation ("good" for Chinese orphans; "good" for adoptive parents), they have not explained why the central government did not adopt a less coercive or less destructive family-planning policy, which greatly disadvantaged girls; why it did not allow Chinese families with one child to adopt orphaned girls; and why it did not do more to assist underfunded orphanages that were struggling to cope with the consequences of the government's one-child policies.

Trafficked Women

Women migrate to work as dancers, bar girls, hostesses, lap dancers, masseuses, and prostitutes in the sex industry, which is located in the North (Western Europe, the United States and Japan), in the Middle East (Israel and Persian Gulf states), and in some countries in the South: Thailand and some Caribbean countries (Jamaica, Cuba, the Dominican Republic) that have become sex-tourist destinations for men from the North.[82] (In Jamaica, some of the sex tourists are females from the North, and also gay men.[83])

Some of the women and girls who work in the sex industry are drawn from local settings and have been described as "voluntary" participants, which means that they participate for economic reasons: they are poor and they need the money. But many are forced to participate by parents or relatives who ask women and girls to become prostitutes in exchange for payments to the parents. In Thailand and South Asia, one-third of the females in the sex industry are girls under seventeen, and many were forced into the industry by their parents, who received loans, grants, or payments from sex-industry operators.[84] Around the world, many adult women are also trafficked in to the industry by criminal recruiters (sometimes by boyfriends or other trafficked women) who use trickery and coercion to force women to work as prostitutes in foreign countries.[85]

Criminal recruiters typically promise women in poor countries that they will find employment in a legal service-sector job—as a waitress or entertainer—in another country, and offer to pay the women's airfare. But when the women arrive, the recruiters steal their passports, insist that the women "repay" the recruiter for the cost of the airfare by working as prostitutes, and beat them if they refuse. It is difficult for the women to go to the police, who are often in league with the traffickers, because policemen often threaten to jail them as illegal immigrants or return them to the custody of the traffickers, who beat the women for seeking help.[86] The U.S. government estimates that there are "between 800,000 and 900,000 such victims of human trafficking every year."[87]

Trafficking is more common in Western Europe, Asia, and the Middle East than it is in either the United States or the Caribbean, where most of the women who work in the sex industry are adults and participate "voluntarily" for economic reasons. In the United States, most of the adult sex workers are "independent entrepreneurs" who work as call girls; or, in Nevada's brothels, as independent contractors; or, in one San Francisco sex emporium, as members of a sex-worker union.

Although prostitution has long existed in many countries, the trafficking of women is a relatively new development associated with the emergence of the global sex industry in the 1980s and 1990s. But it is important to recognize that the global sex industry first emerged in the 1950s and 1960s.

After World War II, the U.S. government established 1,500 military bases around the world "to support 340,000 soldiers in Europe, 144,000 in the Pacific and Asia, and thousands more on land and sea in the Caribbean, the Middle East, Africa, and Latin America."[88] Sex and entertainment industries often grew up alongside these bases; the most prominent or notorious among them were located next to naval ports in the Philippines and Marine bases in Okinawa, Japan.[89] In Olongapo City, next to the Subic Bay naval facility in the Philippines, 25,000 women worked in the entertainment and sex industry in 1987, and 5,000 more women joined the workforce when American aircraft carriers docked in port.[90] U.S. servicemen not only patronized the sex industry in large numbers but also fathered, and then left, children conceived with women prostitutes or romantic partners. In Okinawa, 4,000 children of these liaisons were abandoned by U.S. servicemen.[91] This has been a problem because their mothers could not obtain financial support from departed fathers and because biracial children were shunned and tormented by Japanese children at school.[92] And when they grew up, many outcast biracial girls joined the sex industry to survive.

During the late 1960s, the escalation of war in Vietnam expanded the sex industry in Vietnam and contributed to its rise in other countries, particularly Thailand and Taiwan, where soldiers were sent for rest and recreation.[93] The sex industry in Thailand became the biggest in the world. After the war in Vietnam ended in 1975, Thailand became the principal global destination for sex tourists, mostly businessmen from East Asia, but also from Western Europe and North America. The number of sex trade establishments in Bangkok grew from 977 in 1980 to 5,754 in 1994, and the number of women and girls working in the sex industry grew from 400,000 to perhaps 1.5 million, producing about $5 billion in annual revenues.[94] Many Thai women also migrated to work in the sex industries of other countries, where they were regarded as "exotic" by male patrons. As one analyst observed, "One thing that stands out but stands unexplained [in the literature] is that a large percentage of sex customers seek sex workers whose racial, national or class identities are different from their own. . . . Sex industries today depend on the erotization of the ethnic and cultural 'Other.'"[95] During the 1980s, for example, "There was a switch from white Dutch women to Asian, African, and Latin American women in the red-light districts [in Amsterdam]."[96]

Although the U.S. military withdrew from Vietnam and closed some overseas bases—Subic Bay was closed in 1992—U.S. servicemen overseas continued to patronize sex industries around the world. The war in Iraq (see chapter 12) and the deployment of male servicemen and military contractors to the region has encouraged the growth of the sex industry in the Persian Gulf states, where the trafficking of women is common.[97]

The collapse of communist dictatorships in Eastern Europe and the Soviet Union led to the rise of mafias who trafficked women (as well as guns and drugs) into the sex industry, and to widespread poverty and unemployment, which created a large supply of women eager to obtain jobs in the West. Mafia recruiters are particularly active in very poor, post-communist countries like Albania, Romania, and Moldova, and in very poor South Asian countries: Nepal, Burma, Cambodia, Thailand, and the Philippines. In the 1990s, when a series of economic crises struck the Philippines, the number of women migrating or trafficked to work in the Japanese sex industry increased from 17,000 in 1996 to 70,000 in 2004.[98] After the U.S. government put Japan on a traffic watch list in 2004, the number of Filipina women traveling to Japan on "entertainment visas" fell 95 percent.[99]

Although military deployments, democratization, and economic crises have played an important role in the emergence of the global sex industry and the large-scale trafficking of female migrants, it is not clear why the male demand for sexual services has increased so dramatically in recent years. Several different explanations have been suggested. Some scholars have argued that the increased global mobility of male workers—soldiers, contractors, and businessmen who attend meetings and conferences around the world—has created places (military bases, hotels, and convention centers) where men, traveling alone or in groups (without their wives), congregate in large numbers. The sex industry has grown up around these male-heavy locales to "entertain" the men during their "off-duty" hours.[100] Although the purchase of sexual services is illegal in many places, it is nevertheless condoned by many governments and the hospitality-hotel-convention-tourist industry because it generates billions of dollars in revenue, and it is underwritten by corporations that allow their traveling employees to claim "entertainment" as a business expense.[101]

Whatever the reason, the migrant women who are trafficked to work in the sex industry risk violence and disease, particularly AIDS. Where they are coerced, it is extremely difficult for women in the industry to refuse "unprotected" sex and insist on condom use, which helps protect them from infection. A 2001 U.N. report on AIDS concluded that "the gender dynamics of the epidemic are far reaching due to women's weaker ability to negotiate safer sex."[102]

Although the United States and other countries in Western Europe have passed anti-trafficking laws, and Sweden has decriminalized prostitution for women but increased the penalties for solicitation by men, the expansion of the global sex industry, which has been fueled by growing male demand for the sexual services of migrant women, has meant that this coercive, gender-specific migration has increased.

ECONOMIC CONSEQUENCES IN THE SOUTH

Contemporary migrations have different economic consequences for people in the South and in the North. For people in the South, the remittances sent home by migrant workers in the North have made a significant economic contribution to governments and households in the South. Worldwide, remittances have grown from $2 billion in 1970, to $70 billion in 1995, to $300 billion in 2007.[103] In many countries, remittances are a more important source of income than either foreign investment or foreign aid.[104] In the Philippines, for example, migrant women working as nurses and maids sent home remittances worth $15 billion in 2006, six times the amount the country received in foreign aid ($2.4 billion).[105]

For many years, economists and demographers regarded remittances as unimportant. They argued that remittances were small in size and that the people who received them simply "wasted" the money on consumer goods, rather than saving or investing it, and this did little to promote economic development.[106] But this view changed after a 2003 World Bank study showed that remittances were much bigger than economists thought and that they made positive contributions to governments, households, and businesses.

Governments

When migrants earn money in the North, they are paid in dollars, pounds, or Euros. They then "remit" or send the money home to relatives, who change dollars or other hard currencies for the local currency. The government collects the hard currency and uses it to repay debts and pay for imported oil (oil is bought and sold in dollars), medicine, food, and machinery, and repay debts (see chapter 4). In many countries, the hard currencies that governments need to purchase international goods are difficult to obtain. Remittances provide governments with the foreign exchange they need to keep their economies solvent.

But migrants do not always remit their earnings through formal banking channels. Many migrants use personal couriers to carry their money home, and their relatives use dollars to conduct their business, without ever changing it into the local currency. In some countries, between 50 and 80 percent of migrant remittances are used in informal markets, where people use hard currencies as a hedge against domestic monetary problems (inflation or deflation) and keep their savings in dollars, which they regard as safer.[107] However, this practice deprives governments of hard currency and weakens the value of the domestic currency. In El Salvador, the government in 2001 adopted the dollar as its official currency because most of the country's business was already being conducted in dollars and the local currency had become nearly worthless.[108] Adopting the dollar enabled the government to collect dollars more easily, which was a good thing, though the subsequent decline of the dollar on international markets made some imported goods more expensive, which was a bad thing.[109]

Migrant Households and Communities

Remittances from migrant workers in the North provide a valuable source of income for households in the South, which has helped reduce poverty. Households spend the money they receive on food and other necessities, but they also use it to pay for their children's education or build or improve their home, which is an important source of capital because people can then use their homes as collateral for loans that they can use to invest in business.[110] They invest remittances in agriculture—buying land and improving the productivity of their farms—and they invest in businesses and shops to generate new income for household members.[111] In many respects, households make much more effective use of remittance income than government bureaucracies make of foreign aid or private corporations make of foreign investment. Moreover, households use their remittances not only to improve their own economic circumstances, but also to improve their communities.

For example, in 1945, two brothers from Chinantlan, Mexico, hitchhiked to New York City, where they took jobs mopping floors.[112] Pedro and Fermin Simon invited other relatives to join them, who traveled singly or in groups to join them in New York over the next fifty years. When they arrived, the brothers helped them find work and housing. They pooled their income and, importantly, sent money home to other relatives. They tithed a portion of their earnings and formed a committee to raise money for specific development projects in Chinantlan. They bought bricks for the town square and spent $100,000 for a new potable water system. "Our priority is to give this little town its most basic needs," explained Abel Alonso, the New York committee's president. By the late 1990s, the New York–based migrant community sent Chinantlan $2 million each year and its leaders functioned like a government in exile. "We have no other source of income besides New York," Chinantlan's mayor, Dr. Francisco R. Calixto, said. The migrants returned regularly to the town for vacations and many built homes and retired there. Their Social Security checks and pensions provided an ongoing source of hard currency for the local economy and helped create jobs in local small businesses. Could the World Bank, foreign investors, or the indebted Mexican government have done as much for the development of this small town? Would they have even tried?

If governments in the North really wanted to contribute to economic development in the South and improve the lives of people living there, allowing migrants to work in the North and send money home would be a good way to do it.

Of course, migration North has some adverse consequences in the South. When skilled workers and students migrate, they contribute to a "brain drain" that can deprive countries of their skills and "waste" the money these countries spent to educate them. One study estimated that in 1990, developing countries lost about $7,400 for each of the 90,000 highly skilled workers who migrated to the United States.[113] This suggests that if the North wanted to promote economic development, they would encourage the migration of unskilled workers but discourage the migration of skilled workers, which is the opposite of current policy in the North.

Migration is also hard on households because it separates family members. In the Philippines, 8 million children, 30 percent of all Filipino children, live in households where one parent has migrated overseas.[114] It is difficult to say what impact "wild geese" family structures have on children. Some experts say it is good for them economically, but bad for them emotionally.[115]

ECONOMIC CONSEQUENCES IN THE NORTH

Migration has had positive and negative economic consequences in the North. Generally, migration has been good for businesses, households that employ domestic labor, consumers, and older workers and governments.

Businesses

Many businesses in agriculture, construction, manufacturing and service industries rely on migrant workers who work for low wages at jobs that domestic workers shun. (Although businesses often argue that domestic workers "refuse" to do some jobs, it is often the case that domestic workers shun some jobs because employers "refuse" to pay decent wages.) For example, ABM Industries, a janitorial service company with 73,000 employees and annual revenues of $2.6 billion, relies heavily on migrant workers.[116] Nearly 20 percent of the 262,000 workers employed by the custodial-service industry are migrant workers.[117]

Businesses benefit from migrant workers because they can pay them minimum wage and because it is difficult for migrants to organize unions or go on strike, particularly if they are illegal migrants. In the United States, migrant labor is particularly important to businesses in Midwestern states, where the unemployment rate is half the national average, where workers are scarce, and where the economy is growing. In Nebraska and Iowa, the demand for workers in construction and meatpacking has been met, in large part, by migrants.[118] Migrants in the Midwest have been credited with reviving small towns, increasing school enrollments, and keeping schools from closing.

Where migrant workers are hard to find, businesses can suffer. In upstate New York, Jim Rittner cut down forty acres of thirty-year-old cherry and peach trees on his farm because he could not find migrant workers to pick the fruit, and the Department of Agriculture estimated that eight hundred farms in the state, "with sales in excess of $700 million, were highly vulnerable to going out of business or forced to severely cut back their farm operations due to shortages of migrant workers."[119]

Some businesses profit from services provided to migrant workers. Western Union, a 150-year-old firm that went bankrupt in 1992 when the Internet ruined its telegram business, now earns $1 billion a year by helping poor migrants send remittances home to their families, and it has "five times as many locations worldwide as McDonald's, Starbucks, Burger King, and Wal-Mart combined."[120] Travel agencies, legal service providers, and tax-preparation businesses also profit from the sale of services to migrant workers in the United States.

High-tech firms benefit from the migration of skilled migrant workers. Dick Ward, an Intel manager, said, "Our whole business is predicated on inventing the next generation of computer technologies. The engine that drives that quest is brainpower. And here at Intel, much of that brainpower comes from immigrants."[121]

Robert Kelley Jr., the president of a high-tech industry association, shares this view: "Without the influx of Asians in the 1980s, we would not have had the entrepreneurial explosion we've seen in California," which created new businesses and jobs.[122]

Households that employ workers as maids and nannies also benefit economically from immigration, in part because their work frees women to find jobs outside the home, which can substantially increase household income.[123]

Consumers

In the North, consumers pay less for fresh fruits and vegetables, packaged meat, homes and apartments, food at schools and restaurants, and care at daycare centers, hospitals, and nursing homes than they would if there was not a large supply of unskilled immigrant workers. Consumers also benefit from immigrant culture. For example, migrant entrepreneurs have diversified the kind of food offered to consumers: Mexican burritos, Thai pad thai, Chinese takeout, Japanese sushi, Spanish tapas, Middle Eastern gyros, and Indian curry. These cuisines have complemented the culinary contributions of earlier immigrants: Italian, French, Jewish, and German.

Older Workers and Governments

Most of the workers who migrate North are young and pay taxes on their incomes. Studies have shown that illegal immigrants, even those with false papers, pay taxes on their incomes, and they typically pay more in taxes than they receive in benefits.[124] The taxes they pay contribute to transfer-payment programs like Social Security, unemployment insurance, and Medicare, which benefit older workers, the unemployed, and disabled workers.

For governments in the North, the taxes paid by migrant workers are important because their populations are aging and the number of young workers who contribute money to support social security programs for older workers is shrinking. The young population is shrinking because fertility rates are dropping—young families have fewer children and they have them later in life then they did in the past. At the same time, the older population is growing because mortality rates are falling—people are healthier, receive better medical care, and live longer than they did in the past. These two developments threaten to create a "crisis" for transfer-payment systems like Social Security. Eventually, there will not be enough active young workers to support the large number of retired workers. When this happens, some time in the next thirty to fifty years (depending on the demographic structure in countries in the North), governments will either be forced to raise taxes on young workers or cut benefits for older workers (or delay the retirement age), or both, which will be politically difficult to do.

Migration can help solve the "crisis" of Social Security because migrant workers increase the number of young, active workers and reshape the demographic structure so it does not become too top-heavy (see chapter 1). In effect, migrant workers can "rescue" social security programs for domestic workers.

Japan provides a good example of these developments. In Japan, the population of young workers is shrinking and the population of old workers is growing rapidly. In 2050, "Japan will have 30 percent fewer people, and one-million 100-year-olds."[125] This will lead not only to a "scarcity of workers and falling demand, but also a collapse of the pension system as the tax base shrinks and the elderly population booms."[126] According to the United Nations, "To stave off such a disaster, Japan would need 17 million new immigrants," or about 400,000 new immigrants each year (there are now only 1.5 million immigrants in Japan).[127]

Disadvantaged Workers

Although migration has provided important economic benefits for businesses, consumers, and older workers, numerous studies have shown that it has *disadvantaged* young, unskilled domestic workers, particularly those without a high-school diploma in the North.[128] In the United States, immigration disadvantages young black men.[129] In Georgia, "the jobless rate for black men is nearly triple that of Hispanic men," many of whom are immigrants.[130]"If you have 10 factory . . . openings," Joyce Taylor, a county clerk observed, "I would say Hispanics would get the majority of the jobs now."[131] Naturally, this has led to tension between the two communities.

Migration disadvantages young, unskilled, and minority domestic workers because it increases the supply of labor and depresses wages. And when it pits workers from different ethnic groups against each other, it makes it difficult for them to organize and use collective bargaining to raise wages for both groups. This is particularly a problem in the United States, where real wages for unskilled workers have fallen since 1970. But while migration contributes to stagnant wages in the United States, it is not the only factor. Other developments have also contributed to falling wages for this group. Widespread deindustrialization, which was a product of growing competition with Western Europe and Japan (see chapter 1), the outsourcing of jobs to China (see chapter 7), the decline of unions, the entry of women into the workforce (women are typically paid less than men in the same job), and the consolidation of businesses as a result of mergers have also undermined wages for unskilled domestic workers.

POLITICAL CONSEQUENCES

Workers from the South have been migrating North for decades. But in recent years, this migration has triggered a variety of political responses in the North and in the not-so-poor countries of the South were migrants have congregated. People opposed to migration have engaged in anti-immigrant violence, organized volunteer militias to police immigrants, passed anti-immigration laws, and worked to change public attitudes towards immigrants.

In 2008, anti-immigrant violence erupted in South Africa, a not-so-poor country that is a destination for migrants from Zimbabwe and Mozambique. In Johannesburg, "hundreds of poor black South Africans turned on their even poorer neighbors, killing at least 22 people and driving many more from their homes. Shouting, 'Who are you? Where are you from?' the mobs shot, stabbed, beat, and burned their victims to death."[132]

In Naples, Italy, a mob attacked and burned a gypsy camp in May 2008 after hearing an unfounded rumor that gypsies had stolen a baby from an Italian woman's home.[133] In the same month, the mayor of Rome vowed to bulldoze 20 immigrant shantytowns in the city.[134]

In Malaysia, the three million immigrants from Myanmar have been the "targets of an expanding campaign of harassment, arrest, whippings, imprisonment, and deportation."[135] The Malaysian government organized and armed a volunteer militia and encouraged it to raid the homes of suspected illegal immigrants. According to a Human Rights Watch report, militia members "break into migrant lodgings in the middle of the night without warrants, brutalize the inhabitants, extort money and confiscate cell phones, clothing, jewelry and household goods, before handcuffing migrants and transporting them to detention cells for illegal immigrants."[136] If tried and convicted, illegal immigrants face up to five years in jail, and a whipping, then deportation.

In the United States, like Malaysia, anti-immigrant groups have organized, and sometimes armed, volunteer militias such as the Minutemen to protect the border and prevent immigrants from entering the country illegally. Elsewhere, individual citizens, anti-immigrant groups, and public officials have moved to enforce housing codes and other local ordinances as a way to drive immigrants out of town. In Elmont, New York, Patrick Nicolosi, "a self-appointed watchdog, tries to get local officials to investigate houses that he and his allies suspect of [housing code] violations, to crack down on day laborers," and enforce laws against loitering.[137] Some of his neighbors, who were not illegal immigrants but U.S. citizens, complained that Nicolosi and his anti-immigrant allies had harassed them because they were Hispanic and noted that Mr. Nicolosi was himself an illegal immigrant until he received citizenship under the 1986 amnesty program.[138]

Public officials have joined the effort, citing legal and illegal immigrant families for health and housing code violations.[139] In some cases, they have arrested migrants carrying false papers and prosecuted them for identity theft.[140] Around the United States, state legislatures introduced 1,404 anti-immigration laws in 2007 and enacted 170 of them, while local governments have adopted many more.[141] Some new laws made it a crime for businesses to hire illegal immigrants, made it more difficult for illegal immigrants to obtain a driver's license, or denied them public benefits and medical aid.[142] In Canada, a small town near Montreal adopted a code of conduct for immigrants (it has none) that prohibited, among other things, "the covering of women's faces except for on Halloween and the use of public stoning as a form of punishment."[143]

Still, officials in other cities support immigrant communities. In California, officials in San Francisco, Richmond, and San Raphael passed resolutions that denounced immigration raids, expressed their determination to provide "sanctuaries" for immigrants, and ordered local officials not to cooperate with federal immigration authorities.[144]

In the United States, state and local anti-immigrant legislation has led to the exodus of migrants from some communities, particularly in Arizona, which has passed strict anti-immigrant legislation. Experts say that the rising vacancy rates in the state's rental housing market and declining school enrollments in some districts are proof that immigrants have moved elsewhere.[145] Some business leaders, landlords, and public school officials argue that the exodus of migrant workers will worsen the economic problems facing the state.

State and local politicians have used widespread public opposition to immigration to justify new anti-immigrant legislation. A 2002 poll found that 65 percent of people in the United States, and 77 percent in Germany, thought that immigration should be reduced "a lot or a little."[146] In 2007, 51 percent of African Americans in the United States thought that "Latin American immigrants are taking away jobs, housing, and political power from the black community."[147]

People have been migrating from the South to the North for decades. So why has migration produced a wide-ranging political reaction only in recent years? And why have individuals, militias, and state and local governments in North America and Western Europe expressed political opposition to immigration?

Economic, social, and political developments contributed to the emergence of anti-immigrant politics in the 2000s. In the United States, stagnant wages, rising unemployment, rising household debt, falling home prices and the collapse of some financial institutions, and rising energy and food prices contributed to economic difficulties and to worker anxieties about immigration. Economic anxieties were joined by concerns about security issues raised by the attacks of 9/11 (see chapter 12), particularly the need to "secure the borders" against infiltrating terrorists.

Socially, immigrants' unwillingness or inability to "assimilate" by adopting prevailing linguistic and religious norms contributed to anxiety about immigrants, particularly in communities with *small* foreign-born populations. (Anti-immigrant sentiment is much lower in big cities with large and diverse foreign-born populations than it is in small towns, particularly ones in rural areas, with small immigrant populations.)

Why have grassroots groups and local-state government officials taken steps to protect borders and curb illegal immigration?

Local anti-immigrant politics in the United States has emerged in response to the federal government's "failure" to control migration and curb illegal immigration. In 1986, the Reagan administration and Congress passed legislation that fined employers who hired illegal immigrants as a way to reduce the demand for migrant workers and choke off the supply, but also provided amnesty and eventual citizenship to some of the illegal immigrants already living in the United States. But while the federal government subsequently legalized 2.7 million immigrants, it did little to enforce employer sanctions, largely because the Republican administration did not want to burden businesses with "restrictive" government regulations and intrusive supervision.[148] As a result, the legislation did not reduce the demand for migrant labor or choke off the supply, and illegal immigration continued. In 2006, President George W. Bush introduced legislation to reform the government's immigration policy by tightening employer sanctions, providing a more limited amnesty, and introducing a "guest worker" program (see below) that allowed substantial immigration but under controlled conditions, much like the earlier Bracero Program. But anti-immigration critics complained that the proposed law was not a reform but simply a reprise of the 1986 legislation, which they regarded as a failure. They argued that it would not substantially curb immigration and campaigned against it. The reform bill failed and, in its absence, state and local authorities took matters into their own hands and introduced anti-immigrant legislation of their own.

WESTERN EUROPE

Western European states have taken a different approach to migration. During the 1950s and 1960s, they admitted millions of migrants from South Europe as "guest workers," which allowed them to find work but not obtain citizenship in host countries. They then ended these programs and expelled most migrants in the early 1970s, when an economic recession reduced the demand for foreign labor.

After dictators fell in Spain, Portugal, and Greece, the European Community admitted their democratic successors as members and promoted economic development in these countries. Economic growth created jobs and made it possible for workers in these countries to earn a living at home. The fall of communist dictatorships in Eastern Europe led to another wave of incorporation in the 1990s, when East Germany was united with West Germany and the European Union (the collective successor to the European Community) admitted Poland, Hungary, the Czech Republic, Slovakia, Slovenia, Latvia, Lithuania, and Estonia as members. As before, the European Union promoted economic development in new member states.

An important part of the European Union's approach to enlargement and common economic development has been to permit the migration of workers from member states to other member states. The 1991 Maastricht Treaty, which set up the European Union in 1993, created European citizenship and gave people moving between member states "virtually all citizenship rights, except [the right] to vote in national elections."[149] The 1995 Schengen agreement subsequently removed border controls for EU citizens. But neither Maastricht nor Schengen established a common policy on immigrants from *outside* Europe. (By contrast, the United States extends these rights only to migrants from Puerto Rico, which is a U.S. commonwealth, but not to migrants from other *countries*.)

As a result of the Maastricht/Schengen agreements, cross-border migration increased across Europe. An estimated 800,000 Belgians work outside of Belgium and 86,000 French migrants work in the United Kingdom.[150] After Poland became a member, several million Polish workers migrated to work in Germany and the United Kingdom.[151] Spain has gone from being a country that exported workers in the 1960s to a country that imports workers from the European Union, Romania, and North Africa. Migrant workers in Spain, many employed in agriculture, are credited with reviving rural villages.[152] Although the foreign-born population in Spain has grown from 2 percent to 10 percent since 1985, domestic workers are more tolerant of migrants than their counterparts in the United States or Germany.[153] Since 1985, Spain has carried six amnesty programs for illegal immigrants.[154]

The European Union is one of the few places where migration has done what economists expected it to do. It has contributed to economic development in both sender and receiver countries and, over time, reduced wage disparities between them. In the four years after Poland became an EU member, economic development and rising wages in Poland began to slow the pace of migration abroad, persuaded some earlier migrants to return home, and closed the wage gap between workers in Poland and workers in other EU countries.[155]

Still, migration in the European Union is complicated by conflicts between different waves of migrants. First, there are large numbers of immigrants from outside Europe and the former colonies who came and stayed during the postwar "guest worker" period: Turks in Germany, Algerians in France, Indonesians in the Netherlands, Indians and Pakistanis in the United Kingdom. Although many have lived in Europe for decades, they may not have been treated as citizens, or accorded rights given more recent immigrants from EU states, or assimilated as much as other immigrant groups.[156] This has caused friction between immigrant groups and, in France, riots in the immigrant neighborhoods of Paris.

Second, a wave of migrants from outside the European Union have made their way, legally and illegally, into Europe. They come from very poor countries in Africa and the Middle East and voyage to Spain, Portugal, Sicily, and the Greek islands, where they hope to gain entry into the European Union. In Greece, authorities "detained 112,364 illegal migrants in 2007, three times the number in 2004."[157] Like the United States, the European Union has found it difficult to curb illegal immigration. In 2008, Europe housed an estimated eight million illegal immigrants, compared with about 12 million in the United States. The entry of legal and illegal immigrants from outside Europe has created problems, in part because they have non-European ethnic, linguistic, and religious backgrounds and have been unwilling or unable to assimilate to European cultural norms. The acrimonious public debate over whether girls should be permitted to wear headscarves in public schools is a particular expression of this general problem.

In Germany, the situation is even more complex. Recall that ethnic Germans living in Eastern Europe and the Soviet Union "returned" to Germany after the war. They were then joined by East Germans who fled into West Germany, and by guest workers from southern Europe. After the guest worker program ended, many Turkish migrants stayed on, though they were not granted citizenship. A second wave of East German refugees migrated to West Germany in the fall of 1989, an exodus that toppled the regime and led to German reunification (see below). After reunification, East Germans migrated West and West Germans migrated East. During the 1990s, refugees from wars in Yugoslavia fled to Germany. After 2004, migrants from Poland and other new EU states came to work in Germany, and these legal migrants were joined by illegal immigrants from outside Europe. The foreign-born population in Germany grew from half a million in 1950 to three million in 1970, and seven million in 1995.[158] These successive waves of migration have created a complicated set of relations between different immigrant-resident groups in Germany.

POLITICAL MIGRATIONS

Most contemporary migration has been driven by economic forces. But large-scale migrations have also been triggered by political developments in the postwar period. To appreciate some of the historical causes and social consequences of politically generated migrations, it is useful to examine the migrations related to the partition of states, migrations that were promoted by dictatorships as part of a political strategy, and migrations triggered by ethnic conflict and war.

Partition

In the postwar period, the United States, the Soviet Union, Great Britain, and other "great powers" partitioned a number of countries so that competing political parties could obtain states of their "own."[159] The superpowers divided Korea, China, Vietnam, and Germany along ideological lines and transferred power either to "capitalist" political parties (in South Korea, Taiwan, South Vietnam, and West Germany) or to "communist" parties (in North Korea, China, North Vietnam, and East Germany). By contrast, Great Britain, with some assistance from the superpowers, divided India, Palestine, and later Cyprus between competing ethnic-religious groups: Hindus and Muslims in India and Pakistan, Jews and Muslim Arabs in Palestine, Orthodox Greeks and Muslim Turks in Cyprus (see chapters 8 and 9). Pakistan was later subdivided between different ethnic groups, though both groups were Muslim. The superpowers divided these countries because they thought that partition might avoid or end conflict between competing political parties and create states where they could separately practice their own ideological/ethnic politics.

In the 1990s, another group of countries were partitioned. Czechoslovakia and Ethiopia were divided into two parts, Yugoslavia into five states (maybe six in coming years), and the Soviet Union into fifteen successor states. Unlike the great power partitions of the postwar period, these countries were divided by people living in these countries, though the great powers played a role in the breakup of Yugoslavia and, more recently, of Serbia-Kosovo. Today, political conflicts in Belgium, Iraq, and Israel and the occupied territories may eventually result in partition. (In 1947, the United Nations partitioned Palestine into two states, but only one successor state—Israel—emerged after the war. A Palestinian state was never created, though there have been efforts to create one in the West Bank and Gaza Strip.)

Although the history of partition in each country is complex and diverse, the division of countries led to a common problem: large-scale migration across newly drawn borders. In Korea, 1.8 million people left the North for the South before the outbreak of the Korean War, and another 1.2 million moved south during the war.[160] During the last stage of the Chinese civil war, 1.5 million Nationalist troops and refugees from the mainland fled to Taiwan.[161] In Vietnam, 900,000, most of them Catholics, left the North for the South. Although states in Korea, China, and Vietnam sealed their borders soon after partition, migration in Germany continued after partition until Berlin was subdivided and the border sealed in 1961. During that period, between two and three million people, roughly one-sixth of the East German population, migrated to West Germany.[162] During the war that accompanied partition in Palestine in 1947, about 726,000 Arab Palestinian refugees left Israeli-controlled lands, according to the United Nations.[163] In India, 17.2 million moved across Indo-Pakistani borders in the first six months after partition in 1947, the largest, fastest migration in human history.[164] Muslims and Hindus joined miles-long refugee columns. Lord Mountbatten observed one column that was fifty miles long.[165] The exodus of Hindus and Muslims, in roughly equal numbers, was stimulated by anxieties about how the new governments would treat "minority" residents (Hindus in Pakistan; Muslims in India) and by the widespread violence that erupted around the country. Nearly one million people were killed in the three-month carnage that followed partition.[166] One office in Bengal reported, "This country has become a battlefield since the 16th of August. One village attacks another village and one community another community. Nobody could sleep for a week. Villages are being destroyed and thousands are being killed or wounded. Smoke fires are seen everywhere all around my village."[167]

Officials in India, as in every other divided state, were completely unprepared for the migration that accompanied partition. Pendrel Moon, a government minister in Pakistan admitted, "I foresaw, of course, a terrific upheaval. But I quite failed to grasp the speed with which disturbances and displacements of population . . . would resolve themselves into a vast movement of mass migration."[168]

People in divided states migrated in large numbers because they feared that the governments that took power would not respect their religious-ethnic identities or political affiliations; because they wanted to avoid the wars that threatened between sibling states; and in some cases, because they were driven from their homes by neighbors, militias, or government forces.

More recent partitions have also triggered disruptive migrations. The breakup of Yugoslavia created 1.4 million refugees. Tens of thousands migrated or were forced across newly drawn borders between Eritrea and Ethiopia. Several million Russians living in non-Russian republics migrated to Russia after the breakup of the Soviet Union. But while many migrants left, many also stayed. Governments in the Baltic passed laws depriving Russians of various rights as a way to encourage them to depart. But these policies have strained relations between indigenous groups and between the Baltic states and the Russian Federation.[169] Czechoslovakia has been an exception to the general rule. Partition in 1992 produced very little migration between the successor states and relations between the two countries have been amiable.[170]

Although partition triggered large migrations, many people decided to stay and take their chances. There are as many Muslims living in "Hindu" India as live in Pakistan, which is an "Islamic state." A sizable Muslim-Arab population lives in Israel, and Arab-Palestinians make up a majority of the population in the Israeli-occupied territories. There are still "Serbs" living in Kosovo and "Croats" in Bosnia. These "residual" populations are often seen by government as a political or social problem, which can lead to conflicts between these groups and the state (see chapters 8 and 9).

Partition not only spurred migration, it led in many cases to conflicts between divided states. This in turn triggered new waves of refugees, as it did during the long wars in Korea and Vietnam, the three Indo-Pakistani wars (see chapter 8), and five Arab-Israeli wars (see chapter 9). So, for example, millions of Bengali refugees fled into India during the 1974 Pakistani civil war. India then entered the war to end the fighting, return the refugees to their homes, and set up an independent state in Bangladesh.[171] The problem for divided states has been that the socially disruptive migrations associated with partition have created immediate difficulties and persistent, ongoing, long-term problems that have triggered further migrations.

Migration as a Political Strategy

Capitalist dictatorships and communist regimes have sometimes promoted migration as a political strategy. Although migration produced
some economic gains, in some cases it also resulted in political disaster
and regime change, as it did for capitalist dictatorships in southern Europe during the 1970s and for the East German communist regime in 1989.

In the 1950s and 1960s, capitalist dictatorships in Spain, Portugal, and
Greece encouraged young men to migrate to Western Europe and work
there as guest workers. (The Marcos dictatorship in the Philippines also
promoted migration, though it did so later—in the 1970s—and generally
promoted the migration of women, not men.) These regimes promoted
migration because they thought it would reduce unemployment and export dissent. Dictators worried that if young, unemployed, and unmarried men sat around in cafes and complained about their economic circumstances, they might become critical of their governments and take
steps to challenge them. By exporting workers, regimes could curb political dissent and secure the economic benefits associated with remittances.

This strategy worked for a time. But in the early 1970s, when European
states ended their guest worker programs and sent migrants home, the
strategy collapsed and an economic crisis ensued (see chapter 6). The economic crisis, which was triggered by return migration, joined with different
political crises—the death of Franco in Spain, Greek military misadventure
in Cyprus, and a military uprising in Portugal—to force fascist dictatorships from power in all three countries. Democratic governments in Spain,
Portugal, and Greece then joined the European Community, which promoted economic development and created jobs for former migrants. Today,
these countries import migrant workers from other countries.

Communist regimes in East Germany, Cuba, and Vietnam also developed migration policies that were designed to vent dissent and secure
economic benefits. But the strategies they adopted differed in important
ways from capitalist dictatorships in Southern Europe or the Philippines.

When communist regimes took power in East Germany (1949), Vietnam
(1956 in the North, 1975 in the South), and Cuba (1959), hundreds of thousands of people fled to neighboring, non-communist countries. The exodus of skilled, educated, and professional workers and entrepreneurs
threatened to cripple their economies, so they sealed their borders and
curtailed most migration.

In East Germany, the regime closed the border with West Germany. But
East Germans continued to migrate into West Berlin as a way to head
west. In 1961, the regime built the Berlin Wall, which it called the "Antifaschistischer Schutzwall" ("anti-fascist bulwark") to close this escape
route and stem the exodus, which had created labor shortages and threat-

ened to bring the economy to a halt. But after it sealed the border, the regime adopted a migration policy that allowed 616,006 East Germans to leave the country between 1962 and 1988. It did so both to vent dissent—to get rid of people opposed to the regime—and to secure the remittances migrants sent home to relatives. The regime also "ransomed" 29,670 politically active dissidents to West Germany, which paid the East German government about DM 100,000 for each refugee.[172]

The regime's carefully controlled migration-ransom policy was undone in 1989 when the communist regime in Hungary collapsed and the new government opened its border with Austria. East Germans took advantage of this back-door route and, during the fall of 1989, 343,854 of them made their way to Hungary and then to the West. By December, 50,000 were leaving each week. This rapid migration brought the East German economy to a standstill. Demonstrators in Leipzig chanted, "We want to leave!" The resulting political crisis led to the fall of the regime, the destruction of the Berlin Wall, and the reunification of Germany. The politics of migration played a crucial role in both the survival and the demise of the communist regime in East Germany (see chapter 6).

In Cuba, the Castro regime generally restricted the exodus of Cubans to the United States after it took power. But in 1980, the regime allowed 125,000 migrants to take boats and voyage from Mariel, Cuba, across the straits to Florida. The regime did so to vent dissent and also to discharge criminals from Cuban jails. As elsewhere, the regime expected migrant Cubans to send money home to relatives in Cuba, just as earlier Cuban refugees had done.

The communist regime in Vietnam adopted a similar strategy. After communist forces from the North took control of South Vietnam in 1975, two million South Vietnamese "boat people" fled to other southeast Asian countries and many migrated eventually to the United States.[173] Like Cuban immigrants, they sent remittances back to relatives in Vietnam. The irony is that by doing so, anti-communist migrants from Vietnam and Cuba help keep communist regimes in power.

Ethnic Conflict and War

In many parts of the world—Sudan, Somalia, Sri Lanka, Rwanda—migration has been driven by ethnic conflict and war. Rwanda, a small country in central Africa, illustrates how ethnic conflict and migration interact.

During the 1930s, when Rwanda was a Belgian colony, authorities introduced an identification system that assigned residents to one of two "ethnic" identities: "Tutsi" and "Hutu."[174] In fact, people in both groups belonged to the Banyariwanda people, spoke the same language, and practiced the same faith: Catholicism. Rwanda is the most Christian country in Africa. The only real difference was economic. The Tutsi minority (15 percent) were generally cattle herders; the Hutu majority (85 percent) were farmers. The Belgians emphasized their economic differences as part of a divide-and-rule colonial policy. The Belgians gave preferential treatment to the Tutsi minority, allowing them some access to education and government jobs, a practice that antagonized the Hutu majority.

When Rwanda won its independence from Belgium in 1961, representatives of the Hutu majority took power, turned Tutsi out of government jobs, and, in 1963 and 1973, permitted Hutu attacks on Tutsi across the country. During the 1973 pogroms, one million Tutsi fled the country (another one million stayed behind), and many of them took refuge in Uganda, where they organized an army that waged a guerrilla war against the Hutu government in Rwanda. In the early 1990s, Tutsi rebels made significant military gains against government troops and forced President Habyarimana into negotiations, which resulted in a 1993 peace treaty.[175] But on April 6, 1994, President Habyarimana was killed when his plane was blown up as it approached the capital. Hutu officials opposed to the government's peace treaty with Tutsi rebels used the president's death as a pretext to attack the resident Tutsi population. The Hutu army, together with an irregular militia called the Interahamwe ("Let us strike together"), which had been organized around male soccer teams, attacked and killed Tutsi men, women, and children across the country.[176] During the next 100 days, the *genocidaires* killed 800,000 Tutsi, or five every minute, at a rate that was faster than Nazis killed Jews and other captives during the Holocaust.[177]

Rebel Tutsi forces advanced on the capital to stop the genocide and defeated the Hutu army. The Hutu army, Interahamwe, and hundreds of thousands of Hutu refugees, many of them *genocidaires*, fled into the Congo and Tanzania, where they gathered in U.N.-sponsored refugee camps.[178] They then used these camps as a base for cross-border attacks on Tutsi in Rwanda.

In 1996, the Tutsi government in Rwanda joined forces with the Ugandan government and sent their armies into the Congo to attack Hutu rebel bases and repatriate the Hutu civilians living in the refugee camps. Hutu refugees living in the Congo and in Tanzania returned to their homes in Rwanda, where they now live alongside the Tutsi survivors and the 800,000 Tutsi who have returned to Rwanda after the genocide.

But the fighting between forces from Rwanda and Uganda and Hutu militias triggered a wider war in the Congo. Rebel forces led by Laurent Kabila, who was fighting to overthrow the Mobutu dictatorship in the Congo, joined with armies from Rwanda and Uganda and these forces together launched a campaign to overthrow Mobutu Sese Seko, who had ruled the Congo since 1965. They soon captured the capital, forced Mobuto into exile, and installed Laurent Kabila as president of the Congo.[179] One year later, Kabila broke with his Rwandan backers, "who then sponsored another rebellion, this time against Mr. Kabila," that ignited a multisided civil war in the Congo. The civil war ended in 2003, but flared up again in 2007. About four million people died of hunger and disease during the civil war, which displaced millions of people across the region. The most recent outbreak in 2007 forced 800,000 people from their homes.[180] "Running, running, we are always running," Simwirayi Byenda, a refugee with two children told reporters. "It is always the civilians who suffer."

NOTES

1. Stephen Castles and Mark J. Miller, *The Age of Migration: International Population Movements in the Modern World* (New York: Guilford Press, 1998).

2. Jason DeParle, "A Tiny Staff, Tracking People across the Globe," *New York Times*, 4 February 2008. See www.migrationinformation.com.

3. There are also about 764 million tourists, temporary cross-border migrants, but they will not be included in this discussion. Robin Cohen, *Migration and Its Enemies: Global Capital, Migrant Labour and the Nation-State* (Aldershot, UK: Ashgate, 2006), 188.

4. Ian Goldin and Kenneth Reinert, *Globalization for Development: Trade, Finance, Aid, Migration and Policy* (Washington, D.C.: World Bank and Palgrave Macmillan, 2007), 157; Deutsche Bank Research, "International Migration: Who, Where and Why?" in *The Migration Reader: Exploring Politics and Policies*, ed. Anthony M. Messina and Gallya Lahav (Boulder, Colo.: Lynne Rienner, 2006), 17.

5. Castles and Miller, *The Age of Migration*, 4–5; Michael Parfit, "Human Migration," *National Geographic* (October 1998), 16; Nick Cumming-Bruce, "World's Refugee Count in 2007 Exceeded 11 Million, UN Says," *New York Times*, 18 June 2008.

6. Castles and Miller, *The Age of Migration*, 9.

7. Immanuel Wallerstein, *The Modern World-System: Capitalist Agriculture and the Origins of the European World-Economy in the Sixteenth Century* (New York: Academic Press, 1974), 92–94.

8. Marcus Rediker, *The Slave Ship: A Human History* (New York: Viking, 2007), 5.

9. Castles and Miller, *The Age of Migration*, 53; Rediker, *The Slave Ship*, 10.

10. Robert J. Flanagan, *Globalization and Labor Conditions: Working Conditions and Worker Rights in a Global Economy* (Oxford: Oxford University Press, 2006), 155–56; Daniel Franklin, "Migration: New Demands and Approaches for Europe," in *Migration Policies in Europe and the United States*, ed. Giacomo Luciani (Dordrecht: Kluwer Academic, 1993), 17.

11. Flanagan, *Globalization and Labor Conditions*, 94.

12. Shigeto Tsuru, *Japan's Capitalism: Creative Defeat and Beyond* (Cambridge: Cambridge University Press, 1993), 68.

13. Richard B. Craig, *The Bracero Program: Interest Groups and Foreign Policy* (Austin: University of Texas Press, 1971), 102–3.

14. Lant Pritchett, *Let Their People Come: Breaking the Gridlock on International Labor Mobility* (Washington, D.C.: Center for Global Development, 2006), 69.

15. Cohen, *Migration and Its Enemies*, 49–50.

16. Pritchett, *Let Their People Come*, 69.

17. Jason DeParle, "Rising Breed of Migrant Worker: Skilled, Salaried and Welcome," *New York Times*, 20 August 2007.

18. Robert Pear, "High-Tech Titans Strike Out on Immigration Bill," *New York Times*, 25 June 2007.

19. See discussion of different theories of international migration in Douglas S. Massey, Juaquin Arango, Graeme Hugo, Ali Kouaouci, Adela Pellegrino, and J. Edward Taylor, "Theories of International Migration: A Review and Appraisal," in *The Migration Reader: Exploring Politics and Policies*, ed. Anthony M. Messina and Gallya Lahav (Boulder, Colo.: Lynne Rienner, 2006), 34–48.

20. Pritchett, *Let Their People Come*, 5–6; Peter Stalker, *Workers without Frontiers: The Impact of Globalization on International Migration* (Boulder, Colo.: Lynne Rienner, 2000), 17.

21. Stalker, *Workers without Frontiers*, 23.

22. Stalker, *Workers without Frontiers*, 23.

23. www.bls.gov/fls/ichccreport.pdf.

24. Eduardo Porter, "Here Illegally, Working Hard and Paying Taxes," *New York Times*, 19 June 2006; Pritchett, *Let Their People Come*, 35.

25. Pear, "High-Tech Titans Strike Out."

26. Tamar Lewin, "Study Says Foreign Students Added $14 Billion to Economy," *New York Times*, 12 November 2007.

27. Bill Orr, *The Global Economy in the '90s: A User's Guide* (New York: New York University Press, 1992), 65–66.

28. Lester Brown, ed., *The State of the World 1992* (New York: Norton, 1992), 350.

29. Philip Shenon, "Rearranging the Population: Indonesia Weighs the Pluses and Minuses," *New York Times*, 8 October 1992.

30. Saskia Sassen, "Foreign Investment: A Neglected Variable," in *The Migration Reader: Exploring Politics and Policies*, ed. Anthony M. Messina and Gallya Lahav (Boulder, Colo.: Lynne Rienner, 2006), 598.

31. Flanagan, *Globalization and Labor Conditions*, 120.

32. See Matthew R. Sanderson, "The Political Economy of International Migration: Demographic Aspects of Global Stratification" (Ph.D. dissertation, University of Utah, 2008); Sassen, "Foreign Investment," 601.

33. Sharon Lafraniere, "Europe Takes Africa's Fish, and Migrants Follow," *New York Times*, 14 January 2008.

34. Lafraniere, "Europe Takes Africa's Fish."

35. Taras Grescoe, "Sardines with Your Bagel?" *New York Times*, 9 June 2008.

36. John Tagliabue, "The Gauls at Home in Erin," *New York Times*, 2 June 2006.

37. Tagliabue, "The Gauls at Home in Erin."

38. Norimitsu Onishi, "For Studies in English, Koreans Learn to Say Goodbye to Dad," *New York Times*, 8 June 2008.

39. Onishi, "For Studies in English."

40. Onishi, "For Studies in English."

41. Sarah J. Mahler and Patricia Pessar, "Gender Matters: Ethnographers Bring Gender from the Periphery toward the Core of Migration Studies," *International Migration Review* 40, no. 1 (Spring 2006): 48.

42. Jason DeParle, "A Good Provider Is One Who Leaves," *New York Times Magazine*, 22 April 2007.

43. Geertje Lycklama à Nijeholt, "The Changing Division of Labor and Domestic Workers," in *The Trade in Domestic Workers: Causes, Mechanisms and Consequences of International Migration*, ed. Noeleen Heyzer, Geertje Lycklama à Nijeholt, and Nedra Weerakoon (London: Zed Books, 1994), 15–16; Castles and Miller, *The Age of Migration*, 152; Cohen, *Migration and Its Enemies*, 169.

44. DeParle, "A Good Provider"; Joseph Berger, "From Philippines with Scrubs: How One Ethnic Group Came to Dominate Nursing," *New York Times*, 24 November 2003.

45. Berger, "From Philippines with Scrubs."

46. Berger, "From Philippines with Scrubs."

47. Tamar Lewin, "Recruiters for 'Sisters' Colleges See a Bounty in the Middle East," *New York Times*, 3 June 2008.

48. Lewin, "Recruiters for 'Sisters' Colleges."

49. Nicole Constable, ed., *Cross-Border Marriages: Gender and Mobility in Transnational Asia* (Philadelphia: University of Pennsylvania Press, 2005), 4.

50. Constable, *Cross-Border Marriages*, 4.

51. Castles and Miller, *The Age of Migration*, 173.

52. Hung Cam Thai, "Clashing Dreams in the Vietnamese Diaspora: Highly Educated Overseas Brides and Low-Wage U.S. Husbands," in *Cross-Border Marriages: Gender and Mobility in Transnational Asia*, ed. Nicole Constable (Philadelphia: University of Pennsylvania Press, 2005), 171.

53. Thai, "Clashing Dreams," 170.

54. Thai, "Clashing Dreams," 171–73.

55. Thai, "Clashing Dreams," 147.

56. Constable, *Cross-Border Marriages*, 10.

57. Thai, "Clashing Dreams," 148.

58. Constable, *Cross-Border Marriages*, 6–7.

59. Caren Freeman, "Marrying Up and Marrying Down: The Paradoxes of Marital Mobility for Chosonjok Brides in South Korea," in *Cross-Border Marriages: Gender and Mobility in Transnational Asia*, ed. Nicole Constable (Philadelphia: University of Pennsylvania Press, 2005), 80–90; Norimitsu Onishi, "Wed to Strangers, Vietnamese Wives Build Korean Lives," *New York Times*, 30 March 2008.

60. Eleonore Kofman, Annie Phizacklea, Parvati Raghuram, and Rosemary Sales, *Gender and International Migration in Europe: Employment, Welfare and Politics* (London: Routledge, 2000), 67.

61. Rita J. Simon and Howard Altstein, *Adoption across Borders: Serving the Children in Transracial and Intercountry Adoption* (Lanham, Md.: Rowman & Littlefield, 2000), 4–5.

62. Ron Nixon, "De-emphasis on Race in Adoption Is Criticized," *New York Times*, 27 May 2008; Simon and Altstein, *Adoption across Borders*, 2, 4.

63. Jane Gross, "U.S. Joins Overseas Adoption Overhaul Plan," *New York Times,* 11 December 2007.

64. Gross, "U.S. Joins Overseas Adoption Overhaul Plan."

65. Simon and Altstein, *Adoption across Borders,* 6; Gross, "U.S. Joins Overseas Adoption Overhaul Plan."

66. Gross, "U.S. Joins Overseas Adoption Overhaul Plan."

67. Dan Frosch, "New Rules and Economy Strain Adoption Agencies," *New York Times,* 11 May 2008; Mireya Navarro, "To Adopt, Please Press Hold," *New York Times,* 5 June 2008.

68. Simon and Altstein, *Adoption across Borders,* 8.

69. Simon and Altstein, *Adoption across Borders,* 6, 8, 14, 16, 19.

70. Gross, "U.S. Joins Overseas Adoption Overhaul Plan"; Marc Lacey, "Guatemala: U.S. Adoptions to Go Through," *New York Times,* 12 December 2007.

71. Lacey, "Guatemala."

72. Karin Evans, *The Lost Daughters of China* (New York: Tarcher/Putnam, 2000), 105.

73. Evans, *The Lost Daughters of China,* 111.

74. Evans, *The Lost Daughters of China,* 114.

75. Evans, *The Lost Daughters of China,* 113.

76. Kay Ann Johnson, *Wanting a Daughter, Needing a Son: Abandonment, Adoption, and Orphanage Care in China* (St. Paul, Minn.: Young and Young, 2004), 60.

77. Johnson, *Wanting a Daughter,* 6.

78. Johnson, *Wanting a Daughter,* 6.

79. Johnson, *Wanting a Daughter,* 11.

80. Johnson, *Wanting a Daughter,* 70.

81. Johnson, *Wanting a Daughter,* 38, 22.

82. Giovanna Campani, "Trafficking for Sexual Exploitation and the Sex Business in the New Context of International Migration: The Case of Italy," in Martin Baldwin-Edwards and Joaquin Arango, *Immigrants and the Informal Economy of Southern Europe* (London: Frank Cass, 1999), 241.

83. Beverly Mullings, "Globalization, Tourism, and the International Sex Trade," in *Global Sex Workers: Rights, Resistance, and Redefinition,* ed. Kamala Kempadoo and Jo Doezema (Lanham, Md.: Rowman & Littlefield, 1999), 66.

84. Sipiporn Skrobanek, Nataya Boonpakdee, and Chutima Jantateero, *The Traffic in Women: Human Realities of the International Sex Trade* (London: Zed, 1997), 66.

85. Campani, "Trafficking for Sexual Exploitation," 240–41.

86. Campani, "Trafficking for Sexual Exploitation," 240.

87. Goldin and Reinert, *Globalizaton for Development,* 166.

88. Cynthia Enloe, *Bananas, Beaches and Bases: Making Feminist Sense of International Politics* (Berkeley: University of California, 1989), 66.

89. Edward A. Gargan, "Traffic in Children Is Brisk (Legacy of the Navy?)," *New York Times,* 11 December 1997.

90. Enloe, *Bananas, Beaches and Bases,* 86–87.

91. Calvin Sims, "A Hard Life for Amerasian Children," *New York Times,* 23 July 2000.

92. Sims, "A Hard Life."

93. Skrobanek, Boonpakdee, and Jantateero, *The Traffic in Women*, 8.

94. Skrobanek, Boonpakdee, and Jantateero, *The Traffic in Women*, 57; Enloe, *Bananas, Beaches and Bases*, 35–36; Kamala Kempadoo, "Introduction," in *Global Sex Workers: Rights, Resistance, and Redefinition*, ed. Kamala Kempadoo and Jo Doezema (Lanham, Md.: Rowman & Littlefield, 1999), 6.

95. Julia O'Connell Davidson and Jacqueline Sanchez Taylor, "Fantasy Islands: Exploring the Demand for Sex Tourism," in *Global Sex Workers: Rights, Resistance, and Redefinition*, ed. Kamala Kempadoo and Jo Doezema (Lanham, Md.: Rowman & Littlefield, 1999), 37.

96. Skrobanek, Boonpakdee, and Jantateero, *The Traffic in Women*, 14.

97. William Finnegan, "The Counter Traffickers," *The New Yorker*, 5 May 2008.

98. DeParle, "A Good Provider."

99. DeParle, "A Good Provider."

100. Kofman et al., *Gender and International Migration*, 115.

101. Kofman et al., *Gender and International Migration*, 115–16.

102. Barbara Crossette, "In India and Africa, Women's Low Status Worsens Their Risk of AIDS," *New York Times*, 26 February 2001.

103. Sharon Stanton Russell, "Remittances from International Migration: A Review in Perspective," in *The Sociology of Migration*, ed. Robin Cohen (Cheltenham, UK: Edward Elgar, 1996), 79–80; Jason DeParle, "Migrant Money Flow: A $300 Billion Current," *New York Times*, 18 November 2007.

104. Elisabeth Malkin, "Mexicans Miss Money from Relatives Up North," *New York Times*, 26 October 2007.

105. DeParle, "A Good Provider"; Parfit, "Human Migration," 31; Goldin and Reinert, *Globalization for Development*, 177.

106. DeParle, "A Good Provider."

107. Stalker, *Workers without Frontiers*, 80; Russell, "Remittances from International Migration," 682.

108. Marla Dickerson, "'Dollarization' Buys Optimism, Little Else," *Seattle Times*, 5 August 2007.

109. Dickerson, "'Dollarization.'"

110. Russell, "Remittances from International Migration," 686, 678.

111. Stalker, *Workers without Frontiers*, 82.

112. Deborah Sontag, "A Mexican Town That Transcends All Borders," *New York Times*, 21 July 1998.

113. Russell, "Remittances from International Migration," 78.

114. Arlie Russell Hochschild, "Love and Gold," in *Global Woman: Nannies, Maids, and Sex Workers in the New Economy*, ed. Barbara Ehrenreich and Arlie Russell Hochschild (New York: Henry Holt, 2002), 22.

115. DeParle, "A Good Provider."

116. Eduardo Porter, "Here Illegally, Working Hard and Paying Taxes," *New York Times*, 19 June 2006.

117. Porter, "Here Illegally."

118. Richard Dooling, "Immigration Beefs Up Nebraska," *New York Times*, 11 June 2006.

119. Joshua Brustein, "Farm Machines Replace Shrinking Migrant Work Force," *New York Times*, 27 May 2008.

120. Jason DeParle, "A Western Union Empire Moves Migrant Cash Home," *New York Times*, 22 November 2007.

121. Stephen Moore, "Give Me Your Best, Your Brightest: Immigration Policy Benefits U.S. Society despite Increasing Problems," in *The Migration Reader: Exploring Politics and Policies*, ed. Anthony M. Messina and Gallya Lahav (Boulder, Colo.: Lynne Rienner, 2006), 330.

122. Moore, "Give Me Your Best," 330.

123. *The Economist*, "A Modest Contribution," in *The Migration Reader: Exploring Politics and Policies*, ed. Anthony M. Messina and Gallya Lahav (Boulder, Colo.: Lynne Rienner, 2006), 337.

124. Porter, "Here Illegally"; Nina Bernstein, "Tax Returns Rise for Immigrants in U.S. Illegally," *New York Times*, 16 April 2007.

125. Howard W. French, "Insular Japan Needs, but Resists Immigration," *New York Times*, 24 July 2003.

126. French, "Insular Japan."

127. French, "Insular Japan."

128. Flanagan, *Globalization and Labor Conditions*, 102; George J. Borjas, "The New Economics of Immigration: Affluent Americans Gain, Poor Americans Lose," in *The Migration Reader: Exploring Politics and Policies*, ed. Anthony M. Messina and Gallya Lahav (Boulder, Colo.: Lynne Rienner, 2006), 320–25; Stalker, *Workers without Frontiers*, 86–87; James Surowiecki, "Be Our Guest!" *The New Yorker*, 11 & 18 June 2007.

129. Rachel L. Swarns, "Bridging a Racial Rift That Isn't Black and White," *New York Times*, 3 October 2006.

130. Swarns, "Bridging a Racial Rift."

131. Swarns, "Bridging a Racial Rift."

132. Barry Bearak and Celia W. Dugger, "South Africans Take Out Rage on Immigrants," *New York Times*, 20 May 2008.

133. Jeffrey Stinson, "Immigrants Feel Squeeze in Europe," *USA Today*, 10 June 2008.

134. Stinson, "Immigrants Feel Squeeze."

135. Seth Mydans, "A Growing Source of Fear for Migrants in Malaysia," *New York Times*, 10 December 2007.

136. Mydans, "A Growing Source of Fear."

137. Nina Bernstein, "On Lucille Avenue, the Immigration Debate," *New York Times*, 26 June 2006.

138. Bernstein, "On Lucille Avenue."

139. Kareen Fahim, "Indian Immigrants Hit by Housing Crackdown," *New York Times*, 6 August 2006.

140. Damien Cave, "Local Officials Skirting Federal Rules in a Bid to Snare Illegal Aliens," *New York Times*, 9 June 2008.

141. Djulia Preston, "Immigration Is at Center of New Laws around U.S.," *New York Times*, 6 August 2007.

142. Preston, "Immigration Is at Center."

143. Ian Austin, "Immigration in Quebec: A Hornet's Nest," *New York Times*, 23 December 2007.

144. Jessie McKinley, "San Francisco Bay Area Reacts Angrily to Series of Immigration Raids," *New York Times*, 28 April 2007.

145. Randal C. Archibold, "Arizona Seeing Signs of Flight by Immigrants," *New York Times*, 12 January 2008.

146. Pritchett, *Let Their People Come*, 74.

147. Julia Preston, "Poll Surveys Ethnic Views among Chief Minorities," *New York Times*, 13 December 2007.

148. Jason DeParle, "Spain, Grappling with Illegal Immigrants, Tries Forgiveness," *New York Times*, 10 June 2008.

149. Castles and Miller, *The Age of Migration*, 99.

150. Tagliabue, "The Gauls at Home in Erin."

151. Judy Dempsey, "Polish Labor Crunch as Workers Go West," *New York Times*, 19 November 2006.

152. Emma Daly, "By Enticing Foreigners, Villages Grow Young Again," *New York Times*, 31 July 2003.

153. DeParle, "Spain, Grappling with Illegal Immigrants"; Pritchett, *Let Their People Come*, 74.

154. DeParle, "Spain, Grappling with Illegal Immigrants."

155. Julia Weriger, "As the Poles Get Richer, Fewer Seek British Jobs," *New York Times*, 19 October 2007.

156. Castles and Miller, *The Age of Migration*, 72, 81.

157. Anthee Curassava, "Greek Islands, Overwhelmed by Refugees, Seek Help," *New York Times*, 7 May 2008.

158. Castles and Miller, *The Age of Migration*, 72, 80.

159. See Robert K. Schaeffer, *Warpaths: The Politics of Partition* (New York: Hill and Wang, 1990); Robert K. Schaeffer, *Severed States: Dilemmas of Democracy in a Divided World* (Lanham, Md.: Rowman & Littlefield, 1999).

160. John Sullivan, *Two Koreas—One Future?* (Lanham, Md.: University of America Press, 1987), 100; Gregory Henderson, R. N. Lebow, and J. G. Stroessinger, *Divided Nations in a Divided World* (New York: David McKay, 1974), 60–61.

161. Henderson, Lebow, and Stroessinger, *Divided Nations*, 100.

162. Henderson, Lebow, and Stroessinger, *Divided Nations*, 28.

163. Benny Morris, *The Birth of the Palestinian Refugee Problem, 1947–1949* (Cambridge: Cambridge University Press, 1988), 297–98. Israeli sources put the figure at 520,000; Arab sources claim 900,000.

164. R. F. Holland, *European Decolonization, 1918–1981* (New York: St. Martin's Press, 1985), 80.

165. H. V. Hodson, *The Great Divide* (Oxford: Oxford University Press, 1985), 411.

166. Holland, *European Decolonization*, 80.

167. Hodson, *The Great Divide*, 404.

168. Hodson, *The Great Divide*, 404.

169. Dan Bilefshy, "Latvia Fears New 'Occupation' by Russians but Needs the Labor," *New York Times*, 16 November 2006.

170. For a discussion of Czechoslovakian exceptionalism, see Schaeffer, *Severed States*, 224–29.

171. Schaeffer, *Severed States*, 177–83.

172. Robert K. Schaeffer, *Power to the People: Democratization around the World* (Boulder, Colo.: Westview, 1997), 182–83.

173. Thai, "Clashing Dreams," 145–46.

174. Philip Gourevitch, *We Wish to Inform You That Tomorrow We Will Be Killed with Our Families* (New York: Farrar, Straus & Giroux, 1998), 56–57.

175. Gourevitch, *We Wish to Inform You*, 99.

176. Gourevitch, *We Wish to Inform You*, 93

177. Gourevitch, *We Wish to Inform You*, 133.

178. Parfit, "Human Migration," 11.

179. Lydia Polgreen, "Fear of New War As Clashes Erupt on Congo's Edge," *New York Times*, 13 December 2007.

180. Polgreen, "Fear of New War."

6

⑨

Dictatorship and Democracy

In late 1989, communist dictatorships in seven Eastern European countries suddenly fell, most of them peacefully, and civilian democrats assumed power for the first time in more than forty years. The simultaneous collapse of seven communist dictatorships, capped by the dramatic
opening and then destruction of the Berlin Wall, the most visible symbol
of dictatorship in one-party states, was the high-water mark of contemporary democratization. But democratization in Poland, East Germany,
Czechoslovakia, Hungary, Bulgaria, Romania, and Albania was not the
first nor the last episode of democratization in recent years. Events in
Eastern Europe were only one act in a global drama spanning the past
twenty years, a process that first toppled capitalist dictatorships in southern Europe, Latin America, and East Asia, then destroyed communist
regimes in Eastern Europe and the Soviet Union, and, finally, brought an
end to whites-only rule in South Africa.[1]

The rise of democracy in at least thirty countries around the world since
1974 was not only welcome but remarkable for two reasons.[2] First, democratization or the transfer of political power from dictators or one-
party regimes to civilian democracies occurred peacefully. With the exception of some violence in Romania and considerable violence in South
Africa, the transfer of power was achieved without bloodshed. Second,
dictators themselves often initiated the process, taking steps that made
possible open elections and a return to civilian authority. Few analysts or
dissident movements opposed to dictatorship expected dictators to propose dramatic reforms or to surrender power without a fight. But many
of them did just that, opening negotiations with their opponents, quitting

their offices, and retiring from public life. Of course, dissident movements and popular protests played an important role in some countries, particularly in South Korea, Poland, and South Africa. But in most cases, mass movements organized protests only after dictators initiated reform. Democratization might therefore be described as a "devolutionary" process because the transfer of political power was peacefully achieved and because it did not result in the kind of violent revolutionary change associated with the American, French, or Russian revolutions.

Contemporary democratization occurred in great regional waves, moving westward around the world. It began in southern Europe in the mid-1970s, then moved west across the Atlantic to Latin America. The dictatorships in major Latin American countries began folding in the mid-1980s, and most of the remaining dictatorships disappeared by the end of the decade. In the late 1980s, several countries in East Asia democratized, followed in 1989 by the collapse of communist regimes in Eastern Europe. A few years later, the Soviet Union democratized and dissolved into separate republics, as did Yugoslavia and Czechoslovakia. Then South Africa abandoned apartheid, adopted majority rule, and in 1994, inaugurated a black president.

In each of these geographic regions, democratization had rather different causes. For example, the debt crisis contributed to democratization in most Latin American countries, while in East Asia it was rapid economic growth that helped trigger democratization. Economic stagnation was a primary cause of democratization in the Soviet Union and Eastern Europe, while trade sanctions, embargo, and divestment played a major role in South Africa. In these different regions, economic crises of one sort or another, compounded by problems associated with military defeat, the illness or death of an aging dictator, popular uprising, or changed superpower policy forced dictators to surrender power.

Although democratization had very different causes in each region, its political and economic consequences were everywhere much the same. In political terms, democratizing countries drafted new constitutions, held elections, and allowed numerous and diverse political parties to participate. In economic terms, most democratizing countries opened their economies to foreign investment and trade, sold state-owned assets and industries to private investors and entrepreneurs—processes associated with globalization—and reduced military spending.

People expected democratic politics and economic policies to solve the problems that first confronted dictators and helped force them from power. Unfortunately, civilian democrats in many countries have been unable to solve their country's economic problems. Where democratic politics have been corrupted or new economic policies have failed, new dictators have threatened a return to power.

To understand the prospects of democratizing states, we will look first at the problems that led to the collapse of capitalist dictatorships and communist regimes in different regions during the past twenty years. Then we will examine the common political and economic strategies adopted by most civilian democrats after they assumed power. In this context we will discuss the problems and prospects of democratizing states around the world.

SOUTHERN EUROPE: FALLING BEHIND

Although fascist dictatorships in Italy and Germany were crushed by Allied forces during World War II, dictatorships in Portugal and Spain survived the war, largely because they stayed neutral in the conflict. In Portugal, António de Oliveira Salazar, a former economics professor, assumed power in 1930; in Spain, General Francisco Franco defeated republican forces during a bitter civil war and became dictator in 1939. The Iberian dictatorships, which would survive until the early 1970s, were joined by Greece in 1967, when a military coup established a dictatorship on the Ionian Peninsula.

By Western European standards, Portugal, Spain, and Greece were poor countries. After World War II, the dictatorships in Portugal and Spain, and the civilian and later military government in Greece, made some economic gains, but their improvements in per capita incomes and literacy were based on relatively weak economies. Because none of them had substantial industries that could compete in overseas markets, they all imported more than they exported and posted trade deficits for every year between 1946 and 1974. As the Turkish economist Caglar Keyder noted, "None of the southern European countries (except Spain in 1951 and [again] in 1960) ran a commodity surplus in all the years between 1946 and 1974. This is to say that none of [them] ever reached a point when their economies generated sufficient exports to pay for their imports."[3]

Most economies and the dictators who run them cannot long survive persistent trade deficits of this sort. But Iberian and Ionian dictatorships were able to make up for their industrial deficiencies and grow somewhat during this period (Spain more than Portugal or Greece) because they exported workers, not industrial goods, imported free-spending Western European tourists, and received significant infusions of cash from external sources.[4] As Raymond Carr and Juan Aizpurua note:

> The economy was refueled from abroad: by tourist earnings, by the remittances of emigres working abroad, and by foreign loans. Only these invisible earnings and loans made it possible to realize . . . plans for rapid growth without running up against the balance of payments problems that bring growth to a grinding halt in most poor economies.[5]

After World War II, Western European countries, with the assistance of the U.S. Marshall Plan, recovered and began to grow. They soon experienced labor shortages, in part because the war had killed or crippled many potential workers. So they began hiring workers from Europe's periphery, particularly from Spain, Portugal, southern Italy, Greece, and Turkey. Workers in those countries, particularly in rural areas, were eager to seek work abroad because they faced poverty and unemployment at home. During the 1960s, nearly one million Portuguese workers left the country to work in France and West Germany, and in the early 1970s, 100,000 were emigrating annually.[6] In Spain, half a million workers had left the province of Andalusia during the 1950s. This emigration emptied rural villages. The Castilian novelist Miguel Dlibes wrote of one village: "Cartiguera is a dying village, in agony. Its winding streets, invaded by weeds and nettles, without a dog's bark or a child's laugh to break the silence, enclosed pathetic gravity, the lugubrious air of the cemetery."[7]

But while this was a difficult and sometimes painful process, it greatly reduced domestic unemployment, providing a safety valve for dictatorships, and helped boost their economies because emigrants sent remittances, or money they earned abroad, back to families who stayed behind. In Spain, for example, emigrants' remittances made up about one-half of the annual trade deficit in the 1960s.[8]

While they exported workers, governments in Spain, Portugal, and Greece also imported tourists and spent much of their budgets to build the hotels, resorts, and infrastructure needed to develop this industry. For example, by 1973, the number of tourists visiting Spain exceeded the number of people living there.[9] These tourists spent money and created jobs, and this, too, helped what the economist Paul Samuelson called "market fascism" in Spain to survive.[10]

All three countries also received large infusions of economic aid from external sources. Although the United States had fought to destroy dictatorships in Europe during World War II, it provided financial aid to Iberian dictatorships after the war in return for the establishment of U.S. military bases there in the 1950s and gave economic and military assistance to Greece, a country bordering communist dictatorships in Eastern Europe.[11] Portugal also relied on another source of external wealth during the postwar period: its colonies in Africa. Portugal derived considerable income from its exploitation of Mozambique, Angola, and Guinea-Bissau. The Salazar dictatorship refused to relinquish its colonies, despite armed rebellion by independence movements and the withdrawal of other European states from Africa during the period of decolonization in the 1950s and 1960s.

Buoyed by income from emigrant workers, tourists, and superpower aid, the dictatorships in Portugal, Spain, and later Greece managed to keep their economies afloat during the 1950s and 1960s. Of the three, Spain was the most successful, recording rapid rates of growth, which some economists then described as a "miracle." But much of Spain's postwar growth was simply recovering economic ground lost during the Depression, Civil War, and World War II.[12] By the end of the 1960s, it was clear that despite some modest gains, the rest of Western Europe was leaving these countries behind economically. In the 1970s, conditions that had made possible modest economic growth changed, confronting dictatorships with serious economic crises.

The crisis began when the United States devalued the dollar in 1971, making the value of Portuguese, Spanish, and Greek currencies rise. This made it more difficult for them to export goods and it increased their already large trade deficits. This problem was compounded by rising oil prices. When OPEC countries cut oil supplies and raised prices following the 1973 Yom Kippur War, the economic crisis deepened. Portugal was particularly hard hit because OPEC countries refused to sell oil to the dictatorship, now headed by Marcello Caetano, because it had allowed U.S. forces to use Portuguese bases to assist the Israeli war effort.[13]

The oil crisis of 1973–1974 raised the cost of oil for countries with little oil of their own and increased their trade deficits. It also triggered a global recession. In response, Western European countries laid off immigrant workers and sent them home. This led to rising domestic unemployment and discontent and greatly reduced the money received from emigrant workers' earnings, which further increased trade deficits. And during the recession, fewer European families vacationed abroad, reducing tourist receipts.[14]

Of the three, the Portuguese dictatorship found itself in the most difficult straits. This was because its decade-long wars to prevent the independence of its African colonies had drained its treasury. "By 1974, a population of less than 9 million was sustaining a 200,000-man army in Africa and spending over 45 percent of its annual budget on the military," notes Kenneth Maxwell.[15] "The burdens of the African campaigns on a small, poor nation with limited resources and retarded economic and social infrastructures proved unsustainable."[16] With tongue in cheek, *The Economist* then described Portugal as "Africa's only colony in Europe."[17]

The economic crisis triggered by new conditions in the 1970s was compounded in Portugal, Spain, and Greece by military and political problems. For Portugal, massive military spending on wars in Africa did not help the military defeat armed independence movements. In 1973, António de Spínola, a leading Portuguese general, published a book arguing for a political solution. His book, *Portugal and the Future*, convinced many, particularly in the military, that Portuguese defeat was imminent.[18] Military officers who were radicalized by their experience in Africa then organized the

Armed Forces Movement (AFM), which sought to overthrow the Caetano dictatorship, democratize the political system, and end Portugal's anti-colonial wars in Africa. On April 25, 1974, AFM units in Portugal overthrew the Caetano regime.

Military defeat also played an important role a few months later in Greece. To deflect growing discontent at home, the "Colonels," as the Greek military dictatorship was known, supported a coup in Cyprus during the summer of 1974. The coup was led by Greek-speaking Cypriots who wanted to overthrow the government of Archbishop Makarios III so that the Mediterranean island could then be united with Greece.[19] But the coup prompted fighting between the island's Greek- and Turkish-speaking residents, a development that triggered an invasion of the island by Turkish forces and brought Greece and Turkey to the brink of war. Faced with humiliation in Cyprus and a potentially disastrous war with its more powerful neighbor, military leaders in Greece refused to wage war and on July 24, 1974, demanded that the junta surrender power to a civilian government.[20]

In both Portugal and Greece, military defeat in wars abroad either turned elements of the military against the dictatorship (Portugal) or discredited it completely (Greece). Neither dictatorship could survive the loss of legitimacy associated with military defeat.

In Spain, the economic crisis triggered by the dollar devaluation and the OPEC embargo was compounded not by military defeat but by a crisis of succession. Franco was eighty-three years old in 1974. Increasing age and bouts of illness had forced him to designate King Juan Carlos as his successor in 1969. In the years before his death, Franco tried to create political institutions that would survive him. (This was also a problem in Portugal, where Salazar appointed Caetano as his successor in 1968, shortly before he died.) But dictators have found it difficult to transfer power to successors without interruption, largely because dictatorship requires aggregating, not delegating power. So when Franco died in November 1975, Juan Carlos appointed a prime minister who began to dismantle the political institutions of the dictatorship and move the country toward civilian, democratic government, a protracted process that took three more years. Elites in Spain moved toward democracy because they wanted to join the Spanish economy with the rest of Western Europe. But the European Community (EC) would not let Spain share the substantial economic benefits of EC membership so long as it remained under fascist rule. By abandoning dictatorship, elites hoped to join the EC, which would help Spain overcome both the immediate economic crisis and chronic economic backwardness and promote new economic development. Franco's death gave them the political opportunity to pursue a democratization policy that many regarded as an economic necessity.[21]

Spanish elites who pressed for democratization (called *aperturistas*) and those arguing for continued dictatorship (*immobilistas*) were also mindful of the social and political turmoil prompted by the coup and collapse of the Caetano regime in neighboring Portugal, a revolution that brought to power communists and then socialists. They were also aware of the trials and convictions of military leaders and torturers in Greece following democratization there. For Spanish elites, a carefully managed devolution of power from Francoist dictatorship to civilian democracy seemed a good alternative to events in either Portugal or Greece.

In all three countries, political intermediaries played important roles in the devolution of power. In Portugal, a movement of Marxist military officers served as intermediaries; in Greece, the military president and a former prime minister living in exile in Paris managed the government before full-scale elections were held; and in Spain, a king and his technocratic allies in government oversaw the devolutionary process.[22] All of them made democratization and entry into the EC their primary political and economic objective.[23] When elections were held, socialist governments assumed power in all three countries, a development that would have been unimaginable in the early 1970s, because socialist and communist parties were banned or severely restricted by dictatorships in all three countries.[24] All three soon joined the EC: Greece in 1981 and Spain and Portugal in 1986, a development that proved to be a great boon to their economic fortunes in the 1980s.

The transition to democracy on the Iberian Peninsula did not go unnoticed in Latin American dictatorships, where most countries were once colonies of Portugal and Spain. In 1978, seventeen of the twenty Latin American countries were governed by military or authoritarian governments, according to political scientist Robert Pastor.[25] Unlike dictators in Portugal and Spain, most Latin American dictators would survive the 1970s. But the debt crisis that emerged in the early 1980s would create problems that would sweep most of them from power and establish democratic civilian governments by the end of the decade. By 1990, "17 of the 20 countries and over 90 percent of its population [could be] said to live under democratic governments," Pastor observed. "More of Latin America is now democratic . . . than at any time in the previous 160-year period of the continent [since] the struggle for separation from Spain and . . . Portugal."[26]

LATIN AMERICA: DEBT AND DEVOLUTION

Most Latin American dictatorships had come to power in the 1950s and 1960s, though a few, like Chile's general Augusto Pinochet Ugarte, seized power in the 1970s. Modest economic growth throughout the

1960s provided economic credibility to dictators and the bureaucracies associated with them. The alliance between military rulers and bureaucratic elites formed the social basis of what political scientists called "bureaucratic authoritarian regimes" in the region.

The 1973 oil crisis and the recession associated with it created many of the same economic problems for Latin American dictatorships that it did for regimes in southern Europe. It increased the cost of imported oil and food, which raised prices and contributed to inflation, and created trade deficits, which weakened their currencies. But while it created immediate problems for some Latin American countries, the oil crisis also presented dictatorships with an opportunity. Rising oil prices were good for oil-producing countries (which in Latin America included Mexico and Venezuela) and new revenues from higher oil prices found their way into financial pools that were then made available to dictatorships by lenders in the North.

Because Latin American dictators had access to low-cost credit through private U.S. banks and international lending agencies, they could borrow money to cover trade deficits and invest in their economies. By borrowing money, dictators were able to create jobs, build dams and roads, increase exports, raise military spending, cover budget deficits, and create the kind of economic growth that enhanced their legitimacy. Dictators across the continent borrowed about $350 billion between 1970 and 1983, funds that helped them achieve rapid rates of economic growth. The Brazilian dictatorship, which borrowed the largest amount, also recorded the highest rates of growth, a development that many economists described as a "miracle." Dictators then pointed to this miraculous economic development as proof of their competence, in much the same way that Benito Mussolini boasted of having made Italy's trains run on time.

Massive borrowing enabled Latin American countries to avert the problems that undermined southern European dictatorships in the early 1970s. But having used borrowed money to address their economic problems in the 1970s, Latin American dictatorships created economies that were vulnerable to changed conditions in the 1980s. This vulnerability increased when northern lenders insisted in the late 1970s that loans be tied to floating, not fixed, interest rates, rates that subsequently rose dramatically.

Economic conditions changed when the U.S. Federal Reserve raised interest rates to fight inflation in 1979 and 1980 (see chapter 3). Because most of their loans were tied to U.S. interest rates, dictators had to make higher interest payments on money they had borrowed in the 1970s. Moreover, high U.S. interest rates attracted Latin American investors, who withdrew their money from domestic bank accounts and invested in U.S. government securities. This capital flight from Latin America made it more difficult for countries to purchase imports and repay debt. It also eroded the tax base, depriving Latin American dictatorships of money at a time when they needed it most.

While higher interest rates increased costs, falling commodity prices reduced the income of Latin American countries. The prices Latin American countries could get for their beef, food, timber, minerals, and oil began to fall after 1980 for two reasons. First, borrowed money had been used by dictatorships and private industry to increase their production of exports, and increased production glutted markets. Second, high U.S. interest rates triggered a recession in the United States and around the world, which reduced the demand for goods produced in Latin America and other third world countries. Rising supplies and falling demand led to falling prices. This meant that Latin American dictatorships were earning less while being asked by lenders to pay more.

By 1982, when Mexico announced that it could no longer make payments on its $90 billion foreign debt, most of the Latin American dictatorships faced a serious economic crisis. One World Bank official described Argentina's economy, then saddled with $40 billion in foreign debts, as a "financial Hiroshima."[27]

Although the debt crisis undermined the economic legitimacy of dictators throughout Latin America, the crisis did not alone lead to the collapse of dictatorship. But economic crisis was compounded by political disaster in Argentina, and events there soon led to the collapse of the regime, a development that helped trigger democratization elsewhere.

On April 2, 1982, Argentine troops crossed 300 miles of South Atlantic Ocean and landed in Port Stanley, the capital of an island group the British call the Falklands and the Argentines call the Malvinas.[28] General Leopoldo Galtieri and the rest of Argentina's military junta apparently believed that the British would surrender the islands without a fight and that capture of the islands would bolster domestic support for the regime. Instead, the invasion triggered a ruinous war with the United Kingdom, which destroyed the regime's credibility at home.

Within days of the invasion, a British fleet had shipped south, landing troops on the islands on May 21. They quickly defeated Argentina's ill-equipped army, a force cut off from Argentina by the British navy, and forced its surrender on June 14.

As it turned out, the war in the Falklands/Malvinas was for Argentina's dictatorship what war in Cyprus was for the Greek junta: a humiliating defeat that forced them both from power. In Argentina, power was first assumed by another military government. But deteriorating economic and political conditions soon forced this government to call elections and they transferred power to a civilian government the following year.

The collapse of dictatorship in Argentina, a large and relatively wealthy country in Latin American terms, reverberated across the continent. Dictators in other countries recognized that they shared many of the same economic problems. They also noted that U.S. policy toward dictatorship had undergone an important shift.

For many years, U.S. officials permitted, condoned, or even encouraged dictatorship throughout much of Latin America. But during the Falklands war, U.S. officials did not support or defend the Argentine dictatorship. In fact, in a phone call with Galtieri just one day before the Argentines seized the islands, President Reagan told him, "I do not want to fail to emphasize pointedly that the relationship between our two countries will suffer seriously."[29] Galtieri evidently expected the United States to ignore the conflict, and the junta even suggested to the United States and other Latin American governments that they were obligated to come to Argentina's "defense" under the mutual-security provisions of the 1947 Rio Treaty. They were greatly disappointed when they received no diplomatic assistance from the United States or other dictatorships and felt betrayed after they learned that the U.S. government had provided important military assistance to the United Kingdom during the war. Dictators across the continent regarded these developments as an indication that U.S. policy had shifted and that they could no longer count on U.S. support in the event of a crisis.

In Brazil, the debt had grown from $12.6 billion to $90 billion between 1973 and 1982, inflation had soared into quadruple digits, and the economy had become, from the government's perspective, unmanageable. For more than a decade, the dictatorship had said it was moving toward eventual democratization, but it moved at a glacial pace. Economic crisis and changed political conditions sped up the process after the collapse of the dictatorship in neighboring Argentina in 1983 and in Uruguay in 1984. Brazil's dictators returned power to civilian authority in 1985.

Events then shifted to the Philippines, a country that more closely resembles Latin American countries in political and economic terms than its Asian neighbors. Like many Latin American countries, the Philippines were colonized and Catholicized by Spain, then brought into the U.S. sphere of influence during the Spanish-American War. In 1972, President Ferdinand Marcos declared martial law and assumed dictatorial powers. Like the Latin American dictatorships, the Marcos regime borrowed heavily in the 1970s, the country's foreign debt growing to $26 billion by 1985, an amount equal to the country's annual gross national product (GNP).[30] For the Marcos regime, profound economic crisis was joined by political turmoil when government soldiers assassinated opposition leader Benigno Aquino in 1983. His murder triggered widespread protest and massive capital flight, leading to a moratorium on debt payments and a real decline in the economy.[31] During 1984 and 1985, the economy registered "negative growth," shrinking by nearly 9 percent in those two years.[32] Faced with deteriorating economic and political conditions, Marcos suddenly called a presidential election for early 1986, an election he expected to win over the disorganized opposition.[33] But opposition candidate Corazon (Cory) Aquino, widow of the slain Benigno Aquino, won despite massive election fraud. When Marcos refused to surrender power, civilians and some army units took to the streets, while U.S. officials invited Marcos to find exile in Hawaii. Foreign pull and domestic shove soon forced Marcos from office and into exile in 1986.

In Argentina, Brazil, the Philippines, and throughout Latin America, dictatorships confronted debt-related economic crisis. Because first world lenders insisted that dictators institute austerity programs to ensure repayment of debt, dictators found themselves in the difficult position of introducing extremely unpopular economic policies. To make these policies succeed, dictators needed the acquiescence or cooperation of other social groups, particularly wealthy elites and middle classes. But these groups demanded that they obtain political power and a return to democracy if they were to assume responsibility for managing the economic crisis. Under these circumstances, the dictatorships could not easily refuse. Moreover, they recognized that U.S. support for dictatorship had eroded and worried about the alternatives to a managed or controlled democratization process. They feared popular revolts of the kind that had emerged in Portugal and later threatened in the Philippines. They also worried about events in Greece, where dictators and torturers were tried and jailed for their criminal conduct in office. Many dictators had reason to fear the Greek example. In Argentina, the military had waged a "dirty war" against civilian dissidents during the 1970s and 1980s, kidnapping, torturing, jailing, and murdering its opponents. As one army general explained, "We are going to kill 50,000 people: 25,000 subversives, 20,000 sympathizers, and we will make 5,000 mistakes."[34]

Many officials in dictatorial regimes had also illegally profited from office, siphoning off borrowed public money into private bank accounts. When civilian government returned in Argentina, many people demanded that officials in the dictatorship be tried and punished for economic and political crimes. Military officials lobbied desperately to prevent this, while army units staged mutinies and attempted coups to discourage the civilian government from prosecuting those responsible for the dirty wars.

Throughout Latin America, the generals concluded that a "managed" devolution of power would be preferable to the alternatives. So they devised constitutions that would transfer power while retaining some prerogatives and protections from legal proceedings after they surrendered power. As a result, dictators in Argentina, Uruguay, Brazil, Peru, Ecuador, El Salvador, Panama, Honduras, Bolivia, and Paraguay transferred power, returned to the barracks, or retired from public life.[35] In 1990, even General Pinochet returned power to civilian government in Chile.

EAST ASIA: GROWING PAINS

In 1950, South Korea was the economic equal of Kenya or Nigeria, while Taiwan was comparable to Egypt. But during the next thirty years, the economies of South Korea and Taiwan grew by leaps and bounds. Between 1962 and 1980, South Korea's GNP increased 452 percent, growing from $12.7 billion to $57.4 billion, a development that distanced it from most other poor countries. By 1983, *The Economist* noted, the 18 million people in Taiwan exported more goods than 130 million Brazilians or 75 million Mexicans.[36]

Economic growth in South Korea and Taiwan, which was the envy of poor countries everywhere, was made possible by three important developments. First, both South Korea and Taiwan had been colonized by Japan at the beginning of the century. Japanese colonial administrators developed important economic infrastructures and tied the colonial economies to Japan. Close economic relations survived the war, and when Japan began its remarkable economic ascent, it pulled them along. It is important to note, however, that Japanese economic growth always exceeded that of its former colonies, which meant that their rise, while significant, was not as spectacular as that of Japan.[37]

Second, after the Korean War began in 1950, South Korea and Taiwan received substantial economic and military benefits from the United States, largely because U.S. policymakers regarded them as front-line states in the battle against communism. To shore up Chiang Kai-shek's one-party dictatorship in Taiwan and the military governments that ruled South Korea after 1962, the U.S. government gave $5.6 billion in economic and military aid to Taiwan and $13 billion to South Korea between 1945 and 1978. U.S. aid to Korea was greater than aid to all of Africa ($6.89 billion) and India ($9.6 billion) and nearly as much as to all of Latin America ($14.8 billion) in the same period.[38] The United States also opened its markets to importers from Korea and Taiwan, giving them preferential trading privileges enjoyed by few other countries, while allowing them to erect formidable trade barriers against U.S. imports so they could protect and nurture domestic industry.

Third, the dictatorships in South Korea and Taiwan instituted land reform, thereby creating an urban workforce that could work in their growing export industries. They banned labor unions and strikes to keep wages low, a policy designed to give their export industries a competitive advantage in U.S. and world markets. They also used high tariff barriers to protect domestic industry from foreign competition and fostered the growth of large, export-oriented monopoly firms, what the Koreans call *chaebols* and the Taiwanese call *caifa*, to create businesses that could compete with large transnational firms in Japan and the United States.[39]

These three developments—Japanese colonialism before 1945, U.S. assistance after 1945, and domestic economic policy that capitalized on economic opportunities—enabled the South Korean and Taiwanese economies to grow rapidly, providing considerable economic legitimacy for the dictators that directed them. But changing economic and political conditions in the 1970s and 1980s caught South Korea and Taiwan in an economic squeeze that made it increasingly difficult for the dictatorships to maintain rapid rates of growth.

For Taiwan, problems began in 1972, when President Nixon suddenly recognized communist China and cut off much of U.S. economic, military, and diplomatic aid to the nationalist government. Although these developments shocked the dictatorship, which viewed changed U.S. policy as a betrayal (much as dictators in Greece, Argentina, and the Philippines viewed other U.S. policies), the U.S. government did not curtail Taiwan's trading privileges, so it could still export goods to the United States.

Then, in the early 1980s, both South Korea and Taiwan began to experience growing competition from other Asian countries—China, Thailand, and Indonesia—that adopted their model of economic growth and developed export industries that relied on workers who were paid even lower wages. Workers received $643 a month in Taiwan and $610 in South Korea in 1988; workers received $209 in Indonesia, $132 in Thailand, and $129 in Malaysia.[40] Lower wages enabled businesses in other Asian countries to manufacture goods at prices that undercut firms in South Korea and Taiwan.

While new Asian competitors pushed from below, the United States began to push from above. Because East Asian countries were so successful at selling their goods in U.S. markets, while blocking the sale of U.S. goods in their countries, the United States began running large trade deficits with South Korea, Taiwan, and most importantly, Japan. Persistent trade deficits strained U.S. support for East Asian dictatorships. To reduce its trade deficits with Asian countries and improve its competitiveness in overseas markets, the U.S. government in 1985 devalued the dollar (see chapter 2). This forced up the value of the South Korean currency by 30 percent. "We can absorb wage increases," a South Korean executive explained, "but we can't take any more [currency] appreciation."[41]

In 1987, U.S. officials also began restricting its preferential trading relations with East Asian countries, making it more difficult and more expensive for them to sell goods in U.S. markets.[42] U.S. officials also demanded that East Asian countries lower their trade barriers so that U.S. firms could sell more goods there, using bilateral and multilateral trade talks to press their case.[43]

At the same time, domestic social groups began pressing for change. Rapid urbanization had moved many people off the land and into the cities, but housing shortages and real estate speculation had pushed up housing costs for workers and the middle class. In Taipei, land was more expensive than in Manhattan, but workers earned only half as much as their counterparts in New York City.[44] Workers grew weary of working long hours with little pay—South Koreans worked 54.3 hours a week on average.[45] Although the economy grew and per capita income increased, this did not mean that wages also increased. The dictatorships restrained wage increases and kept them from growing as fast as the economy. Workers who labored hard for their country became increasingly unhappy about their inability to share in its rewards. As a result, legal and illegal labor disputes rose sharply in the mid-1980s. In South Korea, for example, the number of disputes rose from 276 in 1986 to 3,749 in 1987, a thirteenfold increase.[46] Legal and illegal strike activity forced up wages, despite the dictatorships' determined efforts to arrest labor leaders and curb wage increases, and this made South Korean goods more expensive and less competitive on world markets. Moreover, industrial workers, and increasingly, middle-class workers, joined student demonstrators in demanding an end to dictatorship and a return to democracy. Although the dictatorship had been able to contain and isolate student radicals for some years, it became more difficult to do so when economic demands echoed political demands and workers and white-collar employees joined student protests.

For South Korea and Taiwan, economic success invited lower-wage Asian countries to emulate them, invited retaliation by the United States, and triggered domestic protest against them.

In Taiwan, these economic problems were compounded by a succession crisis. The Republic of China was created in 1948 when nationalist armies under General Chiang Kai-shek fled from the Chinese mainland after being defeated by communist armies and took refuge on Taiwan, creating a dictatorship that ran the country under martial law for the next thirty-nine years. When Chiang Kai-shek died in 1975, power invested in the one-party regime passed to his son, Chiang Ching-kuo. By the time he inherited his father's power, Chiang Ching-kuo was sixty-five, already an old man. Near the end of his life, in 1986, he began casting around for a successor. Chiang began to consider reforms that could eventually transform Taiwan into a multiparty democracy and finally lifted martial law in 1987. When he died in 1988, power passed to Lee Teng-hui, who, like Juan

Carlos in Spain, initiated a leisurely reform process that eventually led to open elections in 1992.[47]

In both South Korea and Taiwan, dictators closely observed the hapless demise of the Marcos dictatorship in 1986. Although the Philippines was more like Latin American countries in economic and political terms, it was an Asian society in geographic terms. And its proximity to South Korea and Taiwan made dictators fear the spread of Cory Aquino's "People Power" to restive populations in their own countries.

By 1986–1987, as economic growth slowed from double- to single-digit rates, the dictatorships in South Korea and Taiwan became extremely anxious about economic and political developments. Cho Soon, South Korea's minister of economic planning, warned that without economic reform, "our country will collapse like some of the Latin American countries," such as Argentina.[48] Military leaders like Roh Tae Woo concluded that greater democratic participation was necessary if the country was to move ahead economically. During the late 1980s, military regimes in South Korea and Taiwan came to view democratization as a way to share power with the middle class, create a multiparty system dominated by a center-right party as a way to restore the government's legitimacy, deflect popular protest, and get the economy moving again.

In South Korea, it was Roh Tae Woo who initiated reform in 1987, arguing that "this country should develop a more mature democracy," transforming himself from military leader to presidential candidate.[49] After he won election as president in 1987 when dissident leaders split the opposition vote, he took steps that resulted in the 1992 election of dissident civilian Kim Young-sam. "Now we have finally created a truly civilian-led government in our country," Kim said after the vote.[50]

In 1992, Thailand joined South Korea and Taiwan when the military government stepped down and transferred power to civilians. Like them, Thailand experienced many of the problems associated with rapid growth. When government troops massacred student and middle-class demonstrators demanding political reform in May 1992, the military regime was discredited. Acting as an intermediary, King Bhumibol Adulyadei then assumed the role that Juan Carlos had played in Spain and managed a transfer of power to civilian government by September.[51]

During the late 1980s, communist dictatorships in China and Vietnam also initiated some economic and political reforms to address the economic problems that confronted them (see chapter 7). The aging political leadership of communist parties in China and Vietnam faced the problem of choosing more youthful political successors, a problem shared by rulers in Taiwan. In 1989, students and workers in Beijing occupied Tiananmen Square and demanded political reform, much as demonstrators in Seoul had been doing for some years. It appeared, during the spring of 1989, that these developments in China might lead to democratization there, just as it was then doing in South Korea and Taiwan. But the massacre of protesters camped in Tiananmen Square and the arrest of dissidents throughout the country aborted reforms that seemed in the offing. Events in China did not lead to the collapse of dictatorship. Nor did the Vietnamese regime devolve power, though it, too, was experiencing a deep economic crisis.

Communist regimes in East Asia did not democratize like capitalist dictatorships in southern Europe, Latin America, and East Asia, or like communist dictatorships that would subsequently democratize in Eastern Europe and the Soviet Union. They did, however, reform their economies and adopt many of the same economic policies that civilian democrats deployed in many democratizing states, an approach that in China has been called "market-Leninism."[52] Communist regimes in Asia survived, at least for the present, because they possessed considerable political legitimacy derived from having fought and won wars against colonial powers and domestic rivals. The Chinese communists fought the Japanese in China, U.S. and U.N. forces in Korea, and nationalist rivals during a long civil war. The Vietnamese communists had fought for independence against France, Japan, and the United States, and in 1975 defeated the South Vietnamese regime after U.S. troops withdrew. Their ability to defeat formidable opponents, and their programs designed to improve conditions for the rural population, gave them a reservoir of support they could draw on during the economic and political crises of the 1980s. (Fidel Castro's communist regime in Cuba survives for many of the same reasons.) By contrast, communist dictatorships in Eastern Europe (with the exception of Yugoslavia), assumed power not as a result of their own efforts but because they were installed by the Soviet Union and backed by the Red Army. By the 1980s, dictators in Eastern Europe and the Soviet Union had long since exhausted the patience of domestic populations. So when crisis struck, they found themselves isolated and vulnerable.

THE SOVIET UNION AND EASTERN EUROPE: DECLINE AND DEMOCRATIZATION

Democratization in the Soviet Union and Eastern Europe was a product of an economic crisis originating in the Soviet Union. This crisis was rooted in the stagnation of collectivized agriculture and the heavy burden of military spending in the postwar period. "By the beginning of the 1980s," Soviet leader Mikhail Gorbachev observed, "the [Soviet Union] found itself in a state of severe crisis which has embraced all spheres of life."[53]

During the 1920s and 1930s, the Soviet dictatorship under Joseph Stalin forced small and medium-sized farmers to join large-scale, state-owned collectives. By increasing farm size and mechanizing agriculture under state authority, the government hoped to increase agricultural output and use much of the wealth it produced to finance industrial development. But collectivization was an extremely disruptive process. The regime killed farmers who opposed these measures or sent them to Siberia, and agricultural output faltered, contributing to famine in some regions.

After World War II, agriculture revived and the large-scale collective farms increased production, for a time. But because the communist regime continued siphoning off agricultural resources to finance industrial growth and, during the Cold War, the military, there was little investment in agriculture and few incentives for farmers to increase yields or produce more food. Crop yields in the Soviet Union were only about half those obtained in the United States.[54] These problems reached crisis proportions in the mid-1970s, when Soviet grain harvests failed, a development that sent world grain prices soaring. Because farm production failed to keep pace with the growing Soviet demand for food, the Soviet Union had to import more food. The cost of food imports doubled from $5.1 billion to $10.2 billion between 1974 and 1978, a development that increased the Soviet trade deficit.[55] And because farmers in the 1970s and 1980s began selling a growing share of their crops through unofficial black markets that offered higher prices, there was less food available in state stores. Consumers found it difficult to obtain food using ration coupons, though they could purchase higher-priced food on black markets.[56] The result was long lines, frayed tempers, and considerable resentment.

The Soviet regime might have invested more heavily in agriculture if it had not been preoccupied with military spending. But during the Cold War, the Soviets spent heavily to develop and maintain its new status as a military and political superpower. According to military analyst Ruth Sivard, the Soviets spent $4.6 trillion between 1960 and 1987, or between 12 and 15 percent of its annual GNP, on the military.[57] Other economists argued that the Soviets spent even more, as much as 20 to 28 percent of GNP.[58] (By comparison, the United States spent only 6 to 8 percent of its GNP on the military during the Cold War.)

Massive military spending enabled the Soviet regime to expand its political influence, maintain an occupying army in Eastern Europe, assist socialist movements and communist governments abroad, and obtain substantial income from arms sales to other countries—$64 billion worth of arms between 1973 and 1981.[59] But it did not provide substantial economic benefits in the Soviet Union or secure military advantages abroad.

Domestically, heavy military spending absorbed scarce supplies of capital, skilled labor, and natural resources, diverting resources from other sectors of the economy, particularly from agriculture and consumer industries. As a result, it retarded economic growth and contributed to stagnation and decline.[60] The Soviet invasion of Afghanistan in 1979 greatly increased military spending and stimulated an arms race with the United States in the early 1980s, a development that further increased military spending (see chapter 11). Gorbachev later said that increased military spending during the Afghan war and the arms race of the 1980s had "exhausted our economy."[61]

It also became apparent in the 1980s that massive military spending had not enabled the Soviets to produce weapons that could compete with U.S. and Western European arms on battlefields in Afghanistan and the Middle East (see chapters 9 and 11). In 1982, for example, during an air battle over Lebanon, Israeli pilots flying U.S. and French jets "shot down 80 Soviet-made planes [flown by Syrians] while losing none of their own."[62] This kind of lopsided battlefield performance was dramatically underscored during the 1991 Persian Gulf War, when the U.S.-led coalition crushed Iraqi forces, which were supplied with Soviet arms, destroying 4,000 Soviet-built tanks in the process. The failure of Soviet weaponry in battlefield competition led Soviet military planners to conclude that the entire Soviet military model was "obsolete."[63] Soviet Marshal Dmitry Yazov admitted, "What happened in Kuwait necessitates a review of our attitude toward the [Soviet Union's] entire defense system."[64]

Soviet economic problems worsened after 1985. Falling world oil prices reduced income from oil exports, one of its major sources of foreign currency, and increased its trade deficit. To cover the trade deficit so that it could import the food and technology it needed from the West, the Soviets began borrowing heavily. The Soviet debt to Western European governments and banks nearly quadrupled between 1984 and 1989, growing "from $10.2 billion at the end of 1984 to $37.3 billion at the end of 1989."[65] And the regime's attempts to stimulate the stagnating economy led to growing budget deficits. "In 1981–85, the budget deficit only averaged 18 billion rubles per year, [but] in 1986–89, it averaged 67 billion rubles," a more than threefold increase.[66]

The Soviet economic crisis was compounded by two political crises. First, the communist regime faced not one but three crises of succession between 1982 and 1985. When Leonid Brezhnev died suddenly after reviewing the annual parade celebrating the anniversary of the Bolshevik Revolution on November 10, 1982, a political battle to choose his successor ensued. Brezhnev's chosen successor, Konstantin Chernenko, was passed over and Yuri Andropov was selected as the new Soviet leader. Andropov died after a long illness on February 9, 1984. A second succession crisis then ensued. Andropov's choice, Mikhail Gorbachev, who represented the young, reform-minded wing of the Communist Party, was passed over and Brezhnev's old protégé, Chernenko, was chosen instead. He died a year later, on March 10, 1985, and Mikhail Gorbachev, then fifty-four, assumed power.[67]

Second, by the time the protracted succession crisis had been sorted out, it had become evident that the Soviet Union faced military defeat in Afghanistan. The looming defeat by anticommunist mujahadeen rebels supplied with U.S. arms undermined the legitimacy of a regime that relied on its military standing as a political cornerstone, much as military defeats undermined the legitimacy of dictatorships in Portugal, Greece, and Argentina.

After he assumed power in 1985, Gorbachev took steps to address the economic crisis. By reducing military spending, transferring these resources to other sectors of the economy, and allowing some privatization of state-owned farms and industry to give farmers and workers incentives to increase the quantity and quality of food and consumer goods, Gorbachev hoped to jump-start the languishing economy. But dramatic economic restructuring, called *perestroika*, antagonized the military and the Communist Party bureaucracy that directed agriculture and industry. To overcome opposition within the Communist Party and push ahead with economic reform, Gorbachev needed to develop a wider social constituency that could provide political support for reform outside the party. So Gorbachev promoted political reform and limited democratization, what he called *glastnost* (openness), as a way to rally wider popular support for *perestroika*, demilitarization, and reform. As Gorbachev explained, democratization was "a guarantee against the repetition of past errors, and consequently a guarantee that the restructuring process is irreversible." There was no choice, he said; it was "either democracy or social inertia and conservatism."[68]

Because reduced military spending was a central part of *perestroika*, Gorbachev devoted considerable attention to Soviet military and foreign policy. He withdrew the Soviet army from Afghanistan, stopped aid to the communist regime in Ethiopia, and initiated arms control agreements with the United States and its NATO allies in Western Europe. The 1987 Intermediate-Range Nuclear Forces Treaty, which removed intermediate-range nuclear missiles from Europe, was the first arms control agreement to reduce the size of superpower arsenals. Gorbachev also initiated talks that led to a reduction of conventional troops in Europe. Gorbachev pursued détente with China and sought an end to the long Cold War with the United States. And he renounced long-standing Soviet claims that it had a right to use military force in Eastern Europe to support communist dictatorships there, a "right" that the Soviets exercised in East Germany in 1953, in Hungary in 1956, and in Czechoslovakia in 1968.

By renouncing the right to intervene in Eastern Europe, Gorbachev undercut client dictatorships that had first been installed by the Soviets in the late 1940s. At a press conference on October 25, 1989, Soviet foreign minister Gennady Gerasimov was asked whether the Soviet Union still adhered to the Brezhnev Doctrine, which was used to justify Soviet intervention in Eastern Europe. He said it did not. Instead, he said, new Soviet policy would be called the "Sinatra Doctrine," because the American singer Frank Sinatra "had a song, 'I did it my way.' So every country decides in its own way which [economic and political] road to take."[69]

The Soviet adoption of a new foreign policy toward Eastern Europe fatally weakened communist dictatorships and fueled the rise of dissident opposition movements in Poland, East Germany, Hungary, Czechoslovakia, Romania, Bulgaria, and Albania. By the end of 1989, just two months after Gerasimov's press conference, communist dictatorships in all these countries had been swept from power. Some tried to initiate and manage a devolution of power to dissident democrats so that they could retain some power. But an economic crisis rooted in stagnation and debt—recall that Eastern European countries were the first casualties of the debt crisis in the early 1980s—left them without any economic credibility, while the new Soviet policy deprived them of any remaining political legitimacy or military power.

Because they possessed a very narrow social base, with little economic credibility or political legitimacy, communist dictatorships in Eastern Europe collapsed like a house of cards at the first appearance of organized dissent or concerted civic action. In East Germany, it was the flood of migrants from East Germany to Hungary and then to West Germany that brought down the dictatorship, while in Poland it was an organized labor movement. In Hungary and Czechoslovakia, newly organized dissident movements negotiated an end to dictatorship. In Romania, street demonstrations and mob action brought a bloody end to the brutal regime of Nicolae Ceaușescu.

The dramatic events of 1989 had important consequences for the Soviet Union. The collapse of communism in Eastern Europe, the emergence of protest movements in some Soviet republics, and the extension of reforms in response to a deepening economic crisis led, in August 1992, to a coup by hard-liners in the military and Communist Party who were determined to restore one-party dictatorship. But whereas Chinese hard-liners had been able to crush dissent and maintain dictatorship, Soviet hard-liners found few allies. Instead, people in Moscow and important elements of the army rallied to Russian president Boris Yeltsin, forcing the coup to collapse. These developments led, by the end of 1992, to democratization but also to the division of the Soviet Union into fifteen independent countries.

SOUTH AFRICA: EMBARGO, DEFEAT, AND DEMOCRATIZATION

On February 2, 1990, South Africa's president Frederik W. de Klerk told the country's whites-only parliament that he was legalizing the outlawed African National Congress (ANC) and other black political organizations and would soon release ANC leader Nelson Mandela, who had been imprisoned for twenty-seven years. Although he told one Western diplomat, "Don't expect me to negotiate myself out of power," de Klerk did just that.[70] One year later, he introduced changes that removed "the remnants of racially discriminatory legislation which have become known as the cornerstones of apartheid," and began a process of negotiation and constitutional reform that would lead to elections in April 1993 that swept Mandela to power as the country's first black president. When Mandela was inaugurated on May 10, 1994, joyous black crowds chanted in Xhosa, "Amandla! Ngawethu!" ("Power! It is ours!").[71]

De Klerk initiated the devolution of power in response to a worsening economic and political crisis in South Africa. During the 1950s and 1960s, "the apartheid system probably aided economic growth in South Africa" because it locked in low wages for black South African workers.[72] But by the late 1960s, "this position began to change"; low wages discouraged businesses from spending money to introduce new technology that would increase productivity and make South African goods more competitive on world markets, so economic growth began to slow.[73]

In 1976, South African police killed numerous black demonstrators protesting apartheid in Soweto, an action that triggered a wave of protests and strikes that continued during the 1970s and 1980s. During the second half of the 1980s, "5,000 people died and more than 30,000 were jailed without charge."[74] Black protests called international attention to and condemnation of apartheid. Spurred by protests in South Africa and in Western countries, companies and countries began to withdraw investment from South Africa, levy economy sanctions, and eventually reduce much of their trade with South Africa. By the end of the 1980s, sanctions and disinvestment had brought the South African economy to a standstill. Economists estimated in 1989 that average income had fallen 15 percent from 1980 and that economic sanctions were costing the economy $2 billion a year.[75]

While the economy stagnated, military spending grew, both because the government was waging an internal war against domestic black protesters and because it deployed troops to fight a communist guerrilla movement in Namibia and intervened in neighboring Angola to overthrow the communist government there. These costly external wars led to the South African army's defeat by Angolan and Cuban troops at the battle of Cuito Cuanavale in 1988. As Kevin Danaher and Medea Benjamin noted, "The defeat forced the South Africans to sign a peace treaty [with Angola] on December 22, 1988, requiring them to withdraw from Angola [after a thirteen-year occupation] and a phase-out of their decades-long control of Namibia."[76]

When de Klerk became president on August 15, 1989, after the illness and resignation of hard-line president P. W. Botha, known by associates as the "Old Crocodile," he confronted a worsening economic, military, and political crisis. During the next six months, de Klerk came to view an end to apartheid as the key to ending the crisis. By ending apartheid, he hoped that domestic turmoil would wane, military expenditures could be reduced, and foreign investors and Western governments could be persuaded to end economic sanctions and reinvest in the economy. (As in southern Europe, improved economic relations depended on ending dictatorship.) The collapse of communism during the six months between his appointment as president and his dramatic February 2 speech made de Klerk's decision easier, he said, "because it created a scenario where the communist threat . . . lost its sting."[77]

By initiating a devolution of power, de Klerk sought to manage events so that the white minority could reserve their economic power and retain some political power in a post-apartheid state. And democratization in South Africa has encouraged democratization in other African countries. About half of the continent's forty-eight countries have since held multiparty elections, though this process has had, as yet, only limited results.[78]

DEMOCRACY AND DEVELOPMENT

Contemporary democratization is largely a product of economic crisis. The problems associated with different kinds of economic crises were compounded by political problems related to the death or illness of dictators, defeat in war, and public protest by dissident groups. Faced with difficult economic and political problems, dictators and one-party regimes realized they needed to take drastic steps to resolve their problems. But to be successful, radical economic and political reform needed broad public support. And other social groups refused to accept responsibility for solving economic problems they did not create unless they could obtain real political power. As Walden Bello and Stephanie Rosenfeld have written, "Economic policies that are not supported by a rough consensus forced by democratic means are likely to founder over the long run. Democracy, one might say, has become a factor of production."[79]

Under these circumstances, dictators devolved power to civilian democrats, trying to manage the process so that they could retain some residual political power and possibly protect themselves from prosecution for economic crimes (corruption) or violations of human rights (illegal arrests, torture, murder of dissidents). As we have seen, the devolution of power was abrupt in some places and protracted in others, and dictators had only limited success in managing the democratization process.

Once they assumed political power, civilian democrats had to address difficult economic problems. Although the origin and character of economic crises differed from one region to the next, and from one country to the next within these regions, civilian democrats everywhere adopted a common economic approach that they hoped would solve their separate problems. In nearly all democratizing states, civilian democrats have (1) opened their economies to foreign investment and trade; (2) sold off or privatized state-owned public assets and industries; and (3) cut military spending. This common set of economic policies, which are associated with economic globalization generally, has produced some economic benefits in some democratizing states. But they have also created economic, social, and political problems of their own.

OPENING THE ECONOMY

Around the world, civilian democrats have reduced tariff barriers, opened their economy to foreign investors, and lifted currency exchange restrictions. Lower tariff barriers are designed to make imported goods more readily available to domestic consumers who have long craved many goods from other countries, and force domestic industries to lower their prices and improve the quality of their goods, or go out of business. Foreign competition and the threat of bankruptcy are supposed to shake up domestic industries that have grown inefficient and wasteful as a result of government protection under dictatorship. By opening the economy to foreign investment, governments hoped that foreign companies would inject new money, management skills, and technology into the economy, thereby providing jobs in industries that can compete on world markets. And by lifting currency restrictions, government economists expected world currency markets to appraise their currency and set exchange rates at realistic levels. As a result, the currencies of most democratizing states were devalued, which made their goods cheaper and easier to sell abroad, a development that helped them increase their exports and reduce their trade deficits. These three measures were designed to achieve what the Brazilians called a "competitive integration" with the world economy.[80]

The problem for many countries, however, was that when tariff barriers were lowered, domestic consumers bought expensive imported goods: Nike shoes, Levi's jeans, foreign cars. This orgy of consumer spending on imports created trade deficits, which forced down the value of their currency. And while a devalued currency should have made their goods easier to sell on overseas markets and improved their trade balance, their goods were often regarded as shoddy by consumers elsewhere. The result was that they found it difficult to sell their goods at any price. Foreign investors did open some new factories and purchased some government-owned industries at bargain-basement prices (currency devaluations made businesses cheaper for foreign buyers). But the large-scale investment that many governments expected did not materialize, largely because global economic activity has been slow and there is excess capacity in many industries, and partly because investors are waiting to see whether governments can control inflation and create a favorable business climate and strong consumer market before risking substantial sums of money. Meanwhile, the widespread sale of public assets and industries has glutted investment markets, which has slowed sales and lowered prices.

SELLING OFF STATE ASSETS

Civilian democrats around the world have sold public assets that had previously been controlled by dictators, bureaucrats, and their friends and families, an economic system that Peruvian economist Hernando de Soto has described as "buddy-buddy capitalism" or "crony capitalism."[81] Civilian democrats hoped that the sale of public assets would raise money that could be used to repay debts, reduce government budget deficits, and cut the cost of subsidizing inefficient industries. The sale of public assets was also supposed to provide new economic opportunities for domestic investors, who were expected to emerge as a new class of energetic entrepreneurs.

As a result, the sale of national banks, airlines, telephone companies, shipping lines, cement factories, port facilities, and land has been widespread. In Eastern Europe, assets worth more than $100 billion have been offered for sale.[82] In Brazil, the sale of ninety-two parastatal firms and a port authority by the end of 1992 had been valued at $62 billion.[83]

Privatization has been less extensive in East Asia, though South Korea sold off seven firms in 1990, because both South Korea and Taiwan have few public assets compared with countries in southern Europe, Latin America, or Eastern Europe, and because the state supports large private monopolies, *chaebols*, which they were unwilling to break up.

The problem with this strategy was that few domestic investors could afford to purchase large companies, while currency devaluations made them cheaper, in real terms, for foreign investors. Foreign firms have snapped up the best offerings at bargain prices, but they have not been interested in purchasing poor-quality companies. The worldwide sale of so many properties has glutted investment markets. With airline companies battling each other for scarce travelers in the North, the last thing they needed to do was to purchase the Argentine airline fleet. In Eastern Europe, Poland has managed to sell only a fraction of the firms it has offered for sale, and Germany has sold only a fraction of the 8,000 or so former East German state firms.[84] In Czechoslovakia and in many former Soviet republics, the governments abandoned plans to sell most of their assets on the open market and instead gave or sold vouchers to residents who used them to buy shares in privatized firms. The idea was to sell the firms to domestic buyers and then use the money raised to introduce technology, improve productivity, and increase competitiveness. But this hasn't always worked out. After a Czech government effort to sell some large businesses failed miserably, the prime minister defended his government from criticism, arguing, "We are not a banana Republic."[85] It is not clear that experiments of this sort will provide sufficient capital, that the money collected from small investors will flow to the right firms, that these firms will use the money wisely, or that newly privatized firms will be able to compete effectively against well-established, large-scale firms from Western Europe, North America, or East Asia. So far, the attrition rate of recently privatized firms has been high, and business failures have contributed to rising unemployment in many democratizing states.

DEMILITARIZING THE ECONOMY

Demilitarization has been most dramatic in Eastern Europe and the Soviet Union. Under Gorbachev, the Soviet Union began withdrawing troops from Afghanistan and Eastern Europe and cut military spending and troop levels. This made possible the devolution of power to democrats in Eastern Europe, who promptly slashed military spending, cut troop levels, disbanded party militias, and withdrew from the Warsaw Pact, causing its demise in 1991.[86]

After the 1992 coup failed and the Soviet Union democratized and divided, military spending fell in most of the former republics. The decline was dramatic. In 1987, the Soviet Union had 3.9 million troops under arms and spent $356 billion on defense. In 1994, the Russian government (the largest of the former Soviet republics) had 2.1 million troops under arms and spent only $29 billion on defense.[87]

In Latin America, successive Argentine presidents have cut military spending in half, scaled back the draft, and cut the army to one-half its Falklands/Malvinas size.[88] Armies across the continent have been scaled back. Julio María Sanguinetti, who became Uruguay's president in 1985, described the changed political atmosphere: "If you get a group of Latin American politicians together in a room and ask, 'Who wants to be foreign minister?' everyone will wave his or her hand in the air. But if you ask, 'Who wants to be defense minister?' everyone stares at the floor."[89]

East Asian states have been the exception to this general rule. After Tiananmen Square, the Chinese increased military spending, so Taiwan and South Korea did too. South Korea increased its military spending both because its disagreements with North Korea remain unresolved and because the United States cut back its military spending on the peninsula, which forced the South Korean government to assume a greater share of defense costs.

Although military spending in East Asia has increased somewhat, the general trend is to demilitarize.[90] Governments have demilitarized for a variety of economic reasons. Most governments believe that heavy military spending did little to contribute to economic growth and may even have put their economies at a disadvantage in global competition with states that devoted a smaller percentage of the GNP to military expenditures, like Germany and Japan. In a study on the relation between military spending and economic growth, A. F. Mullins found that

> in general, those states that did best in GNP growth . . . paid less attention to military capability than others. This relation . . . holds right across the range from poor states to rich states and from weak states to powerful. Those that did poorly in GNP growth . . . paid more attention to military capability.[91]

ECONOMIC CRISIS AND DEMOCRACY

During the 1970s and 1980s, economic crises of different sorts created problems that contributed to the collapse of dictatorships around the world. The civilian democrats who assumed power then attempted to solve their separate crises by opening their economies, selling off state assets, and reducing military spending. Some have had more success than others, but difficult economic problems remain for most countries.

In Spain, for example, the economy boomed during much of the 1980s, largely because membership in the EC (now the EU) provided real benefits. But growth slowed dramatically in the 1990s and unemployment grew to a staggering 21.5 percent in 1993, the highest in the EU and twice the average of other EU countries.[92] The problem was that despite improved economic performance, Spanish industry was still not competitive with other European heavyweights. "We were seduced into believing we were in the major league," explained Spanish business consultant Jaime

Mariategui. "But when you are racing a Spanish SEAT [a car made in Spain] against a Mercedes, eventually you [must] face reality."[93] And pointing to the anemic 1 percent growth in 1992 and 1993, and the collapse of Spain's stock market, journalist Roger Cohen observed, "The danger seems real that Spain, having made a great leap, could slip back."[94] A spokesman for a large Spanish firm complained, "Europe means progress, but right here progress means unemployment, and I don't know if that is acceptable."[95]

In Latin America, many new democracies have also recorded impressive economic growth in recent years, but the number of people in poverty has nonetheless increased. "The resumption of economic growth has been bought at a very high social price, which includes poverty, increased unemployment and income inequality, and this is leading to social problems," observed Louis Emerij, an economist at the Inter-American Development Bank in 1994.[96] By the end of the decade, 192 million people or 37 percent of the population will live in poverty. "Growth has really been on only one end of the spectrum, the wealthy. The rich are getting richer and the poor are getting poorer. And this will generate social conflict," argued UN official Peter Jensen.[97]

In 1992 a Harvard University study found that "the vast majority of people in Eastern Europe live in economic conditions demonstrably worse than those under the inefficiencies of central planning."[98] And in the former Soviet Union, a deepening economic crisis has actually lowered the life expectancy of adult men and led to a decrease in the population, a development that British demographer David Coleman described as "an incredibly clear picture of a society in crisis. A decline in life expectancy this dramatic has never happened in the postwar period. . . . It shows the malaise of society, the lack of public health awareness and the fatigue associated with people who have to fight a pitched battle their whole lives just to survive."[99]

Under these circumstances, it should not be surprising that in a 1994 opinion poll, two-thirds of Russians believed that things were "better" under communism than they are "now."[100] In Eastern Europe, substantial minorities, and sometimes majorities, agreed.[101]

Gender and Democratization

In gender terms, dictatorships were male-dominated political institutions. No postwar dictatorship anywhere was led by a woman. In capitalist dictatorships, this gendered, patriarchial form of government generally provided some jobs in the military and bureaucracy to some men, but disadvantaged women, who were expected to stay at home and raise children. In communist regimes, by contrast, this gendered form of government provided these benefits to men but also extended them to women. So in Eastern Europe and the Soviet Union, many women found work in manufacturing and service industries, in professions, and in government, though they were never assigned top positions in any field. When dictatorships fell, men and women were able to create democratic political institutions that were more open to women and men outside the elites who had controlled patriarchial dictatorships. Now, women more often run for office. In a few cases, women even assumed power as president: Corazon Aquino in the Philippines and Violeta Chamorro in Nicaragua.

In economic terms, democratization has had important, though varied gender consequences. Where democratic governments opened their economies to foreign trade and sold public assets, as they did almost everywhere, many of the domestic manufacturing industries were ruined. In southern Europe, Latin America, East Asia, and South Africa, deindustrialization typically resulted in job loss for men, as we have already seen (see chapter 1). Where commodity prices fell, men and women were adversely affected, depending on the gender character of the labor force in agriculture, mining, and natural resource industries. Where governments demilitarized and reduced the size of standing armies, men lost jobs because few women (even in communist countries) served as soldiers in the military.

But while deindustrialization led to job loss for men in many countries, it led to disproportionate job loss for women in Eastern Europe and the Soviet Union. A 1999 UN study found that women lost jobs more often than men. "In the transition to a market economy," the United Nations observed, "the status of women is eroding further."[102]

The scale of deindustrialization in Russia has been enormous. Economists estimate that its GNP has declined 45 percent since the fall of communism in 1992.[103] And nearly 40 percent of the population in the former Soviet Union now live in poverty.[104] These economic developments have had important consequences for women. As the Russian economy contracted, the government lost tax revenue and was unable to provide many services or pay salaries or pensions. This latter development hurt older women who typically survive their male spouse. It also affected younger women. "With political collapse and economic uncertainty [in the former Soviet Union and in Eastern Europe], many women stopped having children or decided to delay motherhood," a UN study found in 2000. "People have been impoverished and decided that having kids at a time of poverty and misery is not the right thing to do, so they cut back. This is family downsizing . . . a rational economic behavior in some ways."[105] As a result, birth rates have fallen sharply. At current rates, demographers expect the population to shrink by as much as fifty-seven million in Russia during the next fifty years.[106]

A Temporary Crisis?

Economists and policymakers who defend the economic policies of democratizing states have argued that contemporary problems are the product of previous economic policy—the residual effects of discredited dictators—or are temporary problems associated with a difficult transition process. That may be. Nonetheless, many people in these countries associate these problems with the democrats who recently assumed power. And unless democratic governments speedily solve some economic problems, public disenchantment may grow.

In some countries, continuing economic crisis, which differs in important ways from the crises that preceded the fall of dictatorship, has created political problems for democratic governments, and dictatorship again threatens. In some Eastern European and former Soviet states, former communists have returned to power in recent elections, ousting the democrats who took power after communism collapsed. Just as the current economic crisis differs from the one that preceded the collapse of dictatorship, the advocates of a new kind of authoritarianism differ from their dictatorial predecessors. In Brazil, for example, Congressman Jai Bolsonaro argued, "Real democracy is food on the table, the ability to plan your life, the ability to walk on the street without getting mugged." He has said, "I am in favor of dictatorship" because "we will never resolve serious national problems with this irresponsible democracy."[107]

Economic crisis, it seems, can create serious problems for dictators and democrats alike. Although dictatorships have collapsed in countries around the world during the past twenty years, many economic and political problems remain. Unless democratic governments can solve some of these problems and make some demonstrable economic progress, the threat of dictatorship, in some form, will remain.

NOTES

1. Robert K. Schaeffer, "Democratic Devolutions: East Asian Democratization in Comparative Perspective," in *Pacific-Asia and the Future of the World-System*, ed. Ravi Palat (Westport, Conn.: Greenwood, 1993); Robert K. Schaeffer, *Power to the People: Democratization around the World* (Boulder, Colo.: Westview, 1997).

2. Samuel P. Huntington, "Democracy's Third Wave," in *The Global Resurgence of Democracy*, ed. Larry Diamond and Marc F. Plattner (Baltimore: Johns Hopkins University Press, 1993), 3.

3. Caglar Keydar, "The American Recovery of Southern Europe: Aid and Hegemony," in *Semiperipheral Development: The Politics of Southern Europe in the Twentieth Century*, ed. Giovanni Arrighi (Beverly Hills, Calif.: Sage, 1985), 141–42.

4. Giovanni Arrighi, "Fascism to Democratic Socialism: Logic and Limits of a Transition," in *Semiperipheral Development: The Politics of Southern Europe in the Twentieth Century*, ed. Giovanni Arrighi (Beverly Hills, Calif.: Sage, 1985), 26; Giovanni Arrighi, "World Income Inequalities and the Future of Socialism," *New Left Review*, 189 (September–October 1991), 47; Keydar, "The American Recovery of Southern Europe," 145.

5. Raymond Carr and Juan Pable Fusi Aizpurua, *Spain: Dictatorship to Democracy* (London: George Allen and Unwin, 1979), 57.

6. Kenneth Maxwell, "The Emergence of Portuguese Democracy," in *From Dictatorship to Democracy: Coping with the Legacies of Authoritarianism and Totalitarianism*, ed. John H. Herz (Westport, Conn.: Greenwood, 1982), 233.

7. Carr and Aizpurua, *Spain*, 67, 68.

8. Eric N. Baklanoff, "Spain's Emergence as a Middle Industrial Power: The Basis and Structure of Spanish–Latin American Economic Relations," in *The Iberian–Latin American Connection: Implications for U.S. Foreign Policy*, ed. Howard J. Wiarda (Boulder, Colo.: Westview, 1986), 139; John Logan, "Democracy from Above: Limits to Change in Southern Europe," in *Semiperipheral Development: The Politics of Southern Europe in the Twentieth Century*, ed. Giovanni Arrighi (Beverly Hills, Calif.: Sage, 1985), 164.

9. Edward Malefakis, "Spain and Its Francoist Heritage," in *From Dictatorship to Democracy: Coping with the Legacies of Authoritarianism and Totalitarianism*, ed. John H. Herz (Westport, Conn.: Greenwood, 1982), 218.

10. Arrighi, "Fascism to Democratic Socialism," 265.

11. Logan, "Democracy from Above," 163.

12. Arrighi, "World Income Inequalities," 47.

13. Maxwell, "The Emergence of Portuguese Democracy," 235.

14. Baklanoff, "Spain's Emergence as a Middle Industrial Power," 140.

15. Maxwell, "The Emergence of Portuguese Democracy," 235.

16. Maxwell, "The Emergence of Portuguese Democracy," 235.

17. Maxwell, "The Emergence of Portuguese Democracy," 235.

18. Rodney J. Morrison, *Portugal: Revolutionary Change in an Open Economy* (Boston: Auburn House, 1981), 2; Logan, "Democracy from Above," 158.

19. Logan, "Democracy from Above," 270.

20. Harry J. Psomiades, "Greece: From the Colonels' Rule to Democracy," in *From Dictatorship to Democracy: Coping with the Legacies of Authoritarianism and Totalitarianism*, ed. John H. Herz (Westport, Conn.: Greenwood, 1982), 253.

21. Malefakis, "Spain and Its Francoist Heritage," 220.

22. Psomiades, "Greece," 258.

23. Michael Harsgor, *Portugal in Revolution* (Washington, D.C.: Center for Strategic and International Studies, 1976), 28; Logan, "Democracy from Above," 166, 168.

24. Maxwell, "The Emergence of Portuguese Democracy," 238.

25. Robert A. Pastor, *Democracy in the Americas: Stopping the Pendulum* (New York: Holmes and Meier, 1989), xi.

26. Pastor, *Democracy in the Americas*, ix.

27. William C. Smith, *Authoritarianism and the Crisis of the Argentine Political Economy* (Stanford, Calif.: Stanford University Press, 1989), 249.

28. Gary W. Wynia, *Argentina: Illusions and Realities* (New York: Holmes and Meier, 1986), 3.

29. Wynia, *Argentina*, 15.

30. Belinda A. Aquino, "The Philippines: End of an Era," *Current History* (April 1986), 158.

31. John Bresnan, *Crisis in the Philippines: The Marcos Era and Beyond* (Princeton, N.J.: Princeton University Press, 1986), 145.

32. Aquino, "The Philippines," 158.

33. Bresnan, *Crisis in the Philippines*, 142.

34. Smith, *Authoritarianism*, 232.

35. Jonathan Hartlyn and Samuel A. Morley, *Latin American Political Economy: Financial Crisis and Political Change* (Boulder, Colo.: Westview, 1986), 1.

36. Chalmers Johnson, "Political Institutions and Economic Performance: The Government-Business Relation in Japan, South Korea and Taiwan," in *The Political Economy of the New Asian Industrialism*, ed. Frederic C. Deyo (Ithaca, N.Y.: Cornell University Press, 1987), 136.

37. Bruce Cumings, "The Origins and Development of the Northeast Asian Political Economy: Industrial Sectors, Product Cycles and Political Consequences," in *The Political Economy of the New Asian Industrialism*, ed. Frederic C. Deyo (Ithaca, N.Y.: Cornell University Press, 1987).

38. Cumings, "The Origins and Development," 67; Hagen Koo, "The Interplay of State, Social Class, and World System in East Asian Development: The Cases of South Korea and Taiwan," in *The Political Economy of the New Asian Industrialism*, ed. Frederic C. Deyo (Ithaca, N.Y.: Cornell University Press, 1987), 167; Walden Bello and Stephanie Rosenfeld, *Dragons in Distress: Asia's Miracle Economies in Crisis* (San Francisco: Food First Books, 1990), 4.

39. Johnson, "Political Institutions and Economic Performance," 147.

40. Bello and Rosenfeld, *Dragons in Distress*, 15.

41. Bello and Rosenfeld, *Dragons in Distress*, 9.

42. Bello and Rosenfeld, *Dragons in Distress*, 9.

43. John F. Cooper, "Taiwan: A Nation in Transition," *Current History* (April 1989), 174.

44. Cooper, "Taiwan," 174.

45. Frank Gibney, *Korea's Quiet Revolution: From Garrison State to Democracy* (New York: Walker, 1992), 83.

46. Bello and Rosenfeld, *Dragons in Distress*, 43.

47. Cooper, "Taiwan," 176.

48. Bello and Rosenfeld, *Dragons in Distress*, 21.

49. Clyde Haberman, "Korean Declares 'Sweeping' Change Is the 'Only' Way," *New York Times*, 5 July 1987.

50. David E. Sanger, "Korea's Pick: A Pragmatist," *New York Times*, 20 December 1992.

51. Joseph J. Wright Jr., "Thailand's Return to Democracy," *Current History* (December 1992), 421–23.

52. Nicholas D. Kristof, "China Sees 'Market-Leninism' as Way to Future," *New York Times*, 6 September 1993.

53. Marshall I. Goldman, "The Future of Soviet Economic Reform," *Current History* (October 1989), 329.

54. Edward C. Cook, "Agriculture's Role in the Soviet Economic Crisis," in *The Disintegration of the Soviet Economic System*, ed. Michael Ellman and Vladimir Kontorovich (London: Routledge, 1992), 199, 200.

55. Michael Ellman, "Money in the 1980s: From Disequilibrium to Collapse," in *The Disintegration of the Soviet Economic System*, ed. Michael Ellman and Vladimir Kontorovich (London: Routledge, 1992), 196.

56. Michael Ellman and Vladimir Kontorovich, eds., *The Disintegration of the Soviet Economic System* (London: Routledge, 1992), 1; Cook, "Agriculture's Role," 210.

57. Ruth Sivard, *World Military Expenditures, 1987–88* (Washington, D.C.: World Priorities, 1987), 5, 54–55; Bomnath Sen, "The Economics of Conversion: Transforming Swords to Plowshares," in *Economic Reform in Eastern Europe*, ed. Graham Bird (Brookfield, Vt.: Elgar, 1992), 21.

58. David F. Epstein, "The Economic Cost of Soviet Security and Empire," in *The Impoverished Superpower*, ed. Henry S. Rowen and Charles Wolf Jr. (San Francisco: Institute for Contemporary Studies, 1990), 153.

59. Alan Smith, *Russia and the World Economy: Problems of Integration* (London: Routledge, 1993), 74, 88–89; Michael Klare, *American Arms Supermarket* (Austin: University of Texas Press, 1984), 312.

60. David Gold, "Conversion and Industrial Policy," in *Economic Conversion*, ed. Suzanne Gordon and Dave McFadden (Cambridge: Ballinger, 1984), 195.

61. Serge Schememann, "The Sun Has Trouble Setting on the Soviet Empire," *New York Times*, 10 March 1991.

62. Mark Kramer, "Soviet Military Policy," *Current History* (October 1989), 351.

63. James Blitz, "Gloom for the Russians in Gulf Weapons Toll," *Sunday Times* (London), 3 March 1991.

64. Blitz, "Gloom for the Russians."

65. Smith, *Russia and the World Economy*, 158.

66. Ellman and Kontorovich, *The Disintegration of the Soviet Economic System*, 25, 114.

67. Stephen White, *After Gorbachev* (Cambridge: Cambridge University Press, 1993), 1–8.

68. Grigorii Khanin, "Economic Growth in the 1980s," in *The Disintegration of the Soviet Economic System*, ed. Michael Ellman and Vladimir Kontorovich (London: Routledge, 1992), 29.

69. Ralf Dahrendorf, *Reflections on the Revolution in Europe* (New York: Times Books, 1990), 16.

70. Allister Sparks, "Letter from South Africa: The Secret Revolution," *The New Yorker*, 11 April 1994, 59.

71. Christopher S. Wren, "Mandela, Freed, Urges Step-Up in Pressure to End White Rule," *New York Times*, 12 February 1991.

72. T. C. Moll, "'Probably the Best Laager in the World': The Record and Prospects of the South African Economy," in *Can South Africa Survive? Five Minutes to Midnight*, ed. John D. Brewer (New York: St. Martin's Press, 1989), 153.

73. Moll, "Probably the Best Laager in the World," 144, 153.

74. Pauline H. Baker, "South Africa on the Move," *Current History* (May 1990), 197.

75. "How Do South Africa Sanctions Work?" *The Economist*, 14 October 1989, 45; Baker, "South Africa on the Move," 200.

76. Kevin Danaher and Medea Benjamin, "Great White Hope de Klerk Brings Glasnost to Pretoria," *In These Times*, 7–13 February 1990, 10.

77. Baker, "South Africa on the Move," 197.

78. John Darnton, "Africa Tries Democracy, Finding Hope and Peril," *New York Times*, 21 June 1994.

79. Walden Bello and Stephanie Rosenfeld, "Dragons in Distress: The Crisis of NICs," *World Policy Journal*, 7:3 (1990), 460.

80. Riordan Roett, "Brazil's Transition to Democracy," *Current History* (March 1989), 149.

81. James Brooke, "Peru Rises Up against Red Tape's 400-Year Rule," *New York Times*, 8 August 1989.

82. Steven Greenhouse, "East Europe's Sale of the Century," *New York Times*, 22 May 1990.

83. Eul-Soo Pang and Laura Jarnagin, "Brazil's Catatonic Lambada," *Current History* (February 1991), 75.

84. Stephen Engleberg, "First Sale of State Holdings a Disappointment in Poland," *New York Times*, 13 January 1991; Ferdinand Protzman, "Privatization Is Floundering in East Germany," *New York Times*, 12 March 1991; Peter S. Green, "Bonanza or Bust? Czech Sale of Privatized Assets Fizzles," *New York Times*, 18 December 2001.

85. Peter Passell, "A Capitalist Free-for-All in Czechoslovakia," *New York Times*, 12 April 1992.

86. Daniel N. Nelson, "What End of Warsaw Pact Means," *San Francisco Chronicle*, 24 April 1991.

87. *New York Times*, "The World's Shrinking Armies," 30 May 1994.

88. Gary W. Wynia, "Argentina's Economic Reform," *Current History* (February 1991), 59–60.

89. James Brooke, "Latin Armies Are Looking for Work," *New York Times*, 24 March 1991.

90. *New York Times*, "The World's Shrinking Armies."

91. A. F. Mullins Jr., *Born Arming: Development and Military Power in New States* (Stanford, Calif.: Stanford University Press, 1987), 103.

92. Craig R. Whitney, "Western Europe's Dreams Turning into Nightmares," *New York Times*, 8 August 1993.

93. Roger Cohen, "Spain's Progress Turns to Pain," *New York Times*, 17 November 1992.

94. Cohen, "Spain's Progress Turns to Pain."

95. Cohen, "Spain's Progress Turns to Pain."

96. Nathaniel C. Nash, "Latin American Speedup Leaves Poor in the Dust," *New York Times*, 7 September 1994.

97. Nash, "Latin American Speedup."

98. Silvia Brucan, "Shock Therapy Mauls Those Who Unleashed It in Eastern Europe," *World Paper* (June 1994), 3.

99. Michael Specter, "Climb in Russia's Death Rate Sets Off Population Implosion," *New York Times*, 6 March 1994.

100. Michael Burawoy, "Reply," *Contemporary Sociology* 23 (January 1994): 166.

101. Burawoy, "Reply."

102. Lenore B. Goldman, "To Act without 'Isms': Women in East Central Europe and Russia," in *The Gendered New World Order: Militarism, Development, and the Environment*, ed. Jennifer Turpin and Lois Ann Lorentzen (London: Routledge, 1996), 41.

103. Alan B. Krueger, "Economic Scene: Legal Reform Is What the Old Soviet Bloc Needs to Put It on the Path to Growth," *New York Times*, 29 March 2001.

104. *New York Times*, "Study Finds Poverty Deepening in Former Communist Countries," 12 October 2000.

105. Steven Erlanger, "Birthrate Dips in Ex-Communist Countries," *New York Times*, 4 May 2000.

106. Erlanger, "Birthrate Dips."

107. James Brooke, "A Soldier Turned Politician Wants to Give Brazil Back to Army Rule," *New York Times*, 25 July 1993.

7

❧

The Rise of China

China, the world's largest dictatorship, is growing by great leaps and bounds. During the last twenty-five years, Chinese farmers increased grain production from 300 million tons to nearly 500 million tons, making China the world's largest grain producer.[1] In the same period, the value of goods exported by businesses in China grew twenty times, while imports increased fourteenfold, leaving China with a large and growing trade surplus.[2] Overall, China's gross domestic product (GDP) quadrupled during this period, and China became the world's sixth largest economy.[3] Given its rapid rate of growth, many economists, like Nobel Prize winner Kenneth Arrow, believe that "China will become the largest economy in the world" in coming years, surpassing the United States as early as 2015.[4]

These developments have raised living standards and lifted aspirations in China. But China's economic ascent has not been entirely benign. While China has developed economically in recent years, its growth has undermined or compromised economic development in other countries around the world and created new problems for people in China. Put simply, dictatorship in China, which has been assisted politically by the United States and supported economically by investors from around the world, has been responsible both for China's remarkable economic ascent and for the troubling problems associated with it.

To understand the accomplishments and challenges presented by communist dictatorship in China, it is important to look first at developments that contributed to China's rise in the 1970s, 1980s, and 1990s, and then examine the contemporary problems associated with its ascent, both in China and around the world.

STAGNATION, RECOGNITION, AND REFORM

China's rapid economic growth is a recent development, dating back only to the late 1970s. But dictatorship in China is older, dating back to the late 1940s, when Mao Ze Dong led the Communist Party to victory in a civil war with nationalists led by Chaing Kai-shek.

Like other communist dictatorships in the postwar period, the Chinese regime tried to promote agricultural development and industrialization by mobilizing indigenous Chinese natural and human resources. They received little foreign aid from other countries (the Soviet Union briefly provided some assistance in the 1950s) and adopted a set of mercantilist, self-reliant, do-it-alone economic policies. Some economic initiatives, such as the Great Leap Forward in the late 1950s, were conspicuous failures. The regime's attempt to promote rapid industrialization disrupted agricultural production and led to widespread food shortages and acute famine, contributing to the death of between fourteen million and twenty-six million people.[5]

In the late 1960s, Mao launched a political initiative—the Great Proletarian Cultural Revolution (GPCR)—which was intended to remove the bureaucratic and ideological obstacles to development and force the pace of change. But like the Great Leap Forward, the GPCR disrupted the economy without producing substantial benefits.

Still, despite these setbacks, the dictatorship managed to increase grain production from 195 million tons in 1957 to 304 million tons in 1978. This allowed China to feed its population, which grew from 574 million to 962 million in this period, and improve diets slightly.[6] And despite economic disruptions and political distractions, the regime was able to promote substantial industrialization, which laid the foundation for industry in China. "Without that foundation," one economist has argued, "the post-Mao reformers [of the late 1970s] would have had little to reform."[7]

By the early 1970s, China was politically isolated, both by the United States and its allies, and by the Soviet Union and its allies. Its economy was stagnant, having achieved only modest improvements in agriculture and industry, and it conducted virtually no foreign trade. Socially, China was exhausted by the years-long political turmoil of the GPCR. It was at this juncture that the communist dictatorship first received critical assistance from the United States.

In 1972, President Richard Nixon traveled to China and established new diplomatic relations with its communist leaders. Under terms of the

Shanghai Communiqué, the United States agreed to recognize communist China (and its claim to Taiwan as a part of China), admit it to the United Nations (and displace Taiwan as the representative of China in the United Nations), and make it a permanent member, with veto powers, of the UN's Security Council. Nixon's initiative brought an end to China's decades-long political isolation and gave the regime new political legitimacy, both at home and abroad. Recognition also made it possible for the regime to conduct an active diplomatic policy and secure foreign loans and trade deals that could assist its economic development.

Why did Nixon end years of bipartisan U.S. opposition to communist dictatorship and assist Mao's regime in China? Nixon did so because he wanted to divide the two major communist countries (the Soviet Union and China) and use China to put pressure on the Soviet Union and its ally in North Vietnam, which was then engaged in a war with the United States in South Vietnam. Because China bordered North Vietnam, Nixon wanted China to threaten North Vietnam from the rear, making it more tractable in negotiations with the United States. The new U.S.-China alliance helped persuade North Vietnam to agree to a ceasefire in South Vietnam one year later, in 1973. Nixon's dramatic move altered the Cold War balance of power, assisted U.S. policy in Vietnam, and changed political relations and calculations around the world.

In 1976, Mao and his premier, Zhou Enlai, both died, triggering a struggle for political power in China. After a two-year battle, Deng Xiaoping emerged as the regime's new leader. When he came to power, Deng was determined to adopt reforms that would promote agriculture, spur industrial growth, and, importantly, ensure the survival of one-party dictatorship in China. Because the policy reforms that emerged were adopted as a result of trial and error, Deng described his approach as "crossing the river by groping for stepping stones."[8]

Reform in China had three important features in the late 1970s and early 1980s. First, the regime abandoned policies that encouraged population growth and in 1979 adopted a strict, one-child-per-family policy, which was enforced by close surveillance of child-bearing households, widespread sterilization and abortion, and a system of economic penalties and incentives. Although China's population continued to grow, the pace of growth slowed appreciably in subsequent years.

Second, the regime abandoned its collective approach to agriculture and leased public land to farm households for long periods, giving them an incentive to make independent production decisions and reap their benefits. The government increased the prices paid to farmers for produce delivered to the state and allowed farmers to sell food produced in excess of their quotas on the open market for even higher prices. By giving farmers more control over the land, higher prices, and economic incentives, the regime hoped to increase productivity and grow more food. Farmers did just that. In the first six years of reform, grain production in China increased from 305 million tons in 1978 to 407 million tons in 1984.[9] Rising incomes and improved diets secured new political support for the regime in the countryside.

Third, the regime borrowed money from abroad to invest in industry, permitted foreign investors to build factories that could produce goods for export, devalued the currency to make these goods cheaper and easier to sell, and allowed Chinese entrepreneurs to set up businesses that could produce goods for sale on domestic markets. Using foreign loans totaling $40 billion the first decade of reform, and foreign investment amounting to $28 billion in this period, the dictatorship was able to modernize and expand industry, providing jobs for urban and rural workers.[10] The value of China's exports grew fivefold, from $9.7 billion in 1978 to $52.5 billion in 1989.[11]

The income earned from exports enabled the regime to purchase the imports it needed—oil and technology—and invest in developments that would spur the country's economic growth. Deng's package of reforms came to be described as "market-Leninism" because they promoted capitalist or market practices in agriculture and industry, but did so under the direction of a communist or Leninist dictatorship. As Deng explained, his reforms used "communism as the basis, capitalism as the means."[12]

REFORM, CRISIS, AND TIANANMEN SQUARE

While the reforms under Deng slowed population growth, increased food production, expanded exports, and spurred economic growth, they also created problems, inflation chief among them. By using higher prices to increase food production, the government drove up the price of food for urban workers, and rising prices led to inflation. The influx of money and credit from foreign loans, foreign investment, and heavy government spending on development also increased the money supply and contributed to inflation.

In China, as in any country, inflation is a discriminatory economic process (see chapter 3) because some people can keep up with rising prices while others fall behind. In China, inflation, which reached double-digit levels in 1985 and rose 28 percent in 1989, disadvantaged urban residents and people on fixed government salaries or pensions. By 1988, half of all urban households saw their incomes fall.[13] As one Shanghai worker explained, "I wish we could go back to Mao's day. At that time, we had no inflation and were guaranteed a certain living standard. Now I can hardly afford to feed my family."[14]

During the late 1980s, support for economic reform waned as inflation, and the austerity measures designed to curb it, took hold. It was in this context that students assembled in Tiananmen Square in the spring of 1989 to mourn the death of Hu Yaobang (a supporter of more freedom in China, forced to resign as party secretary) and rallied again to welcome Soviet leader Mikhail Gorbachev. At these rallies, they called for political reforms that would yield greater democracy. They were joined by workers upset about the adverse effects of inflation and economic reform. As support for the demonstrators grew, particularly among urban workers around the country, the Communist Party was faced with a serious and direct challenge to its authority.

After a period of negotiations, threats, and internal debates, the dictatorship ordered military forces to assault the peaceful demonstrators camped in the square on June 3–4. They killed about 2,600 demonstrators during the night and morning.[15] A few days later, Deng proclaimed victory: "This was a test [for the regime], and we passed."[16]

The regime was able to survive this test because it had retained the support of the rural peasantry, which had benefited from rising food prices, because the opposition to the dictatorship was relatively weak (urban workers are a minority of the population in China), and because the regime was able to draw on continued political and economic support from the United States and foreign investors after the crisis.

At this crucial juncture, the United States and other countries filed diplomatic protests with the Chinese dictatorship, but they did not sever diplomatic ties, impose economic sanctions or embargoes, or isolate China, as they had done to other countries after similar occasions (for instance, when the Soviet Union invaded Afghanistan in 1979). President George Bush stuck with China during the crisis for several reasons. First, Bush was committed to President Nixon's China policy, having served as U.S. ambassador to China under Nixon in 1974–1975. Bush believed that the United States still needed China as an ally to curb the Soviet Union, which was still a communist dictatorship in 1989. That would soon change, but this was not, evidently, apparent to U.S. leaders in the summer of 1989. Second, Bush wanted to protect U.S. investments in China, which had become substantial during the first decade of reform under Deng, and to reap the future economic benefits of trade with China, which would have been threatened if the United States broke with the regime over the massacre in Tiananmen Square.

Of course, many of these assumptions changed with the fall of communist dictatorships in Eastern Europe a few months later and the subsequent collapse of the Soviet Union a few years later. At it turned out, the fall of communist dictatorships in Europe, combined with the collapse of capitalist dictatorships in Latin America and East Asia, would create an enormous economic opportunity for the dictatorship in China, an opportunity that would enable China to seize a global comparative advantage in low-wage labor.

CHINA'S COMPARATIVE ADVANTAGE

In the late 1980s and early 1990s, capitalist and communist dictatorships around the world collapsed (see chapter 6). The civilian, democratic governments that took power embraced globalization and adopted policies that opened their economies to foreign investment and trade. As a result, hard-working, low-wage workers around the world were offered to employers who produced goods for domestic and global markets. Even though China did not democratize, it was able to obtain a global "comparative advantage" in low-wage workers and convince foreign businesses to invest in the dictatorship in China, not in emerging democracies. China was able to obtain a comparative advantage and secure the lion's share of global foreign investment because it had earlier adopted policies designed to offer vast supplies of labor to prospective employers at very low wages, because it kept wages low as the Chinese economy grew, and because new democratic governments in other countries abandoned policies that dictators had used to coerce labor and suppress wages.

In China, the dictatorship had long used policies designed to control labor and suppress wages. But in the 1980s and 1990s, policies and practices combined to create a low-wage work force that would undercut low-wage workers in other countries and successfully attract foreign investment. The system that emerged consisted of three elements: (1) labor control; (2) migration; and (3) political repression.

Labor Control

In 1955, the dictatorship established control over workers by creating a residential permit system (*hukou*) that assigned households to a particular residence and allowed members to move to a different location—to marry, take a job, care for relatives—only with the government's official permission.[17] Because the government wanted to keep workers in rural areas—collective farms needed their labor and the government could not provide jobs or housing for them in the already-crowded cities—it was very difficult for rural workers to leave the countryside or for urban workers to move from one city to another. The government also sometimes sent urban workers to rural areas as punishment for political "errors" or ideological "crimes," particularly during the GPCR. Under the *hukou* system, individuals who finished their education were assigned by the government to a particular job, where they were expected to stay for their entire career.[18] It was very difficult for individuals or households to change jobs without risking the loss of housing, health care, pensions, future employment, education for their children, food rations, and, for urban residents, permission to continue living in an urban area.[19]

During the 1980s and 1990s, as foreign businesses and domestic entrepreneurs opened factories and new businesses, the government controlled access to these jobs. To meet the demands of new businesses, the regime permitted the creation of some limited labor markets made up of workers that employers could hire. But these markets remained limited and were allowed to operate primarily in the "special economic zones" created for this purpose in cities along the coast.[20]

To provide workers with an incentive to increase productivity in agriculture and industry after 1978, the dictatorship began paying higher prices to rural farmers, which increased their incomes and, as we have seen, contributed to inflation. The regime also allowed businesses to pay higher wages in industry. During the first twenty years of reform (1978–1997), rural and urban wages increased substantially, providing workers with better diets, higher living standards, and increased disposable income. For example, rural and urban wages increased sixteenfold during this period.[21]

This sounds like an impressive achievement, particularly in a dictatorship. But it's impressive only because worker wages were rock-bottom to begin with. It means that rural per capita incomes grew from $16 a year in 1978 to $261 annually in 1997, from 5 cents a day to less than a dollar a day, and urban per capita incomes from 10 cents a day to about $2 a day.[22]

The dictatorship could increase wages and still provide low-wage labor at a bottom-dollar global price because wages were punishingly low to begin with.[23] In China, the "poverty line" is $75 a year, a category into which 85 million people fall, and 203 million more still live on less than $2 a day.[24]

While wages in China have risen, there is little reason to expect that they will continue to rise at the same rate in coming years, because economic growth in the coastal cities has triggered a huge migration that will keep wages in check.

Migration

As jobs were created in coastal cities and wages for urban workers rose, people in rural areas began migrating to the cities, despite legal prohibitions against it. The dictatorship estimates that 114 million workers are migrants in China, a migratory flood without parallel in human history.[25] Shanghai alone is home to three million migrants. "By comparison, the entire Irish migration to America . . . involved perhaps 4.5 million people."[26]

The regime permits this migration because the "blind drifters" (*mangliu*) or "outsiders" (*waidiren*), as they are called, will do unpleasant work that urban workers are unwilling to do, because they can be employed on a temporary or casual basis and can easily be dismissed or replaced without recourse (they can lay no claim to permanent or contractual employment because they have no residency permits), and because they can be forced to work under onerous conditions—twelve to fifteen hours a day, in hazardous or toxic work environments—for little or, in many cases, no pay.[27] Officials recently reported that migrants who labor on government construction projects were owed $43 billion in unpaid wages, and "some have remained unpaid for up to 10 years."[28]

Because migrants are essentially illegal, undocumented workers in China, they cannot lay claim to the legal protection provided workers with government-issued residential permits and work assignments, do not qualify for the minimum wage, and cannot send their children to school or obtain food rations or any of the health and pension benefits associated with legal employment.[29] If they complain about working conditions or demand higher wages, they can be arrested and "deported" (returned to the countryside) because they have no legal standing.[30] If migrants are injured or killed on the job—industrial accidents in China claimed more than 30,000 fingers from workers last year—they receive lower compensation than regular workers. The families of legal urban workers killed in a gas-line explosion received $17,000 for their loss; migrant families only $5,000.[31]

The regime's labor policy—which allows massive migration to occur, but also keeps it illegal—has created a huge reserve army of labor. This illegal army supplies workers to government industries and private domestic employers at less-than-legal wages and helps suppress the wages of legal workers. As this migrant army swells, it will be increasingly difficult for legal workers to demand higher wages because they know that employers might replace them with desperate, more tractable, illegal migrants.

This means that wages, both for legal and illegal workers in China, will not rise in coming years as fast as they did during the first twenty years of reform. Moreover, migration is not the only mechanism the regime uses to keep wages low and prevent them from rising, a development that would erode China's comparative advantage in low-wage labor. It also relies on an extensive repressive apparatus, which is designed to keep economic and political demands in check.

Repression

The dictatorship in China, like most dictatorships, outlaws individual dissent and public political protest, union organizing and collective bargaining, religious practices and proselytizing, and any other activity that questions or challenges its authority.[32] It routinely uses the army, the legal system, and its regular police forces to keep protest in check and the Communist Party in power. But unlike many other dictatorships, it depends on another instrument: the Communist Party cadre dispersed throughout public and private institutions. In 1989, the regime could count on forty-five million party members. But after Tiananmen Square, the party recruited heavily, and by 2000 it could count on sixty-eight million members to monitor the population; report dissent; manage schools, businesses, and government bureaucracies; and disseminate the government's propaganda.

In China, it is illegal even to petition the government for redress. Individuals who traveled to Beijing to submit petitions to the regime were beaten, arrested, and deported to rural prisons, and a human rights organization in China reported that 36,000 had been so treated in 2003.[33] Collective protests in 2003, in which more than three million people took part, were ruthlessly suppressed by government forces.[34] The government imprisons millions of people and executes thousands each year for political and civil crimes that are not regarded as capital crimes in most other countries: fraud, tax evasion, and prostitution.[35] Not surprisingly, 93 percent of Chinese surveyed "did not regard China as a country ruled by law."[36]

In this context, the massacre of peaceful protesters at Tiananmen Square in 1989 was significant because it signaled the regime's determination to keep low-wage worker policies in place and prevent Chinese workers from demanding economic or political reforms that might increase wages or demand safety regulations, job protection, or health care or pension benefits that would raise the indirect costs of doing business in China and thereby undermine the government's ability to sell its workers on the world market for the lowest possible price.

DEMOCRATIZATION ELSEWHERE
ADVANCES DICTATORSHIP IN CHINA

Until the wave of democratization in the late 1980s and early 1990s, China's dictatorship was just one among many. Although the regime in China could offer low-wage workers to business investors, so, too, could many other countries. But that changed with the collapse of capitalist dictatorships in Latin America and East Asia, and the fall of communist regimes in Eastern Europe and the Soviet Union.

When civilian democrats assumed power, they reformed their economies, opened them to foreign investment and trade, and invited businesses from around the world to employ their hard-working, low-wage workers. But while the new democracies could offer workers who were willing to work cheap, they could not promise investors that these workers would remain inexpensive.

In a democracy, workers can move freely, change jobs, and organize collectively to demand higher wages. To keep workers, businesses have to accommodate at least some of their demands. Moreover, workers can vote in elections and choose politicians who will provide workers with social benefits and take steps to protect them from hazardous or onerous working conditions. In democracies, workers can therefore raise the direct cost of doing business (by demanding higher wages) and increase the indirect costs of doing business (by imposing taxes on businesses to pay for unemployment insurance and contribute to health care and pension funds or by adopting laws requiring business to spend money to comply with safety or environmental regulations). Business investors understand that when they hire low-wage workers in democratic countries, they can expect the cost of doing business there to rise over time. But this is not the case in a dictatorship, like China, where the government actively suppresses wage demands and does everything it can to reduce the indirect costs to businesses by allowing them to operate without meaningful worker safety or environmental regulation, without assuming any real responsibility for the health or welfare of their employees.

If China had democratized, as it might have done in 1989, it would not have secured a decisive global, comparative advantage in low-wage workers. But because it remained a dictatorship, while others around it democratized, it could offer a low-wage work force that could not easily demand higher wages or impose higher costs on the firms that employed it.

Obviously, China's global comparative advantage was gained unfairly. When Adam Smith and David Ricardo formulated the idea that countries should specialize at what they did best in the global division of labor, they imagined that a country's comparative advantage would be fairly obtained, not artificially acquired by dictatorships using the coercive powers available to them. Instead of playing on a level playing field, dictatorship in China has rigged the game of providing low-wage labor and tilted the field in its favor, which disadvantages all the other players on the field. This is particularly evident for other democratizing countries that offered low-wage workers to global investors in the 1990s.

South Korea and Taiwan

In the late 1980s, regimes in South Korea and Taiwan began to democratize, holding elections and opening them to participation by opposition parties (see chapter 6). Like China, dictators in these countries had suppressed worker demands and offered compliant, low-wage workers to foreign and domestic employers. But as they democratized, investors began to relocate many businesses to low-wage China, where dictatorship remained intact.

This process was first apparent in industries that produced footwear, toys, games, and sporting goods for export to the United States. In 1987, when the democratization process began, 60 percent of the goods imported by the United States came from South Korea and Taiwan, and only 5 percent from China. But by 1991, as democratization widened and deepened, U.S. imports of these goods from South Korea and Taiwan had fallen to 30 percent, while imports from China had risen to 30 percent. And by 1999, the origin of these goods had completely reversed: China produced 60 percent of all U.S. imports, and only 5 percent originated in South Korea and Taiwan.[37]

Much the same happened with personal computers produced in Taiwan. In 1999, only 28 percent of the PCs manufactured by Taiwanese computer firms were made in China, but by 2000, 42 percent originated from the mainland.[38] During the 1990s, "tens of thousands of firms in Taiwan shifted operations across the strait to the mainland," largely because they were "lured by Chinese wages that [were] a fifth of those in Taiwan."[39]

During the 1990s, businesses based in Taiwan invested an estimated $100 billion in China. Not surprisingly, economic growth in Taiwan slowed dramatically, from 6 percent annual growth in GDP in 1990 to 0 percent in 2001, and unemployment tripled.[40]

Mexico

For decades, Mexico was run by a single political party, the Institutional Revolutionary Party (PRI). During the 1990s, opposition parties began to win local and state elections, challenging its political and economic domination. But as Mexico democratized, electing Vincente Fox, Mexico's first non-PRI president, in 2000, investors abandoned low-wage factories—*maquiladoras*—established along Mexico's border with the United States, and moved to China. During the 1990s, Mexico "lost nearly half a million manufacturing jobs and 500 *maquiladora* manufacturers" to China, where workers earned one-quarter of the wages paid to Mexican workers.[41]

In all three of these cases, it is important to note that it is not just wage levels that affected business decisions to abandon low-wage workers in East Asia and Latin America, but also democratization, which promised to raise wages over time. Essentially, investors voted with their money against democracy in South Korea, Taiwan, and Mexico, and for dictatorship in China.

Some economists argue that investors have shifted factories from democracies to dictatorship only because workers in South Korea, Taiwan, and Mexico were more costly. But that does not explain why investors should choose China over India, where there are just as many bottom-dollar, low-wage workers who would work hard for foreign businesses.

India

Like China, India has a huge population of poor, hard-working people, who are desperate to work for cheap. In India, 350 million people earn less than $1 a day, and 40 million are officially unemployed.[42] Like China, the government in India decided in the early 1990s to abandon its mercantilist policies, reform the economy, and welcome foreign investment and trade as a way to spur growth in its moribund economy.[43] Moreover, India has one "comparative advantage" that China cannot easily match: 34 million Indian workers are extremely proficient in English, which is one of India's official languages and a regular part of the curriculum in its schools. Indeed, many American and Indian firms have taken advantage of low-wage, well-educated, English-speaking Indian workers to outsource U.S. jobs in the service industries—computer software, accounting, customer support, insurance claims—to India, where low-wage workers conduct business and communicate in English with customers in the United States. In the early 2000s, firms providing these services employed about one million workers in India, and observers estimated that this outsourcing work force would grow to three or four million in the next decade.[44]

In strict economic terms, India can match or surpass China's comparative advantage in low-wage labor. But it has not managed to persuade foreign investors to choose India over China, except in the relatively small, English-speaking service sector. Between 1990 and 2000, foreign businesses invested $4 billion in India, but over ten times that amount, $40 billion, in just one year in China, and eighty-five times that amount over the same period.[45]

Businesses chose China over India not because wages are lower in China, but because China is a dictatorship that promises to keep wages low, while India is a democracy, where even low-wage workers can demand more. Workers in India can do so because they can move freely, organize collectively, and shift jobs in search of higher wages or better working conditions. That is exactly what has happened in the English-speaking outsourcing service industry in India. Businesses there compete for the best workers, workers can search for the best opportunities, and, as a result, "wage increases have been the most common method of attracting and retaining employees. For example, Wipro, the big outsourcing company, gave its 24,000 employees in India an average raise of 10 percent [in

2003]."[46] But these conditions, whereby businesses compete for workers who can leave for other jobs, do not exist in China.

Not only do workers in democratic India have the ability, given them by its laws, to bargain effectively with domestic and foreign employers to raise wages, they can use their right to vote to obtain benefits that are unavailable in a dictatorship.

In India, foreign investment in outsourcing services provided jobs, raised wages, and spurred economic growth in cities around the country. But because India is a democracy, voters unhappy with the unequal distribution of opportunities and wealth associated with service industry development threw out the governing political party that had promoted it and in 2004 elected a new government, which promised to redistribute the benefits of growth more widely.[47]

Foreign investors took the election results to mean that the new government would permit workers to demand higher wages and increase the indirect costs of doing business, as democratic governments everywhere do, to provide benefits to workers outside the service industry.[48] As a result, the Indian stock market plunged and foreign investment, which was never strong, slowed.[49] As events in India demonstrate, investors, given a choice, would rather employ low-wage workers in a dictatorship, in China, where there is little upward pressure on wages and indirect business costs, than hire workers in a democracy, where the upward pressure on wages and benefits is stronger. This is true even though employers could pay the same low wages in both countries.

During the 1990s, the global comparative advantage in low-wage labor swung decisively to China as other low-wage countries democratized. This advantage resulted in a flood of foreign investment in China. It came from investors in other low-wage countries like South Korea and Taiwan, from overseas Chinese communities, and from high-wage countries like the United States and Japan. This surge of foreign investment promoted economic growth in China. But it also effectively undermined development elsewhere, particularly in low-wage, democratic countries.

THE FOREIGN INVESTMENT FLOOD

During the first decade of reform, China received $51 billion (about $5 billion annually) in foreign investment, half of it in the form of loans, which China had to pay back, and half of it in the form of direct investment, which it did not have to repay.[50] In this period, investors from Taiwan, Hong Kong, and Macao provided three-quarters of the money.[51] These modest flows were significant for China because they spurred economic growth during the 1990s.

But after Tiananmen Square, when the dictatorship in China secured its global comparative advantage in low-wage labor, foreign investment soared, and modest flows became a torrential flood, a monsoon of money. Overseas Chinese investors were joined by investors from the United States and Japan. In 1991, foreign investment doubled to $12 billion, quadrupled the next year to $58 billion, and doubled again in 1993 to $111 billion.[52] By 1993, China had captured one-third of all global foreign direct investment (FDI), as it is called.[53]

During the 1990s, investment continued to pour into China, primarily as FDI, not loans. Some economists estimate that China received $339 billion during the decade, and by 2000 had captured one-half of all global foreign investment.[54] Although there was a slight downturn in 2000, China received foreign investment worth about $40 billion annually in the early 2000s. By 2004, China had become the world's second largest recipient of foreign investment, after the United States.[55]

These investment flows are unprecedented. They dwarf U.S. spending during the Marshall Plan, which helped all of Europe recover from World War II (see chapter 1). The Marshall Plan was used to assist democratic countries, but today, ironically, foreign investment is being used to assist a communist dictatorship.

Naturally, investment on this scale produced sizzling, double-digit rates of annual economic growth in China, helping it weather the economic storm that flattened economies in other East and South Asian countries during the late 1990s. But there remained one obstacle to continued economic growth in China: the trade barriers and restrictions imposed on Chinese exports by countries belonging to the World Trade Organization (WTO), the United States chief among them. Foreign investors primarily built factories in China to produce goods for export to global markets. But because China did not belong to the WTO, its member countries could tax or restrict goods imported from China, making it difficult for the regime to realize all the possible gains from foreign investment. But that would change with China's admission to the WTO.

ADMISSION TO THE WORLD TRADE ORGANIZATION

During the 1990s, admission to the WTO became a priority for the regime in China, and it began conducting negotiations with the United States and other WTO members to achieve membership and persuade them to lower trade barriers to goods made in China. Officials in many WTO countries opposed China's admission, arguing that if they lowered trade barriers, low-cost imports from China would destroy their domestic industries, primarily because domestic industries paid workers higher wages. This mattered most in labor-intensive industries like textiles, apparel, and shoe manufacture, and less in industries that rely more heavily on machines and technology to do the work.

After long negotiations, the Bush administration in 2001 finalized an agreement admitting China to the WTO and concluded its own bilateral agreement with China, which lowered U.S. trade barriers to Chinese imports. U.S. trade representative Charlene Barshefsky promised that the agreement would "open the world's largest nation to our goods, farm products and services in a way we have not seen in the modern era."[56]

Why did the U.S. government support the dictatorship in China on this issue?

Remember that U.S. firms had invested heavily in China during the 1990s. So it was largely goods they made in China that were being taxed or restricted by U.S. trade barriers. For instance, it helps to know that "12 percent of China's exports to the United States end up on Wal-Mart's shelves . . . and that Wal-Mart's trade with China accounts for 1 percent of that country's gross domestic product."[57]

U.S. businesses—like Wal-Mart, but also General Electric, Target, IBM, Boeing, and Ford—that produce goods in China and import them into the United States wanted to lower U.S. tariffs or taxes on imported goods so they could sell them for lower prices on U.S. markets. Essentially, by reducing U.S. tariffs on goods imported from China, the U.S. government provided a tax break for U.S. corporations, a tax cut not made available to domestic firms that manufacture goods in the United States. The Clinton and then Bush administrations decided to support China's application to the WTO, and assist U.S. firms producing goods in China, because they were persuaded that the benefits to U.S. consumers, and to U.S. businesses in China, outweighed the cost to domestic U.S. businesses, workers, and taxpayers (reducing tariffs means that taxpayers have to make up lost revenue).

Ever since Nixon went to China in 1972 and established diplomatic relations with the dictatorship there, Republican and Democratic presidents have supported expanded investment in and trade with China. They took three steps to accomplish this: first, by establishing diplomatic relations with the dictatorship and admitting it to the United Nations; second, by promoting U.S. investment in China, particularly after the massacre in Tiananmen Square; and third, by sponsoring China's admission to the WTO.

POLICY RATIONALES

To justify these policies, U.S. officials and economists have advanced two arguments. First, they have argued that diplomatic recognition and investment would promote democratization in China. Proponents of this perspective maintain that capitalism has a corrosive effect on dictatorship. By promoting the economic freedoms associated with the market, capitalism would eventually promote the political freedoms associated with democracy.

Second, U.S. officials have argued that by establishing trade relations with China, and admitting it to the WTO, they would open the huge domestic market in China to U.S. goods, thereby providing business opportunities for U.S. corporations and jobs for American workers.

Unfortunately, the first argument is based on illusions about the impact of investment and economic growth on political institutions. And the second fails to appreciate the policies that have been adopted by the dictatorship to prevent access to the huge domestic market in China.

Foreign Investment and Democratization in China?

Economists and politicians in the United States have argued that foreign investment in China would promote democratization in China. But is there any evidence that it has? Although it has been more than thirty years since Nixon recognized China, and twenty-five years since the regime opened China to foreign investment, the Communist Party remains in power, having survived the brief challenge to its authority in 1989. If foreign investment were a powerful force for political change, shouldn't the huge investments made during the 1990s have produced some kind of political change?

They have not. Instead of weakening, the Communist Party has become stronger, making it possible for its leader, Hu Jintao, to declare in 2004 that Western-style, multi-party democracy was a "blind alley."[58]

The adoption of capitalist economic strategies in China and the influx of foreign investment has done little to promote democratization during the last thirty years. This should not be surprising. Imagine, for a moment, that West Germany had invested $1.5 trillion in East Germany during the 1970s and 1980s, when the communist dictatorship in the East still held power. (West Germany did invest this amount of money in the East between 1990 and 2004, but only after the dictatorship fell in 1989.[59]) Would massive West German investment have promoted democracy in the East? Or would it instead have secured a long political future for the dictatorship there? If the dictatorship had received West German investment, wouldn't it have plausibly claimed that economic growth was due to the regime's own sound economic policies and astute political leadership, not due to foreigners? Wouldn't this have been a plausible argument for most of its citizens? It should be obvious that if West Germany had invested in dictatorship, the East German regime would likely still be in power today. And so it is with China. Foreign investment has not undermined dictatorship in China, but instead rescued and strengthened it.

Opening China's Market to Foreign Firms?

For thirty years, foreign investors have hoped that trade with China would enable them to capture markets and sell their goods to domestic consumers in China. But they have generally failed to appreciate the fact that the dictatorship has adopted policies designed to prevent foreign access to China's domestic market.

The regime has allowed foreign investors to use China as a vast *maquiladora*, a place to produce goods for export. This enables the government to secure the hard currency it needs to purchase raw materials, capital goods, and technology that it cannot yet make on its own. But it has discouraged foreign firms from producing or selling consumer goods in domestic markets. The regime wants to build up state and private firms to produce for the domestic market, so it has adopted a number of policies that assist domestic firms and discourage foreign ones.

First, the regime allows state-run and private Chinese firms to hire illegal, undocumented migrant workers and pay them less than the workers assigned to or hired by foreign firms. So the government provides low-wage workers to foreign firms, but even lower-wage workers to domestic businesses, giving them a crucial cost advantage. It also allows domestic firms to evade the restrictions, laws, and benefits required of foreign firms, giving domestic firms an additional savings on their indirect costs.

Second, the regime provides loans at below-market interest rates to domestic firms, and, if they incur losses, provides subsidies to keep them afloat. This enables them to obtain capital more cheaply than foreign firms.[60]

Third, the regime encourages domestic firms to create excess manufacturing capacity and produce more goods than the market requires, driving down prices and making it difficult for foreign firms to profit from the production of these goods.[61]

Fourth, the government demands that foreign firms transfer their technologies in return for permission to do business in China, then passes these technologies on to domestic firms so they can develop the technological capacity to compete with foreign businesses. Or it simply tolerates the theft or piracy of foreign patents and copyrights.

Fifth, the regime discourages retail outlets, many of them state owned, from displaying or selling foreign goods or, if they do, to push domestic goods first.

Finally, and perhaps most importantly, the regime devalued the Chinese currency several times in the late 1980s and then again in the early

1990s. These devaluations made imported goods more expensive and domestic goods cheaper, which gave Chinese firms an important price advantage over foreign firms that imported goods to sell in China.

As a result of these policies, domestic Chinese firms have obtained numerous advantages—cheaper labor and capital, access to technology and
retail outlets, and favorable exchange rates—over foreign firms. This
makes it very difficult for foreign firms to compete effectively with domestic firms or capture the markets they expected to find when China
"opened" its doors to foreigners.[62]

The fact is that instead of foreign firms capturing China's markets,
China has captured foreign markets. Proof of this is simple. The United
States registered its first trade deficit with China in 1983, when it recorded
a $300 million deficit.[63] This deficit has grown every year since, reaching
$83 billion in 2001 and $125 billion in 2004.[64]

It turns out that the hopes of U.S. politicians and economists, who argued that investment would democratize China and open its market to
U.S. goods, were misplaced. All the evidence suggests that these hopes
were illusory.

PROBLEMS AHEAD

China's economic ascent, which was assisted by foreign diplomacy, investment, and trade, has come at the expense of economic development and
democratic government in many countries around the world. But while
economic expansion in China has been dramatic, even spectacular in global
terms, China nonetheless faces a series of important problems in coming
years, which are largely by-products of its success. How these problems are
addressed, both by China and by other countries around the world, will determine how China's economy and its dictatorship fare in the near future.

There are four important problems that could create obstacles to continued economic success and political dictatorship in China: (1) inflation, which
is a by-product of foreign investment; (2) exchange rates, which are a by-
product of China's determination to protect domestic firms and markets; (3)
food supplies, which may grow short as a by-product of China's economic
growth and migration, which is a by-product of efforts to increase food supplies; and (4) Taiwan, which is a by-product of postwar partition in China.

Inflation

Foreign investment has provided enormous benefits for China's economy
and its political leadership. But massive foreign investment has a down-
side: it contributes to rising inflation.

When foreigners invest in China, their dollars and other hard currencies are converted into yuan, the Chinese currency, so it can be spent inside the country. But this increases the supply of money and credit in China, which contributes to inflation.[65] Prices are pushed up not only by foreign investment, but also by growing demand for food, energy, and raw materials, which are consumed by growing industries.[66] Meanwhile, massive migration to the cities drives up urban rents. By the early 2000s, inflation increased annually at a rapid rate.[67]

Inflation is a troublesome problem for businesses, workers, and government officials in China because it is a discriminatory economic process. It is a particularly difficult problem in China because, as we have seen, it is very hard for workers there to obtain higher wages, which would help them keep up with rising prices. As a result, inflation erodes their real incomes and standards of living—some groups (rural workers, migrants, women) more than others (urban workers, men, workers employed by foreign firms).

Inflation is not only an economic problem, it is also a political one. Rising inflation in the late 1980s contributed to widespread dissent and support for protesters at Tiananmen Square. In the early 2000s, inflation could again contribute to dissent and public protest.

To curb inflation and deflect protest, the dictatorship could restrict the influx of foreign investment; raise interest rates (as Volcker did in the United States in the early 1980s; see chapter 3); and restrict the cheap loans it makes to domestic businesses. But doing so would slow China's economic growth, increase unemployment, and disadvantage domestic firms, which rely on cheap capital to maintain their competitive advantage over foreign firms in the domestic market. So the regime faces a difficult dilemma: let inflation continue, and contribute to lower living standards and possible political protest; or curb inflation, and contribute to a recession, higher unemployment, and bankruptcies among domestic firms.

Exchange Rates and Trade

In the mid-1980s, and again in the early 1990s, China devalued the yuan, by about 80 percent.[68] Since 1994, the yuan has been fixed at a rate of 8.27 yuan to the dollar.[69]

These devaluations gave China economic advantages because the exchange rate made goods manufactured in China cheaper and easier to sell in foreign markets (particularly the United States), and made imported goods (particularly from the United States) more expensive and therefore more difficult to sell in China. Changing exchange rates gave China a competitive advantage, much like a generous handicap in golf (see chapter 2).

For the Chinese, currency devaluations produced a desirable outcome. They helped create a trade surplus with the United States. But for the United States, devaluations contributed to a large and growing trade deficit with China, amounting to $125 billion in 2004.

Under normal conditions, China's trade surpluses with the United States would force China to revalue its currency and make it stronger, thereby reducing its golfer's handicap to reflect its improved economic game, and making its exports more expensive and its imports cheaper. But the regime has prevented any readjustment in two ways. First, it has not permitted the yuan to be freely convertible on global currency exchange markets, as other currencies have been since the end of Bretton Woods in 1971 (see chapter 2).[70] Because the yuan is not convertible, it is not subject to market forces, only government dictates.

Second, the regime has used the dollars earned from trade surpluses with the United States to buy U.S. Treasuries, which has helped keep the value of the dollar high, the yuan low, and the exchange rate unchanged. Estimates vary, but economists believe that China has purchased $400–$600 billion in U.S. Treasuries.[71] Essentially, the Chinese have used dollars they earned from the United States to keep U.S. manufactures at a disadvantage, both in U.S. and in Chinese markets.

Growing U.S. trade deficits with China have provoked a reaction by businesses and workers in the United States. They have pointed out that rising trade deficits have contributed to job loss in the United States— three million manufacturing jobs were lost between 2001 and 2004—and made it difficult for U.S. firms to compete.[72] "With this currency manipulation, my customers, my employees are being slaughtered by unfair competition," argued Bill Hickey, president of Lapham-Hickey Steel. "It's like running a 100-yard dash against a team that starts on the 50-yard line."[73]

But the Bush administration rejected complaints that China's currency policies helped subsidize its exports or violated international monetary or trade laws, despite the fact that even a modest, 25 percent rise in the value of the yuan "would ultimately add 500,000 high-paying jobs to the American economy."[74]

Why did the U.S. government reject efforts to change China's exchange rate policy, even though doing so would reduce U.S. trade deficits and create U.S. jobs?

The reason was that many U.S. firms support China's efforts to keep exchange rates as they are. Remember, many of the goods imported from China are made by U.S. manufactures based in China. U.S. corporations opened factories in China so they could export goods to the United States, where they end up on shelves at Target and Wal-Mart. These firms want to keep exchange rates as they are because if the value of the yuan rose, it would increase the price of goods they sell in America.

Another issue is related to this. If the value of the yuan increased, and the dollar fell, the value of China's vast stocks of U.S. Treasuries would decline. The U.S. government needs to sell Treasuries to cover the huge budget deficits created by the Bush administration. If China sold Treasuries in advance of a yuan appreciation, the Federal Reserve would likely have to raise interest rates to persuade investors to buy U.S. bonds. This, of course, could trigger a recession and increase unemployment in the United States. So U.S. budget deficits and fear of rising interest rates prevents or discourages U.S. policymakers from forcing China to alter its exchange-rate policies.

Of course, the Chinese regime could alter exchange rates on their own. But if they did, this would likely reduce China's export earnings and increase imports, which would result in job loss in China. Again, the dictatorship faces a difficult choice: keep exchange rates as they are and risk possible retaliation from the United States or change exchange rates and slow economic growth in China. In all likelihood, they will probably adjust rates a little, to throw U.S. policymakers a bone, but not enough to affect trade patterns significantly. However, this will only defer the problem, both for China and the United States.

Food and Migration

In 1995, Lester Brown, head of the Worldwatch Institute, wrote a book called *Who Will Feed China?* In it he noted that reform in China had encouraged farmers to increase agricultural productivity and grow more food.[75] Annual grain production grew from 276 million tons in 1979 to 461 million tons in 1997.[76]

While this was a remarkable achievement, Brown noted that reforms had also resulted in the conversion of farm land to other uses, as businesses built factories, cities built housing, and the government built new roads and dams.[77] At the same time, rising incomes had increased the demand for food, diets improved, and the population continued to grow, though at a slower pace. These developments, he argued, were not unique to China. They had previously occurred in Japan, South Korea, and Taiwan, where economic growth had led to cropland conversion and growing consumer demand. If China followed their example, it would eventually have to import large quantities of grain to meet its needs. But because China is so large, it would have to import much larger quantities, which could force up world grain prices. This would benefit farmers, but seriously disadvantage the world's poor, who could ill afford more expensive food.

Brown's book triggered a wide-ranging debate.[78] By and large, Brown's critics have argued first that the conversion of arable land to other uses has been slower and less dramatic than he imagined.[79] Second, they argued that the productivity of agriculture in China could still increase substantially. By consolidating farms, mechanizing agriculture, and introducing higher-yielding grain varieties, farmers in China could make up for cropland losses and meet growing consumer demand.[80]

Part of the problem, though, is that Chinese government agricultural statistics are unreliable, which makes it difficult to make accurate assessments and predictions. Because China is so large, a small error can have huge consequences. For example, it takes 370,000 tons of grain just to provide each adult in China with one more bottle of beer each year. So if each adult in China drank three more bottles of beer each year, brewers would need to secure as much additional grain as Norway, the entire country, grows annually.[81]

More recent estimates show that cropland is being converted at a rapid rate: China lost 2 percent of its farmland in 2003, and grain production has not improved. In 2003, China produced 401 million tons, "down 18 percent from the record 486 million tons produced in 1998, according to the USDA."[82] In 2003, China consumed 40 million tons of imported grains, and this has contributed to rising food prices, up 28 percent in China during that year.[83] Unfortunately, the problems Brown identified are still troublesome.

Moreover, if the regime consolidated farms and introduced new machinery to increase productivity and crop yields, they would displace hundreds of millions of rural people. Experts estimate that there are 200 million surplus laborers in agriculture, so any effort to consolidate or mechanize agriculture would force them off the land.[84] Presumably, they would join the tide of illegal, migrant workers in China, effectively doubling or tripling its size, heading into the cities. Once you know that China already has 166 cities of more than one million people—the United States has only nine—you can appreciate how difficult it will be to build housing and provide jobs for the millions of rural people displaced by agriculture.[85]

This creates a serious set of problems for the regime. Continued economic growth may force it to increase food imports, which would drive up prices in China and around the world. But the alternative—increasing food production in China—could unleash a flood of low-wage workers into the cities.

Taiwan

The dictatorship in China has long wanted to reunite with Taiwan. The island was separated from the mainland by the onset of the Korean War in 1950, when the United States sent its fleet into the Taiwan Straits to prevent a communist assault on Chiang Kai-shek's nationalist forces, which had fled to the island during the Chinese civil war (1947–1949).[86]

Because U.S. officials regarded Taiwan as a front-line, anti-communist state during the Cold War, they lavished economic and military aid on the capitalist dictatorship there. U.S. aid promoted rapid economic growth in Taiwan during the 1950s and 1960s. But U.S. assistance evaporated when Nixon recognized communist China and forced Taiwan off the Security Council and out of the United Nations. Despite this political reversal, the regime in Taiwan was able to continue making economic gains, and Taiwan became a relatively prosperous country.

Chiang Kai-shek died in 1975 and passed power to his son, Chiang Ching-kuo, who ruled until his death in 1988. But just before he died, Chiang Ching-kuo began democratizing Taiwan's political institutions, a process that proceeded slowly but eventually took root by the end of the 1990s. During this period of democratization, Taiwanese businesses began investing heavily in communist China (see above). While economic ties between the two countries grew stronger, political differences grew wider because Taiwan became a democracy while China remained a dictatorship. As a result of their growing political differences, Taiwan's leaders have argued that the island should protect its democratic institutions, perhaps even become fully independent.

But Taiwanese independence and democracy is anathema to China's communist leaders, who regard Taiwan as part of China. And they have threatened to invade Taiwan if it moves toward independence.

The problem, of course, is that a Chinese invasion of a small democratic country would wreck China's diplomatic standing in the world, destroy trade relations, disrupt investment patterns, impose enormous suffering on people in both countries, and raise the risk of nuclear war (China possesses nuclear weapons and Taiwan has the ability to make them fairly quickly).

Generally speaking, the United States and members of the United Nations support China's claim on Taiwan, regarding it as a part of China's sovereign territory. But they do not want to see China reclaim Taiwan by force. If China invaded Taiwan, the United States would be forced to make a difficult choice: side with China to protect U.S. economic interests, and thereby support a dictatorship against a democracy; or protect democratic Taiwan from attack by a communist dictatorship, and thereby risk substantial U.S. economic interests in China.

There is some chance that leaders in China and Taiwan might agree to reunify peacefully. They have each raised the possibility of doing so over the years. China's reincorporation of Hong Kong and Macao was, for some years, viewed as a possible model. But China's record in Hong Kong, where the dictatorship has undermined the democratic guarantees

provided in the reunification agreements with Great Britain, has persuaded many in Taiwan that reunification would greatly disadvantage
them, both in economic and political terms.[87] So long as China remains a
dictatorship, peaceful reunification with Taiwan is unlikely, and any attempt to force the issue could be disastrous.

The problems associated with investment, exchange rates, food, and
Taiwan could slow, stall, or wreck continued economic growth in China
and trigger conflict in China and abroad. How these problems will be addressed is unclear, but it is certain that whatever steps are taken, their consequences will reverberate around the globe.

NOTES

1. Jianfa Shen, "Agricultural Growth and Food Supply," in *Changing China: A
Geographical Appraisal*, ed. Chiao-min Hsieh and Max Lu (Boulder, Colo.: Westview,
2004), 52; Lester R. Brown, *Who Will Feed China?* (New York: Norton, 1995), 27.

2. United Nations, *China Human Development Report, 1999: Transition and the
State* (Oxford: Oxford University Press, 1999), 41.

3. Ted C. Fishman, "The Chinese Century," *New York Times Magazine*, 4 July 2004.

4. David Wessel and Marcus Walker, "Good News for the Globe," *Wall Street
Journal*, 3 September 2004; Maurice Meisner, "China's Communist Revolution: A
Half-Century Perspective," *Current History* (September 1998), 247.

5. Elisabeth J. Cross, "The New Peasant Economy," in *Transforming China's
Economy in the Eighties*, ed. Stephan Feuchtwang, Athar Hussain, and Thierry
Pairault (London: Zed Books, 1988), 101.

6. Shen, "Agricultural Growth and Food Supply," 50.

7. Meisner, "China's Communist Revolution," 246.

8. Richard Baum, *Burying Mao: Chinese Politics in the Age of Deng Ziaoping*
(Princeton, N.J.: Princeton University Press, 1994), 17.

9. Mark Selden, *The Political Economy of Chinese Development* (Armonk, N.Y.: M.
E. Sharpe, 1983), 19.

10. Gerald Segal, "The Challenges to Chinese Foreign Policy," in *The Reform
Decade in China: From Hope to Dismay*, ed. Marta Dassu and Tony Saich (New York:
Columbia University Press, 1992), 183; Renssalaer W. Lee III, "Issues in Chinese
Economic Reform," in *Economic Reform in the Three Giants*, ed. John Echeverri-Gent
and Friedemann Muller (Washington, D.C.: Overseas Development Council,
1990), 86; Ian Jeffries, *Socialist Economies and the Transition to the Market: A Guide*
(London: Routledge, 1993), 159.

11. Susan L. Shirk, *The Political Logic of Economic Reform in China* (Berkeley: University of California Press, 1993), 48.

12. Selden, *The Political Economy of Chinese Development*, 8.

13. Barry Naughton, "Inflation and Economic Reform in China," *Current History* (September 1989), 270; Nicholas Kristof, "China Erupts," *New York Times Magazine*, 4 June 1989.

14. Lee, "Issues in Chinese Economic Reform," 80.

15. Baum, *Burying Mao*, 263–64, 276.

16. *New York Times*, "Deng's Speech: 'We Faced a Rebellious Clique' and 'Dregs of Society,'" 30 June 1989.

17. Kam Wing Chan, "Internal Migration," in *Changing China: A Geographical Appraisal*, ed. Chiao-min Hsieh and Max Lu (Boulder, Colo.: Westview, 2004), 229–30; Selden, *The Political Economy of Chinese Development*, 14.

18. Margaret Maurer-Fazio, "Building a Labor Market in China," *Current History* (September 1995), 285.

19. Maurer-Fazio, "Building a Labor Market," 286; Selden, *The Political Economy of Chinese Development*, 14.

20. Jan S. Prybyla, "All That Glitters? The Foreign Investment Boom," *Current History* (September 1995), 278.

21. United Nations, *China Human Development Report 1999*, 68.

22. United Nations, *China Human Development Report 1999*, 68.

23. United Nations, *China Human Development Report 1999*, 32; Selden, *The Political Economy of Chinese Development*, 16.

24. George J. Gilboy and Eric Heginbotham, "The Latin Americanization of China?" *Current History* (September 2004), 257; Jim Yardley, "In a Tidal Wave, China's Masses Pour from Farm to City," *New York Times*, 12 September 2004; Joseph Kahn and Jim Yardley, "Amid China's Boom, No Helping Hand for Young Qingming," *New York Times*, 1 August 2004.

25. Yardley, "In a Tidal Wave"; Gilboy and Heginbotham, "The Latin Americanization of China?" 260.

26. Yardley, "In a Tidal Wave."

27. Ann Tyson and James Tyson, "China's Human Avalanche," *Current History* (September 1996), 277; Maurice Meisner, "The Other China," *Current History* (September 1997), 267.

28. Yardley, "In a Tidal Wave."

29. Gilboy and Heginbotham, "The Latin Americanization of China?" 260.

30. Eric Eckholm, "How's China Doing? Yardsticks You Never Thought Of," *New York Times*, 11 April 2004.

31. Gilboy and Heginbotham, "The Latin Americanization of China?" 260.

32. Joseph Kahn, "When Chinese Workers Unite, the Bosses Often Run the Union," *New York Times*, 29 December 2003; Charles Hutzler, "In Rural China, Religious Groups Face Suppression," *Wall Street Journal*, 27 July 2004.

33. Matt Pottinger, "China Takes Hard Line on Protesters," *Wall Street Journal*, 10 September 2004.

34. Kahn and Yardley, "Amid China's Boom."

35. James D. Seymour, "Human Rights, Repression, and 'Stability,'" *Current History* (September 1999), 283.

36. Yongnian Zheng, *Globalization and State Transformation in China* (Cambridge: Cambridge University Press, 2004), 147.

37. Nicholas R. Lardy, *Integrating China into the Global Economy* (Washington, D.C.: Brookings Institution Press, 2002), 160–61.

38. Lardy, *Integrating China*, 52.

39. Keith Bradsher, "After an Exodus of Jobs, a Recovery in Taiwan," *New York Times*, 19 March 2004.

40. Jason Dean, "China Raises Economic Heat on Taiwan," *Wall Street Journal,* 27 July 2004; Bradsher, "After an Exodus of Jobs."

41. Fishman, "The Chinese Century," 50; Elisabeth Malkin, "A Boom along the Border," *New York Times,* 26 August 2004.

42. Amy Waldman, "What India's Upset Vote Reveals: That High Tech Is Skin Deep," *New York Times,* 15 May 2004.

43. Amy Waldman, "Sikh Who Saved India's Economy Is Named Premier," *New York Times,* 20 May 2004; Pankaj Mishra, "India: The Neglected Majority Wins," *New York Review of Books,* 12 August 2004, 30–31; Alan Heston, "India's Economic Reforms: The Real Thing?" *Current History* (March 1992), 113–16.

44. Katherine Boo, "The Best Job in Town: The Americanization of Chennai," *New Yorker,* 2 August 2004; Saritha Rai, "India Sees Backlash Fading over Boom in Outsourcing," *New York Times,* 14 July 2004; Jonathan D. Glater, "Offshore Services Grow in Lean Times," *New York Times,* 3 January 2004; Steve Lohr, "Many New Causes for Old Problem of Jobs Lost Abroad," *New York Times,* 15 February 2004.

45. Saritha Rai, "India Market Falls on Jitters After Election," *New York Times,* 15 May 2004.

46. Noam Scheiber, "As a Center for Outsourcing, India Could Be Losing Its Edge," *New York Times,* 9 May 2004.

47. Saritha Rai, "Indian Voters Turn a Cold Shoulder to High Technology," *New York Times,* 12 May 2004.

48. Saritha Rai, "India Budget Raises Taxes to Finance Aid to Poor," *New York Times,* 9 July 2004.

49. Saritha Rai, "India's Economic Growth Is Expected to Slow," *New York Times,* 6 August 2004.

50. Prybyla, "All That Glitters?" 275; Selden, *The Political Economy of Chinese Development,* 35; Yasheng Huang, "The Role of Foreign Invested Enterprises in the Chinese Economy: An Institutional Foundation Approach," in *China, the United States, and the Global Economy,* ed. Shuxun Chen and Charles Wolf Jr. (Santa Monica, Calif.: RAND Corp., 2001), 151.

51. Prybyla, "All That Glitters?" 276.

52. Huang, "The Role of Foreign Invested Enterprises," 151; Prybyla, "All That Glitters?" 275.

53. Harry S. Rowan, "China and the World Economy: The Short March from Isolation to Major Player," in *China, the United States, and the Global Economy,* ed. Shuxun Chen and Charles Wolf Jr. (Santa Monica, Calif.: RAND Corp., 2001), 216.

54. Zheng, *Globalization and State Transformation in China,* 4, 6.

55. Matt Pottinger and Phelim Kyne, "Beijing Restrains Growth in Loans but Raises Risks," *Wall Street Journal,* 14 July 2004.

56. Lardy, *Integrating China,* 2–4.

57. Fishman, "The Chinese Century," 22.

58. Joseph Kahn, "For China, One Party Is Enough, Leader Says," *New York Times,* 16 September 2004.

59. Mark Lander, "East Germany Swallows Billions, and Still Stagnates," *New York Times,* 21 July 2004.

60. Lardy, *Integrating China*, 89; Thomas R. Gottschang, "The Economy's Continued Growth," *Current History* (September 1992), 271.

61. Keith Bradsher, "Another Leap by China, with Steel Leading Again," *New York Times*, 1 May 2004.

62. Christopher Rhoads and Charles Hutzler, "China's Telecom Forays Squeeze Struggling Rivals," *Wall Street Journal*, 9 September 2004.

63. Xianquan Xu, "Sino-U.S. Economic and Trade Relations," in *China, the United States, and the Global Economy*, ed. Shuxun Chen and Charles Wolf Jr. (Santa Monica, Calif.: RAND Corp., 2001), 239.

64. Keith Bradsher, "In China, Troubling Signs of an Overheating Economy," *New York Times*, 14 April 2004.

65. Bloomberg News, "China Says Influx of Capital Hurts Efforts to Cool Economy," *New York Times*, 3 September 2004.

66. Keith Bradsher, "Newest Export out of China: Inflation Fears," *New York Times*, 16 April 2004.

67. Associated Press, "China's Prices Gave No Sign of Slowing in August," *New York Times*, 14 September 2004; Keith Bradsher, "As Prices Rise in China, Signs of Inflation," *New York Times*, 24 August 2004.

68. Lardy, *Integrating China*, 49; Gang Xu, "China in the Pacific Rim: Trade and Investment Links," in *Changing China: A Geographical Appraisal*, ed. Chiao-min Hsieh and Max Lu (Boulder, Colo.: Westview, 2004), 176.

69. Chris Budkley, "China Is Said to Consider Revaluing Its Currency," *New York Times*, 10 February 2004.

70. Lardy, *Integrating China*, 8.

71. Louis Uchitelle, "U.S. Trade Partners Maintain Unhealthy Long-Term Relationship," *New York Times*, 18 September 2004; Eduardo Porter, "How a Stronger Yuan Could Hurt the U.S.," *New York Times*, 29 February 2004; Keith Bradsher, "China Anxiously Seeks a Soft Economic Landing," *New York Times*, 7 May 2004; Keith Bradsher, "China Ponders Interest Rise as Economy Heats Up," *New York Times*, 18 January 2004; Eduardo Porter, "Hoping the Yen, If Not the Yuan, Will Show Muscle," *New York Times*, 18 March 2004.

72. Elizabeth Becker, "Staring into the Mouth of the Trade Deficit," *New York Times*, 21 February 2004; Neil King Jr. and Michael Schroeder, "China Is Talk of Campaigns," *Wall Street Journal*, 20 July 2004.

73. Elizabeth Becker, "Industry and Labor Step Up Fight over China's Currency," *New York Times*, 10 September 2004.

74. Eduardo Porter, "Market Place," *New York Times*, 11 February 2004.

75. Brown, *Who Will Feed China?* 78–79.

76. Brown, *Who Will Feed China?* 74.

77. Brown, *Who Will Feed China?* 57–60.

78. Vaclav Smil, "Feeding China," *Current History* (September 1995), 280; Clifton W. Pannell and Runsheng Yin, "Diminishing Cropland and Agricultural Outlook," in *Changing China: A Geographical Appraisal*, ed. Chiao-min Hsieh and Max Lu (Boulder, Colo.: Westview, 2004), 34.

79. Pannell and Yin, "Diminishing Cropland," 43.

80. Vaclav Smil, "Feeding China: From Wanting to Wasting," *Current History* (September 2003), 270.

81. Brown, *Who Will Feed China?* 50–51.

82. Jim Yardley, "China Races to Reverse Its Falling Production of Grain," *New York Times*, 2 May 2004.

83. Yardley, "China Races."

84. Sun Sheng Han, "Agricultural Surplus Labor Transfer," in *Changing China: A Geographical Appraisal*, ed. Chiao-min Hsieh and Max Lu (Boulder, Colo.: Westview, 2004), 66.

85. Gilboy and Heginbotham, "The Latin Americanization of China?" 258.

86. Robert K. Schaeffer, *Power to the People: Democratization Around the World* (Boulder, Colo.: Westview, 1997), 134–35.

87. Joseph Kahn, "Let Freedom Ring? Not So Fast. China's Still China," *New York Times*, 3 May 2004; Joseph Kahn, "China's Offshore Headaches: Is the Leash Short Enough?" *New York Times*, 3 March 2004.

COLLATERAL DAMAGE: THE CAUSES AND CONSEQUENCES OF 9/11

The events of September 11, 2001—the airplane hijackings and attacks on the World Trade Center Towers and the Pentagon and the downing of another plane in Shanksville, Pennsylvania—can be traced to a number of separate historical developments in the Middle East, a region stretching from the Mediterranean to the Bay of Bengal, from the Caspian to the Red Sea. The events of 9/11, in turn, had important consequences for people in the region—invasions of Afghanistan and Iraq by the United States and its allies—and economic and political consequences for people around the world.

But while 9/11 was an important turning point, it was not the first or perhaps even the most important turning point for this region in the period since World War II. Two other important watersheds stand out: 1947–1948 and 1978–1980. Developments in those years led to conflicts that contributed, directly and indirectly, to 9/11 and subsequent events. An account of these different turning points or watersheds will be told in the next four chapters.

Chapter 8 is an analysis of partition in India and Pakistan in 1947–1948 and the myriad conflicts triggered by partition: four Indo-Pakistani wars and an insurgency in the Kashmir. Chapter 9 is an account of the partition of Palestine in 1948 and the five Arab-Israeli wars and two intifadas, or insurgencies in the occupied territories. Chapter 10 is an account of Iran's revolution in 1979 and its invasion by Iraq in 1980 and of the role that oil played in revolution and wars in the Persian Gulf. Chapter 11 is an account of revolution in Afghanistan in 1978, its invasion by the Soviet Union in 1979, and the multi-sided civil wars that followed in its wake.

This section analyzes the roles that the superpowers and neighboring states played in the destruction of Afghanistan and the rise of the Taliban and its Al Qaeda allies, who organized the attacks of 9/11. This is followed, in chapter 12, by a discussion of 9/11: invasions of Afghanistan and Iraq and the impact of 9/11 on political and economic developments around the world.

Events that occurred many years ago have had consequences that reverberate today. An appreciation of distant and sometimes obscure developments is essential to any serious understanding of 9/11. Keep in mind, too, that these events have been associated with important collateral developments: massive migrations by political refugees; conflicts over the role of women in public and private life; and, perhaps most important, the proliferation of nuclear weapons in a region that has for nearly sixty years been host to stubborn and recurrent conflicts.

8

§

The Legacy of Partition in India and Pakistan: 1947–1948

When the British withdrew from their colonies, from India in 1947 and from Palestine in 1948, they also divided political power between competing independence movements and assigned them to separate states.[1] Partition was designed to create sovereign states that would accommodate different ethnic-religious groups: Hindus in India and Muslims in Pakistan; Jews in Israel and Arab Muslims in Palestine. (For various reasons, the state for Arabs in Palestine did not come into being, as we shall see in chapter 9.) The British expected partition to solve conflicts between ethnic-religious groups in their former colonies. But it did not. Instead, partition created social and political problems that triggered war, insurrection, and the proliferation of nuclear weapons in both regions. In India, partition led to four Indo-Pakistani wars, one insurrection in the Kashmir, and the acquisition of nuclear weapons by governments in both countries. In Palestine, partition resulted in five wars between Israel and its neighbors, two insurrections or intifadas in the occupied territories of the West Bank and Gaza, and the acquisition of nuclear weapons by Israel. To appreciate these developments, we will look first at partition in India.

DIVISION AND SUBDIVISION IN INDIA AND PAKISTAN

After World War II, the British began withdrawing from many of their colonies around the world. But when they transferred power in India, they divided it between two competing independence movements: the

Indian National Congress, led by Mahatma Gandhi and Jawaharlal Nehru; and the Muslim League, led by Ali Jinnah.

The British decided to withdraw from India for various reasons. Postwar governments abandoned colonial rule in India because World War II had ruined British finances and they could no longer afford to maintain their far-flung empire; because the new Labor Party government was opposed to colonization in principle; and because the United States, which provided Great Britain with economic and military aid, was determined to pry colonies from European and Asian empires and, through the United Nations, construct a new interstate system based on independent nation-states (not dependent colonies), a project also endorsed by the Soviet Union.[2]

The British decision to partition India and create two successor states— India and Pakistan—was the product of developments in India before and during World War II. Before the war, the leaders of the Indian National Congress and the Muslim League worked together to end British colonial rule. Although they represented different constituencies, they coordinated their political campaigns and shared a common goal: independence in a united India. But the outbreak of war dramatically altered relations between the two movements and drove them apart. In 1939, the British viceroy declared war on India's behalf, without first consulting political parties in India. Infuriated by this decision, leaders of the Congress withdrew from office and campaigned against the British war effort. Many of its leaders were arrested and jailed. The League, by contrast, rallied to support the British and in 1940 demanded that Muslims, who were the largest minority in India, be granted a state of their own after the war, a position they described as "Divide and Quit."[3]

After the war, both movements continued to press for independence, but on different terms. The Congress demanded the British leave behind a single state, in which the two political parties would compete for power based on the democratic principle of majority rule. The League, by contrast, worried that when independence came, the Hindu majority would, "in the name of democracy . . . make use of it to coerce and crush us, its prey, into complete captivity."[4] The League demanded that the British create a separate state for Muslims, where they might rule as a majority, instead of a single state where they would constitute a political minority. Jinnah reminded the British that they "owed" the League for supporting Britain during the war. But he also threatened massive violence if the British did not accede to their demands. "We have forged a pistol," Jinnah said of the League's determination to organize violent protests, "and we are in a position to use it."[5]

Faced with conflicting demands, the British decided that partition promised the best solution. Lord Mountbatten, who negotiated the transfer of power, hoped that partition would avert incipient civil war between Hindus and Muslims; protect the political rights of the Muslim minority; reward Muslims for their service to the Empire during the war; and speed the withdrawal of British troops, who might otherwise be required to stay on in a peacekeeping role. As a concession to the Congress, Mountbatten promised that a referendum would be held at a later date to determine whether people in both countries wanted to reunite in a single state. (Not surprisingly, the postpartition referendum was never held.) Leaders of the Congress reluctantly agreed to accept partition, which they regarded as temporary, because they wanted to curb the violence erupting around the country. But while Nehru agreed that the "Muslim League can have Pakistan if they want," they could do so "only on the condition that they do not take away other parts of India which do not wish to join Pakistan."[6] After partition, this would become an important issue for both countries in Kashmir.

Partition was designed to solve problems in the subcontinent. But it did not. Instead, partition created problems that its proponents did not anticipate and triggered conflicts that have endured to this day.

The first problem partition created was massive migration (see chapter 5). After partition on August 15, 1947, more than seventeen million people picked up and fled across newly created borders in the Punjab and in Bengal.[7] One of the odd features of partition was that it resulted in the creation of two Pakistans, one in the Northwest (Punjab) and one in the East (Bengal), because that is where Muslim populations were concentrated.

Muslims and Hindus fled their homes in roughly equal numbers— Muslims from India to East or West Pakistan, Hindus from East or West Pakistan to India.[8] They migrated because they feared for their future in a country where they would live as a minority, and they fled from the violence visited on them by their neighbors. Nearly one million people were killed in the three-month carnage that followed partition.[9] In many places, whole villages were massacred and left in smoking ruins. Refugee columns, which stretched for miles along roads in the Punjab and Bengal, often passed down the same road, traveling in opposite directions. Where they met, they often fought, leading to slaughter on the open road. Neither the departing British administrators nor the newly installed governments in either country anticipated the violence and they were not prepared to arrest its spread or cope with the huge influx of refugees that flooded across their borders. How could they? This was the largest, fastest, most violent migration in human history.

After the violence abated and the migrations slowed, it became evident that partition had not created states with homogeneous populations. Instead, many Muslims remained in India, either because they were unwilling to leave their homes or because they hoped the new government might accommodate them. So, for example, as many Muslims remained in India, which became a secular and democratic state, as lived in Pakistan. Far fewer Hindus chose to remain in Pakistan, largely because its founders decided it would become an explicitly Islamic state and because Pakistan would be run by military dictators for most of the postpartition period.

Problems associated with partition became particularly acute in Kashmir. After partition, developments there triggered a wider war, the first of many between India and Pakistan.

THE FIRST INDO-PAKISTANI WAR

When the British advanced partition, they weighed the competing interests of the two independence movements in their decision. But they did not consider the interests of other important groups—Sikhs, untouchables, or the princes who presided over 565 states covering about one-third of India, each possessing some measure of autonomy—or include them in the negotiations.[10] When independence came, local potentates were left, more or less, at the mercy of the new central governments, which had made it clear that they would not allow the princes to retain any autonomous authority within the newly independent states.

In Kashmir, a Hindu maharaja governed a Muslim majority and refused, for a time, to join either India or Pakistan. When Muslim rebels moved to overthrow the maharaja and join Kashmir to Pakistan, the Pakistani army moved to assist them. The maharaja's troops could not resist the combined onslaught, so the monarch appealed to India for military assistance. The Indian government agreed to send troops, provided the maharaja transfer his authority to India. The rebellion then developed into a wider, interstate war, and fighting continued until 1949, when both sides agreed to a UN-brokered ceasefire that effectively divided Kashmir between them. (About two-thirds of Kashmir was held by India, the rest by Pakistan.[11])

Partition, then, resulted in the creation of two states on the subcontinent, and one of them, Pakistan, was itself divided into two parts. It also triggered huge, socially disruptive migrations, rebellion, and war, and left in its wake a dispute over Kashmir that would fuel tension and ignite wars in the decades to come.

THE SECOND INDO-PAKISTANI WAR

In August 1965, the Pakistani army infiltrated troops into Kashmir, where they planned to lead an uprising against Indian rule. Leaders in Pakistan evidently hoped that the Indian army, which had been weakened by a disastrous war with China in 1962, would be unable to respond effectively and that the Muslim population would rally to Pakistan's call to arms.[12] They also wanted to prevent India from implementing legislation that would end Kashmir's special constitutional status and integrate it more fully into the Indian polity.[13]

But the Pakistani incursion, which they called "Operation Gibraltar," failed to rouse the populace, and the Indian army moved quickly to contain the incursion. The Pakistani army then moved to assist its advance forces, and the Indian army responded to this escalation by moving its forces against Pakistan outside Kashmir in the Punjab, thereby widening the war.[14]

After fighting reached a stalemate in late September, the two sides once again agreed to a UN ceasefire and later negotiated an agreement that returned both sides to the positions they held before the war.[15]

THE THIRD INDO-PAKISTANI WAR AND THE PARTITION OF PAKISTAN

When the British divided India, they transferred power in Pakistan to the Muslim League, which was strongest in West Pakistan. Jinnah, Pakistan's founder and first president, established Karachi (in the West) as the country's capital. His successors filled civil service and army posts with candidates from the West. Of 741 top civil servants in the mid-1950s, only 51 hailed from Bengal in the East.[16] In 1952, the government adopted Urdu as the country's official language, despite the fact that Bengali was more widely spoken.[17] Fearing that their inability to speak Urdu would bar their entry into the country's universities, civil service, and the army, people in Bengal rioted. These events persuaded politicians in Bengal to break from the Muslim League in 1953 and found the Awami League. Its twenty-one-point program demanded a greater role for Bengalis in Pakistan's affairs.[18]

In 1958, the Pakistani army staged a coup, deposed the civilian government, and installed General Muhammed Ayub Khan as dictator. Civilian opposition to dictatorship soon emerged in both East and West, but parties in each region adopted different strategies toward the regime. In the West, the Pakistan People's Party, led by Zulfikar Ali Bhutto, pressed for an end to military rule, but did so as a kind of loyal opposition, with Bhutto serving for a time as the regime's foreign minister. (It was in that capacity that he conceived and organized Pakistan's incursion into Kashmir in 1965.)

In the East, the Awami League, led by Sheik Mujibur Rahman, demanded both an end to dictatorship and greater autonomy for Bengal, including a new constitution that would give Bengal a stronger voice in Pakistan's parliament. Continued discrimination against Bengalis in the civil service and the army, and the government's policy of disbursing foreign aid primarily in West Pakistan, increased popular support for the Awami League in Bengal.[19]

In 1968, riots in both East and West forced General Ayub to step down. Widespread protest forced his successor, General Yaya Khan, to schedule elections that would install a new parliament charged with drafting a new constitution and restoring democracy.

The elections, originally scheduled for October 1970, were postponed until late December after a ferocious typhoon struck Bengal. The regime's weak response to the disaster assisted the Awami League when elections were finally held. The League captured the overwhelming vote in Bengal and gained an absolute majority—160 of 300 seats—in the new constituent assembly.[20] Although the People's Party won a majority of votes in the West, it secured only half as many seats as the Awami League.

When it became clear that the Awami League was in a position to rewrite the constitution and redistribute political power, the military authorities and civilian parties in the West balked. Arguing that "a majority alone doesn't count in national politics," Bhutto demanded that Mujibur treat the People's Party as one of the "two majority parties," draft a new constitution as equals, and include Bhutto in the cabinet of the new government.[21]

Of course, Mujibur refused to surrender his electoral majority. Bhutto then persuaded the military to postpone the opening of the constituent assembly in an effort to wrest concessions from the Awami League. But this only antagonized League supporters in Bengal, and they began organizing demonstrations, strikes, and widespread civil disobedience. Confronted with this challenge and unwilling to concede power to the electoral majority, the military, with Bhutto's support, decided to crush incipient rebellion in Bengal and "sort this bastard [Mujibur] out."[22] Political leaders in Pakistan could not accept the idea of majority rule in India before partition in 1947, and they could not tolerate the idea of majority rule in Pakistan in 1971.

On March 25, 1971, the army, which was composed primarily of soldiers from the West, arrested Mujibur and Awami League leaders, imposed martial law, and attacked student demonstrators, slaughtering tens of thousands of civilians in Bengal.[23]

The regime's assault in the East ignited civil war. Bengali police officers and soldiers who escaped the army dragnet formed the nucleus of an irregular army—the Mukti Bahini—that waged a guerrilla war against the dictatorship. With the outbreak of civil war, millions of people fled Bengal for India: four million by May, eight million by August, and nearly ten million by mid-November.[24] This massive migration, more than half the size of the 1947 migration—created serious problems for Prime Minister Indira Gandhi's government in India. Cholera swept through refugee camps, migrants taxed government food supplies and resources, and government aid to immigrants fueled resentment among poor Indians in the region.

As the civil war intensified, Gandhi directed the army first to arm and train Mukti Bahini guerrillas and then, in late November, to invade Bengal and defeat Pakistani forces there. The Indian invasion of Bengal triggered a general war between the two countries, not only in Bengal but also in the Punjab and Kashmir.[25] The Indian army quickly routed the Pakistani army in Bengal, restored the Awami League to power, and began returning Bengali refugees to their homes.

As a result of this third war, Pakistan was subdivided and a new state—Bangladesh—was created in the East. Although the Awami League expected independence to promote economic development in Bangladesh, war and migration crippled the economy and the country became one of the poorest in the world. In Pakistan, defeat in Bengal discredited the dictatorship and the army quickly transferred power to a civilian government led by Bhutto. But it did not last long. In 1977, the army again overthrew the civilian government and its new leader, General Muhammad Zia Ul-Haq, had Bhutto arrested and hanged.

NUCLEAR PROLIFERATION

In 1974, India exploded its first atomic bomb. Curiously, India's nuclear weapons program grew out of conflict with China, not Pakistan, but India's acquisition of nuclear weapons triggered the development of nuclear weapons in Pakistan, leading to a nuclear arms race in the subcontinent.

In 1962, India and China fought a brief war over disputed territories in the Himalayas. Chinese forces routed the Indian army during the campaign. Two years later, in 1964, China detonated its first nuclear weapon at Lop Nor. Afterward, Chinese leader Mao Ze Dong said that China's nuclear weapons would "boost our courage and scare others."[26]

And scare others they did. A few days later, Homi Bhabha, head of India's atomic power program, warned that the only defense against nuclear attack by China "appears to be the capability and threat of retaliation."[27] One hundred members of parliament immediately petitioned the government in India to develop nuclear weapons, arguing that China would now grow into a "giant," while India would "remain a dwarf" if it did not follow suit.[28]

At first, India's nuclear weapons program proceeded slowly. But it was accelerated by two events. First, in 1967, China tested a more powerful hydrogen bomb. Then, in 1971, during the third Indo-Pakistani war, U.S. president Richard Nixon sent naval forces, armed with nuclear weapons, into the Bay of Bengal in an effort to deter India from intervening in the civil war in Bengal. The United States did so because it viewed Pakistan, a dictatorship, as its primary ally in the region, not India, even though India was the world's largest democracy.

Government officials in India evidently viewed the U.S. show of force as a real threat, and decided that the possession of nuclear weapons by India might deter such threats in future.[29] But while Indian leaders imagined that their acquisition of the nuclear weapons might deter China and perhaps the United States from threatening India, they did not consider the impact this would have on Pakistan. Defeat in the third Indo-Pakistani war, and India's subsequent test of an atomic bomb in 1974, persuaded Pakistani officials to develop nuclear weapons, presumably to "boost their courage and scare India." The head of Pakistan's nuclear program, Dr. Abdul Quadir Khan, later explained that Pakistan had embarked on a nuclear weapons program "to avoid a nuclear Munich at India's hands . . . which many Pakistanis think very real."[30]

With assistance from China, Pakistan began developing nuclear weapons, and in 1987 President Zia announced that "Pakistan has the capability of building the bomb."[31] The fact that Pakistani leaders referred to their weapon as an "Islamic bomb" antagonized not only India but other countries in the region, particularly Israel.[32]

Still, Pakistan did not actually test nuclear weapons, and India did not conduct further tests, for many years. Then, in May 1998, the Indian government, led by the Hindu nationalist Bharatija Janata Party, conducted a series of nuclear tests. The dictatorship in Pakistan, not to be outdone, responded with tests of its own.[33]

Although Indian leaders celebrated its nuclear tests as representing a "resurgence of India," and Pakistani officials termed their feat a "day of pride," no other country celebrated the proliferation of nuclear weapons in countries that had gone to war three times.[34]

INSURGENCY IN KASHMIR

After disputed elections in 1987, Muslims protesting Indian rule in Kashmir began organizing demonstrations and then violent assaults on Indian politicians, civilians, and military forces across the valley.[35] The uprising, which is what Pakistan had tried but failed to ignite in 1967, had indigenous origins. But Pakistan began providing sanctuary, economic aid, and military assistance to the insurgents, who were led by the Jammu and Kashmir Liberation Front.[36]

Rebellion in the Kashmir very nearly led to a fourth war between India and Pakistan in 1990. Full-scale war was averted, but irregular war continued. Since 1988, 25,000 people, mostly civilians, have lost their lives in the conflict.[37] Eventually, the insurgency triggered yet another war in 1999.

THE FOURTH INDO-PAKISTANI WAR

In May 1999, an Indian army patrol was ambushed and its soldiers killed by Pakistani forces in the mountains along the Line of Control in Kashmir. The Pakistani army then advanced its forces to secure strategic vantages along the ceasefire line.[38] Both armies typically withdrew their troops from high-altitude positions in the mountains in advance of winter, and reinserted them in the spring. Pakistani troops took advantage of this to improve their positions.

Not surprisingly, the Indian army fought back, but conducted fairly limited operations in this incredibly difficult terrain. They did not seek to escalate the war by opening other fronts outside Kashmir. But the fighting was nevertheless sharp and costly. Eventually, Pakistani forces were driven, or retired, from their new positions, and the Kargil War, as it was called, ended as others had: back at the 1949 Line of Control, with the boundary in Kashmir still in dispute.

It is important to note these developments because the events of 9/11 would bring India and Pakistan once again to the brink of war. Pakistan also played an important role in Afghanistan's long wars and contributed to the rise of the Taliban and Al Qaeda. It did so in part because it wanted allies to its "rear" in the event of war with India. And it recruited Taliban and Al Qaeda fighters to assist Muslim insurgents in Kashmir. For its part, India used 9/11 as an opportunity to crack down on insurgents in Kashmir and expose Pakistan as a sponsor of Islamic terrorism. This confrontation can only be understood as part of a long history of enmity, stretching back to partition and the problems associated with it.

As we have seen, the partition of British India created a series of unanticipated problems that contributed to persistent conflict, recurrent war, subdivision, and nuclear proliferation in the subcontinent. But India was not the only place where the British advanced partition as a solution to political problems in postcolonial settings. In the same period, partition was also advanced as a solution to conflict in Palestine. In Palestine, as in India, partition created problems that persist to this day. To understand the myriad Arab-Israeli conflicts, we now turn to the partition of Palestine in 1948.

NOTES

1. See Robert Schaeffer, *Warpaths: The Politics of Partition* (New York: Hill and Wang, 1990).

2. Schaeffer, *Warpaths*, 73–86.

3. Reginald Coupland, *The Indian Problem* (London: Oxford University Press, 1944), 206; P. N. Pandey, *The Breakup of British India* (New York: St. Martin's, 1969), 167.

4. Coupland, *Indian Problem*, 201.

5. H. V. Hodson, *The Great Divide* (Oxford: Oxford University Press, 1985), 166.

6. C. H. Philips and M. D. Wainwright, *The Partition of India* (Cambridge, Mass.: MIT Press, 1970), 219.

7. R. F. Holland, *European Decolonization, 1918–1981* (New York: St. Martin's, 1985), 80.

8. Shahid Javed Burk, "The State and the Political Economy of Redistribution in Pakistan," in *The Politics of Social Transformation in Afghanistan, Iran and Pakistan*, ed. Myron Weiner and Ali Banuazizi (Syracuse, N.Y.: Syracuse University Press, 1994), 300.

9. Holland, *European Decolonization*, 80.

10. Hodson, *The Great Divide*, 367–68; Sumit Ganguly, *Conflict Unending: Indo-Pakistani Tensions since 1947* (New York: Columbia University Press, 2001), 15.

11. Ganguly, *Conflict Unending*, 16–17.

12. Ganguly, *Conflict Unending*, 40–43.

13. Ganguly, *Conflict Unending*, 35–36.

14. Ganguly, *Conflict Unending*, 44–45.

15. Ganguly, *Conflict Unending*, 46–47.

16. Richard Sisson and Leo E. Rose, *War and Secession: Pakistan, India, and the Creation of Bangladesh* (Berkeley: University of California Press, 1990), 10.

17. Sisson and Rose, *War and Secession*, 9.

18. Sisson and Rose, *War and Secession*, 12.

19. Ganguly, *Conflict Unending*, 52–53.

20. Sisson and Rose, *War and Secession*, 33.

21. Sisson and Rose, *War and Secession*, 60–61, 70.

22. J. N. Saxena, *Self-Determination: From Biafra to Bangladesh* (Delhi: University of Delhi Press, 1978), 53; Sisson and Rose, *War and Secession*, 81, 93, 123, 132–33.

23. Lee Burcheit, *Secession: The Legitimacy of Self-Determination* (New Haven, Conn.: Yale University Press, 1978), 206.

24. Burcheit, *Secession*, 207; Sisson and Rose, *War and Secession*, 152.

25. Sisson and Rose, *War and Secession*, 214.

26. John Lewis and L. Zue, *China Builds the Bomb* (Stanford, Calif.: Stanford University Press, 1988), 216.

27. Gerald Segal, *Arms Control in Asia* (New York: St. Martin's, 1987), 103.

28. George Perkovich, "What Makes the Indian Bomb Tick?" in *Nuclear India in the Twenty-First Century*, ed. D. R. Sar Desai and Raju G. C. Thomas (New York: Palgrave-MacMillan, 2002), 28.

29. Raju G. C. Thomas, "Whither Nuclear India?" in *Nuclear India in the Twenty-First Century*, ed. D. R. Sar Desai and Raju G. C. Thomas (New York: Palgrave-MacMillan, 2002), 7.

30. Leonard S. Spector, *Going Nuclear* (Cambridge, Mass.: Ballinger, 1987), 108.

31. *Washington Post*, "Zia Says Pakistan Capable of Building A-Weapon," March 24, 1987.

32. Spector, *Going Nuclear*, 113; Haider K. Nizamani, *The Roots of Rhetoric: Politics of Nuclear Weapons in India and Pakistan* (Westport, Conn.: Praeger, 2000), 95, 108.

33. Nizamani, *Roots of Rhetoric*, 83–84.

34. Nizamani, *Roots of Rhetoric*, 137.

35. Ganguly, *Conflict Unending*, 90–91.

36. Ganguly, *Conflict Unending*, 91–93.

37. Sumit Ganguly, "An Opportunity for Peace in Kashmir," *Current History* (December 1997), 415.

38. Ganguly, *Conflict Unending*, 114–19.

9

֍

Partition in Palestine: 1948

It is commonly said that the Arab-Israeli conflict is "as old as the Bible." Marc Charney, writing in the *New York Times*, argued,

> In the West Bank, Gaza and Jerusalem, the conflict between Israeli and Palestinian is being shaped by a fratricidal agony: competing, centuries-old claims to the same precious strip of land between the Jordan River and the Mediterranean, and the right to exist there as a nation.
>
> The rivalry is as old as the conflict between Moslem and Jew for the legacy of their common father Abraham.[1]

But this view is fundamentally mistaken and historically inaccurate. The present conflict, though serious and persistent, is less than one century old. In fact, Jews and Muslim and Christian Arabs lived together in Jerusalem and its environs for centuries without difficulty. It was only after World War I that conflict between them first emerged, and only after World War II, with the partition of Palestine in 1948, that it assumed its present form.

Before World War I, Palestine, the strip of land between the Jordan River and the Mediterranean, was administered as a territory of the Ottoman Empire. During the war, the British army based in Egypt defeated Ottoman armies in the Middle East, occupied its territories and, after the war, divided them with France. Lebanon and Syria were assigned to France as colonies. Transjordan and Iraq became British colonies, and Palestine was assigned to Great Britain as a mandate by the League of Nations. Under the League's mandate, Britain was supposed to grant self-rule to its inhabitants at some later, unspecified date. Saudi Arabia became an independent kingdom and the Ottoman Empire, shorn of its territories, became Turkey.

World War I was important not only because the British occupied Palestine, but also because the British made political commitments about its future. During the war, on November 2, 1917, British foreign minister Arthur J. Balfour promised Jewish leaders in Europe that Britain would one day provide a "homeland" for Jews in Palestine. Balfour issued similar pronouncements to other constituencies, promising greater self-rule in Ireland (May 16, 1917) and eventual self-rule to nationalists in India (August 20, 1917). He did so to rally various constituents to the imperial war effort when British military fortunes were at a low ebb. After the war, Britain allowed Jews to emigrate into Palestine, joining the small Jewish population that had for centuries inhabited the region, as a step toward fulfilling the Balfour Declaration.

Encouraged by leaders of the Zionist movement, which had been organized in the late 1890s to promote Jewish migration to Palestine and establish a Jewish nation there, Jews from Europe moved to Palestine and purchased land. The pace of immigration accelerated in the 1930s with the rise of the Nazi Party in Germany and the spread of anti-Semitism in many countries across Europe. But growing Jewish immigration, and their purchase of Arab lands, alarmed some Arab Palestinians, and they demanded that the British authorities curb both.

In 1936, Muslim and Christian Arabs organized riots and mounted an irregular war to force the British to transfer power to the Arab Palestinian majority, which would then move to block Zionist land purchases and curtail Jewish immigration.[2]

The British responded to the 1936 Arab revolt first by curbing Jewish immigration to Palestine. This blocked the entry of Jews fleeing Nazi persecution. This measure antagonized Zionists, who saw it as a betrayal of commitments made in the Balfour Declaration, and did little to appease Arab-Palestinian opinion. (Many other countries, including the United States, also curbed or barred Jewish immigration during this period, largely because they argued that it would increase already high levels of unemployment.)

Second, the British commissioned an investigation into the origins of the revolt. In 1937, the Peel Commission issued its report. It found that "Arab nationalism [has become] as intense a force as Jewish [nationalism]" and that "the gulf between the races . . . will continue to widen if the present mandate is maintained."[3] As a solution to the conflict, the commission recommended the partition of Palestine into separate Arab and Jewish states, with Jerusalem remaining under British rule. Partition, the commission argued, would deliver Arabs "from the fear of being 'swamped' by the Jews and from the possibility of ultimate subjection to

Jewish rule," and the Jews would be relieved "from the possibility of . . . being subjected in the future to Arab rule."[4]

When World War II broke out, Zionist and Arab Palestinians took different sides, much as competing independence movements did in India. The Zionist movement supported the British war effort, and many Zionists joined its armies. Arab Palestinian leaders opposed the British and sympathized with Axis forces. By the end of the war, however, as the consequences of restrictive British immigration policy—preventing Jews from escaping the Holocaust—became apparent, radical Zionist militias such as the Stern Gang and Irgun (led by Avraham Stern and Menachim Begin, later a prime minister of Israel), waged a terrorist campaign to force the British out of Palestine. The assassination in Cairo of British minister of state Lord Moyne in 1944 marked the beginning of their campaign to oust the British by force.[5] The bombing by the Irgun of the King David Hotel in Jerusalem in 1946 was the most destructive episode in this campaign.

After World War II, the British made plans to withdraw from India and Palestine in 1948. Lord Mountbatten moved up the date of departure in India to August 1947, but the British kept to their original plan in Palestine, withdrawing on May 14, 1948. When British officials announced this decision on February 18, 1947, they also announced that they were delegating authority over the future of Palestine to the United Nations.[6] The British did so because they did not want to be the target of either Jewish or Palestinian violence in the period before they evacuated, and they hoped that the United Nations, with U.S. and Soviet participation, might broker an acceptable solution to the conflict.

The UN Commission on Palestine recommended partition as a solution to the problem, adopting the plan first advanced by the Peel Commission a decade earlier. The majority report proposed the creation of an Arab and a Jewish state, with Jerusalem assigned to an international trusteeship. The minority report proposed the creation of a single federal government that would consist of two states (states in the American sense of the word, as in North and South Carolina), one Jewish and one Arab.[7] President Harry S. Truman endorsed the majority report recommendation for partition, and the United Nations, with U.S. and Soviet support, ratified the partition of Palestine on November 29, 1947.[8] Although the Soviet Union subsequently sided with Arab states in conflicts with Israel, the Soviets supported the plan because they viewed it as a way to end British imperialism and promote self-determination in the region.[9]

But while the British, the two superpowers, and the United Nations viewed partition as a solution to conflict, the withdrawal of British troops on May 14, 1948, led immediately to war and, in subsequent years, to recurrent war and rebellion.

THE FIRST ARAB-ISRAELI WAR, 1948–1949

Arab Palestinians objected to the UN partition plan, in part because it awarded 55 percent of the land in Palestine to a Jewish state, at a time when "Jews owned less than 10 percent of the land and were one-third of the population."[10] After the UN plan was adopted, fighting erupted between Jewish and Arab militias. Then, when the British departed, Zionist leader David Ben-Gurion declared the creation of the state of Israel, and the fighting immediately escalated into a full-scale war between Israeli forces and armies from neighboring states (Syria, Jordan, Lebanon, Iraq, and Egypt), which moved to support the Palestinians. The Israeli army was able to defeat uncoordinated Arab assaults and forced them to accept separate armistices in February 1949.[11]

The first Arab-Israeli conflict, which the Israelis call their "War of Independence" and which Palestinians call "al-Nakba" or "the Catastrophe," had important social and political consequences.[12]

As in India, partition triggered large, socially disruptive migrations. During the war, fighting forced or persuaded many Palestinians to flee. An Israeli Defense Force study of Arab migration between January 1947 and June 1948 found that most Arabs left their homes as a result of, in order of importance: "1) Direct, hostile Jewish operations against Arab settlements, 2) The effect of our hostile operations on nearby settlements . . . especially the fall of large neighboring cities; and 3) Operations of the [Jewish] dissidents [such as the Irgun and Stern Gang]."[13]

Palestinians did not leave because they were encouraged to do so by Arab state radio broadcasts. Instead, according to Israeli historian Benny Morris, "The Arab Higher Committee decided to impose restrictions and issued threats, punishments and propaganda in the radio and press to *curb* [Arab] migration [emphasis added]."[14] Palestinians left their homes because, like most persons during a war, they sought to escape attack or injury. Israeli demolition teams razed many Palestinian towns, and dynamite supplies ran short, until Israeli officials decided that Arab housing should be kept intact for the postwar influx of Jewish immigrants.[15]

When the fighting ended, the Israeli army sealed the borders and blocked the return of Palestinian refugees. "We must prevent their return at all costs," Ben-Gurion argued.[16] The Israeli government then adopted regulations providing for the seizure of any Arab land or property in an "abandoned area" and transferred seized assets directly to Jewish settlers and to private Jewish agencies, which then leased Arab properties to Jews.[17]

After the war, the Knesset in 1950 adopted the Absentee Property Law, which provided that any Arabs not present and registered during the 1948 census—which was conducted while the war was still in progress, at a time when many Palestinians had fled the fighting—would be regarded as "absentees." This meant that they could not reclaim their "abandoned" property or claim citizenship in the new state. Refugees who returned to their homes were treated by the Israeli government as "illegal immigrants" and were deported to neighboring territories.[18] Foreign Ministry Director General Walter Eytan justified this policy by saying,

> The war that was fought in Palestine was bitter and destructive, and it would be doing the refugees a disservice to let them persist in the belief that if they returned, they would find their homes or shops or fields intact. . . . Generally, it can be said that any Arab house that survived the impact of the war . . . now shelters a Jewish family.[19]

While the Israeli government passed laws depriving many Palestinians of property and rights, they also passed laws—the 1950 Law of Return—granting rights in Israel to Jews living outside the country, largely as a way of encouraging continued Jewish immigration. During the decade that followed partition, nearly one million Jews from around the world—half of them from the Middle East and North Africa, and half from Europe—immigrated to Israel.[20]

Estimates of the number of Palestinian refugees vary, from 520,000 according to Israeli sources to 900,000 according to Arab sources. UN officials estimate a figure in between, about 726,000 refugees.[21] Some Arab Palestinians did not flee and were recorded as present on the 1948 census. After the war, they made up about 11 percent of the population and were allowed to become citizens.

When the war ended, the territory under Israeli control was 21 percent larger than the area allotted to it under the UN partition plan.[22] But while the partition of Palestine resulted in the creation of an Israeli state, it did not result in the formation of a Palestinian state. Because Palestinians relied on neighboring states during the war, they did not create the political institutions that might have formed the nucleus of a separate state. Instead, the lands assigned to the "Palestinian" state were either absorbed by Israel during the fighting or incorporated by neighboring states: the West Bank by Jordan, the Gaza Strip by Egypt.

These two developments—the creation of large Palestinian refugee populations in the West Bank and Gaza (and in Lebanon), and the evaporation of the proposed Palestinian state—are at the heart of the current conflict between Israel and the Palestinians. Although this conflict went into hibernation for the two decades between 1948 and 1967, largely because the Palestinians had been displaced to refugee camps outside Israel, it reemerged as a problem when Israel conquered and occupied territories in the West Bank and Gaza during the Six Days' War in 1967.

Partition also created conflicts between Israel and neighboring Arab states. The armistices of 1949 stopped outstanding conflicts but did not resolve them. Instead, partition and war prepared the ground for a number of wars in subsequent years, just as in India. Unlike the Indo-Pakistani wars, however, which reverberated across the subcontinent but did not have much of an impact outside the region, Arab-Israeli wars have had important global repercussions. It is to these wars and their global consequences that we now turn.

THE SUEZ WAR AND NUCLEAR PROLIFERATION

In 1952, a group of secular military officers led by Gamal Abdel Nasser overthrew Egypt's King Farouk. Nasser wanted to rebuild the Egyptian army, which had been wrecked in the 1948 war with Israel, and promote economic development. To rebuild the army, Nasser needed arms, and to promote development, he needed money to finance the construction of the Aswan Dam across the Nile. U.S. and British officials refused to supply arms that might fuel an arms race and contribute to war, and they decided not to finance construction of the Aswan Dam.[23] Rebuffed by the West, Nasser turned to the Soviet Union for military supplies and on July 26, 1956, one week after the United States turned down his request for dam financing, nationalized the Suez Canal.[24]

Seizure of the canal enabled Nasser to "redress his wounded prestige and also procure the financial assets that would make possible the building of the Aswan Dam."[25] Nasser planned to use the $25 million annual profit from running the canal to pay for the dam.[26]

Although Nasser was within his rights, as a shareholder, to nationalize the canal, his British and French co-owners, who had originally built it, were outraged. They immediately began making plans to invade Egypt, seize the canal, and overthrow Nasser. But they devised an elaborate charade to conceal their intentions.

"Operation Musketeer," as the British called the plan, initially called for a joint British-French assault on Egypt. But they soon enlisted Israel, which wanted to reopen the Straits of Tiran to Israeli shipping (Nasser had closed it), drive Egyptian forces from Gaza and the Sinai, and create a desert buffer between Israel and its primary Arab opponent.[27] Together, the Three Musketeers developed a stratagem. Israel would attack Egypt across the Sinai, using ongoing disputes with Egypt as a pretext for war. The British and French would then invade Egypt and seize the Suez Canal to "protect" it from advancing Israeli forces, their secret partner in this military adventure.

The Israelis opened the fighting on October 29, 1956, and quickly drove Egyptian forces back across the Sinai. As agreed, Britain and France then attacked Egyptian airfields, dropped paratroops into the canal zone on November 5, and landed an invasion force at Port Said the following day.[28] But while military operations proceeded according to plan, the Musketeers failed to consider the wider political consequences of secret collaboration and war.

As soon as war began, U.S. officials mounted a furious campaign to end the fighting and force Britain, France, and Israel to withdraw. President Dwight D. Eisenhower told Secretary of State John Foster Dulles to send a scathing message to Israeli prime minister David Ben-Gurion: "You tell 'em, God damn it, we're going to apply sanctions, we're going to the United Nations, we're going to do everything that there is so we can stop this thing."[29]

The Soviet Union, meanwhile, threatened to intervene on behalf of Egypt, its new ally, and joined the United States in a UN resolution calling for an immediate ceasefire and the withdrawal of Israeli, British, and French forces.[30] In addition, Soviet premier Nikita Khrushchev threatened all three countries with rocket attacks if they did not comply, implying that the Soviets might use nuclear weapons against them. "In what situation would Britain find herself, if she were attacked by a stronger power possessing all types of modern weapons of destruction? . . . for instance rocket equipment," Khrushchev warned.[31]

Alarmed by Soviet nuclear threats, the three allies appealed to the United States for protection, asking Eisenhower to deter Soviet threats. But Eisenhower did not do so because he did not believe that Soviet threats were credible and because he thought that the United States had already extended to its allies sufficient guarantees against nuclear attack by the Soviet Union. But leaders in Britain, France, and Israel did not see it that way. They felt threatened by the Soviet Union and abandoned by the United States. Yaacov Herzog, Ben-Gurion's chief advisor, wrote later that Ben-Gurion was "entirely unprepared by the vehemence of President Eisenhower's backing of the [UN] General Assembly's call for immediate and unconditional Israeli withdrawal. What the U.S. did then was to remove Israel's—as well as Britain's and France's—protective shield against possible [nuclear] retaliation, leaving them all exposed."[32]

Facing determined opposition from both superpowers, the British and French accepted a ceasefire on November 6 and agreed to withdraw one month later. Israel withdrew from the Sinai a few months after this.[33]

But while Operation Musketeer collapsed and the war ended back where it had begun, it had important long-term consequences, nuclear proliferation chief among them.

Nuclear Proliferation

In 1956, Britain, France, and Israel did not possess nuclear weapons. British scientists had participated in the Manhattan Project, which developed nuclear weapons for the United States during World War II. The

British had not pressed ahead with their own weapons program, trusting that the United States would protect it from nuclear threats. But Soviet nuclear threats during Suez, and the U.S. reluctance to deter them, changed British, French, and Israeli views. Instead of relying on the United States, they decided to develop nuclear weapons of their own. According to one historian, the French government's "previous hostility toward nuclear weapons was transformed overnight into a determined and positive interest in national nuclear armament."[34] Angered by U.S. actions, France developed nuclear weapons and later withdrew from NATO. Contemporary disputes between the United States and France—over intervention in Iraq and Middle East policy generally—can be traced, in large part, to events in 1956.

Britain and France rapidly developed nuclear weapons, and the French helped Israel do the same. The French helped Israel build its first nuclear research reactor and provided important technical and material assistance, including weapons-grade plutonium.[35] Although Israeli progress was slow, by 1975 Israel admitted that it had developed the ability to build nuclear weapons.[36] U.S. analysts estimated that Israel had more than ten weapons by 1975 and between 100 and 200 by 1985.[37] But they have never tested any weapons and have adopted an ambiguous policy— "Israel will not be the first country to introduce nuclear weapons into the Middle East"—because "deliberate ambiguity maximizes our deterrence."[38]

The development of Israeli nuclear weapons, in turn, persuaded other countries in the region to try to do the same. Iraq attempted to do so first (with assistance from France), and Iran has tried to do so more recently, in response to both Israeli and Iraqi nuclear programs. (We will examine these developments further in the chapter on Iran and Iraq.) As in the subcontinent, the development of nuclear weapons by countries that participate in recurrent wars has been a worrisome problem.

THE SIX DAYS' WAR AND OCCUPATION

After the 1956 war, Egypt and Syria rebuilt their armies with Soviet assistance. By 1967, Nasser felt strong enough to close the Straits of Tiran to Israeli shipping once again, move troops into the Sinai, and engage in a noisy saber-rattling campaign against Israel.[39] Israel regarded the closure of the Straits as a violation of international law (as did the United States and Great Britain) and an act of war. This, together with the mobilization of Arab armies in Egypt, Syria, and Jordan, persuaded Israeli leaders to make a preemptive strike against Arab forces before they could launch a coordinated attack on Israel.

The third Arab-Israeli war began on June 5, when the Israeli air force attacked and destroyed Arab planes on the ground.[40] Its armies struck across three fronts in Egypt, Jordan, and Syria. In just six days, Israeli forces routed and destroyed Arab armies, taking Gaza and the Sinai from Egypt; East Jerusalem and the West Bank from Jordan; and the Golan Heights from Syria.

But while Israel achieved a stunning, overwhelming, and comprehensive military victory over their many adversaries, the war also created a new set of problems, because its armies occupied territories that were home to large Palestinian populations, many of them displaced or driven from Israel during the 1948 war: 400,000 Palestinians in Gaza; 600,000 in the West Bank; and 100,000 in East Jerusalem.[41]

The occupation of the West Bank and Gaza created several ongoing problems. First, as in 1948, fighting in 1967 also triggered a migration, this time of Palestinians fleeing from the West Bank into Jordan, where they joined Palestinians who had fled in 1948.[42] Second, although a large Palestinian population had been incorporated into territories administered by Israel, the government did not extend to Palestinians any meaningful civil rights. Instead, Palestinians were generally subjected to military, not civilian rule, and treated as illegal immigrants who belonged to another state (Jordan or Egypt).[43] As in 1948, about 36 percent of the land in the West Bank was seized by Israel from "absentee" owners and transferred to Israeli ownership or military control.[44] In East Jerusalem, 4,000 Palestinians were expelled from their homes in the Jewish Quarter to make possible the creation of an all-Jewish Jewish Quarter, and 10,000 residents of villages in the West Bank were expelled from their homes.[45]

Third, the Israeli government began constructing Jewish settlements in the West Bank and Gaza, often on land seized from Palestinians. This process proceeded slowly at first. By 1976, there were 3,176 Jewish settlers in the territories, 10,000 in 1979, 20,000 in 1982, 57,000 in 1987, and 147,000 by 1995.[46] In 2000, 400,000 Jewish settlers lived in the occupied territories, about half of them in neighborhoods around Jerusalem.[47] The immigration of Jewish settlers into the West Bank and Gaza has been a constant source of friction and violence between Israelis and Palestinians since 1967.

Fourth, while Israel occupied territories in East Jerusalem, the West Bank, Gaza, and the Golan, its authority there has not been recognized in international law. Instead, UN resolutions, which have been ratified by the United States and other members of the Security Council, treat Israeli authority in these territories as temporary, pending a negotiated settlement in the region.[48] And while efforts have been made to negotiate a settlement at Camp David and in Oslo (see below), parties to the conflict have been unable to address the issues associated with the Israeli occupation, which began during the Six Days' War.

THE PLO, BLACK SEPTEMBER, AND EXIT TO LEBANON

After the defeat of Arab armies in 1967, Palestinians based in refugee camps in Jordan began launching small-scale guerrilla attacks against Israeli forces in the West Bank. These attacks were organized by Al Fatah, or "the Conquest," which was founded by Yasir Arafat and three others in 1959.[49] Irritated by these attacks, Israel in 1968 launched punitive incursions into Jordan to deter future assaults. In March 1968, Al Fatah fighters stood their ground in the face of an Israeli incursion. After the Israelis withdrew (as they had intended to do), Arafat declared a victory, using the occasion to rally Palestinians to his cause and campaign for leadership of the Palestine Liberation Organization (PLO), an umbrella group created in 1964 by Egypt.[50] In 1969, Arafat assumed the leadership of the PLO, a position he would not relinquish until his death on November 11, 2004.

By 1970, the PLO in Jordan had grown strong enough to challenge King Hussein for power. But Hussein was determined to prevent a PLO rebellion and to remove the cause of Israeli incursions into Jordan. So he mobilized his army to suppress the PLO in September 1970, driving PLO forces out of their bases in refugee camps and into Lebanon, where they established new bases among the 400,000 Palestinian refugees living there.[51] The civil war in Jordan, known to Palestinians as "Black September," shifted the focus of anti-Israeli guerrilla warfare from Jordan to Lebanon, where it contributed later to war and the globalization of violence. Palestinian irregulars, some directly associated with the PLO, and some members of splinter groups and factions, began assaulting non-Israeli targets or attacking Israeli targets outside the region. The hijacking of U.S. and British airplanes and their destruction in the Jordanian desert in 1970, and the capture of Israeli athletes at the 1972 Munich Olympics, which resulted in their deaths, were among the more visible episodes in the global terrorist campaign that emerged after Black September.[52]

September 1970 was also significant because on the 28th, Nasser died. His successor, Anwar Sadat, immediately made plans to renew war with Israel. To this end, he prepared a military, political, and economic plan designed to globalize the Arab-Israeli conflict, engage the attention of the world community, and neutralize the advantages that Israel had secured in the three previous wars. His plan, which came to fruition with Egypt's surprise attack on Israel in October 1973, would have consequences that reverberated across the region and around the world for years to come.

THE 1973 YOM KIPPUR WAR AND THE OIL EMBARGO

On October 6, 1973, the Egyptian army crossed the Suez Canal and attacked Israeli forces in the Sinai. In a coordinated effort, the Syrian army assaulted Israeli positions in the Golan Heights. This combined attack, which came on Yom Kippur, the Jewish Day of Atonement, surprised the Israelis and secured some initial military gains.[53] But in the weeks of heavy fighting that followed, the Israeli army turned the tide and prepared to encircle and destroy one of the Egyptian armies. But before they did, a U.S. and Soviet ceasefire on October 24 brought an end to the fighting.

Although the war was relatively brief, and did not substantially change the boundaries of the Arab-Israeli conflict, it dramatically altered military, political, and economic conditions in the Middle East. First, while Egyptian and Syrian armies did not defeat Israeli armies in the field, they inflicted serious losses, a worrisome development for Israel. The Israelis demonstrated, for a fourth time, that they could defeat their opponents in battle. But they learned that repeated military victories did not secure their forces from attack.

Second, the war engaged both superpowers. The United States was forced to rush military supplies to Israel when it ran short during the war. The Soviet Union was forced to engage in frantic diplomacy to rescue its allies from outright defeat. So the war persuaded both U.S. and Soviet officials more actively to seek a political solution to the conflict.

Third, the war triggered an Arab oil embargo that had serious economic consequences around the world. As he prepared for war, Sadat secretly persuaded Arab oil-producing countries to embargo oil in the event of war. He hoped that this would provide Egypt with important new economic leverage. The oil-producing states agreed to participate in Sadat's plan because they saw war as an opportunity to exercise new political power and viewed embargo as a way to reverse falling oil prices, which had declined as a result of the U.S. dollar devaluation in 1971 (see chapter 2), and increase their revenues. When war broke out, they used U.S. efforts to resupply Israel as the justification for their embargo, though this was a pretext since they were prepared to do so in any event.

The global economic consequences of the oil embargo were vast and varied. The oil embargo swiftly raised the price of oil and contributed to inflation around the world (see chapter 3). Inflation triggered economic stagnation in the rich countries (see chapter 3) and forced many poor countries to borrow heavily (see chapter 4). When the United States moved to curb inflation in the 1980s, high interest rates created a recession at home, which contributed to a crisis in the banking, housing, and agricultural industries (see chapter 3), and created a debt crisis for indebted countries around the world (see chapter 4). In some countries, the debt crisis contributed to democratization (see chapter 6); in others it led to falling commodity prices and growing poverty.

The oil embargo also had important regional consequences. Rising oil prices, and the problems associated with inflation, contributed to revolution in Iran (see chapter 10). Revolution in Iran and battles for control of oil resources in the Persian Gulf contributed to the first and second Gulf wars, which we will examine below.

In short, the 1973 Yom Kippur War had serious and long-lasting global consequences. Ironically, it also led to some positive political developments, particularly for Israel.

CAMP DAVID PEACE ACCORDS

The 1973 Yom Kippur War increased the diplomatic stature and bargaining power of Egyptian president Anwar Sadat. During the next few years, he used it to change political and military relations in the Middle East.

First, in 1975, Sadat jettisoned the Soviet Union as Egypt's superpower ally and turned back to the United States for assistance.[54] U.S. officials welcomed Egypt's return to the West and agreed to provide it with economic aid, which had been terminated in 1956, and use U.S. influence to persuade Israel to open negotiations with Egypt.

Then on November 9, 1977, Sadat made a dramatic announcement. He told the Egyptian parliament that he was ready to travel to the Israeli Knesset (parliament) to demonstrate his willingness to negotiate a permanent peace treaty with Israel. Ten days later, he flew to Israel, appeared in the Knesset, and offered to recognize Israel and make peace if Israel agreed to return Arab territories, including Arab Jerusalem, recognize a Palestinian state, and accept secure boundaries subject to international guarantees.[55]

One year later, President Jimmy Carter persuaded Sadat and Israeli prime minister Menachem Begin to conduct a peace conference at the presidential retreat in Camp David, Maryland. There, in September 1978, Sadat and Begin agreed to conclude a peace treaty that provided for an Israeli withdrawal from the Sinai in return for mutual recognition and the creation of normal diplomatic relations between the two countries. They also agreed, in a separate document, that Egypt, Israel, Jordan, and representatives of the Palestinians would work toward a resolution of the Palestinian problem, leading to the creation of an elected Palestinian authority in the West Bank and Gaza and the eventual withdrawal of Israeli forces from the occupied territories.[56]

The Camp David agreements, and the peace treaty that followed, greatly improved the Israeli position. With the return of the Sinai, Egypt retired from the conflict. Israel no longer had to worry about its strongest, most determined opponent. Moreover, Egypt's withdrawal from the conflict undermined the Palestinians, who had long relied on Egyptian force of arms and on the political and economic support it provided through the PLO. Once the peace treaty was concluded, little effort was made to solve the Palestinian problem, largely because Israel refused to negotiate with the PLO, which had become the political representative of Palestinians throughout the region but which the Israelis regarded as a terrorist organization.

Although Sadat was able to transform relations in the Middle East and secure a partial peace, he was not able to advance a comprehensive peace. He was assassinated in 1981 by members of the Muslim Brotherhood, a militant Islamic group based in Egypt whose ideas greatly influenced radical groups across the region, Hamas and Al Qaeda among them.

INVASION, CIVIL WAR, AND THE DESTRUCTION OF LEBANON

The migration of PLO fighters from Jordan to Lebanon after September 1970 upset the balance of power in Lebanon, the only Arab democracy in the Middle East. In Lebanon, power had been shared between Maronite Christians and Sunni and Shiite Muslims based on a political formula established by the French in 1932. But the growth of Palestinian refugee populations since 1948—400,000 out of a population of three million in 1970—and the influx of new Palestinian groups, including the PLO from Jordan, increased the weight of the Muslim population.[57] However, political power in the government did not reflect these demographic changes. The Christian population, now a minority, retained an edge in government. Skirmishes between the PLO and the Christian-dominated government led to fighting, and Syrian troops entered Lebanon as "peacekeeping" forces to prevent the conflict from escalating into civil war.[58]

In 1978, the PLO launched a series of raids into Israel from its bases in southern Lebanon, much as it had done a decade earlier in Jordan. Israel, much as before, retaliated by launching punitive assaults into Lebanon, and then allied itself with Maronite Christian militias in the South.

Determined to destroy PLO bases in Lebanon, Israel launched a full-scale invasion of Lebanon in June 1982, marking the onset of the fifth Arab-Israeli war. Israeli forces under Defense Minister Ariel Sharon swept PLO and Lebanese army forces into Beirut, which the Israelis then besieged. The two-month siege resulted in 18,000 deaths and 30,000 casualties, most of them civilian.[59] The Israelis lifted the siege only after the Lebanese agreed to evacuate PLO fighters to countries outside the region, and the PLO was forced to move its headquarters to Tunisia.

The Israelis had hoped to install a friendly Maronite leader, Bashir Gemayel, as president of Lebanon. He had secretly agreed to work with Israel to eliminate the Palestinian and Syrian influence in Lebanon and had promised to establish a government friendly to Israel.[60] But Gemayel was assassinated by a huge car bomb at his Phalangist party headquarters. Israel immediately took control of West Beirut, a violation of the ceasefire agreement, and allowed Phalangist militia to assault Palestinian refugee camps at Sabra and Shatila, where they massacred 1,000 to 2,000 civilians.[61]

President Ronald Reagan then landed U.S. marines to enforce the ceasefire and prevent the situation from deteriorating further. But car bomb attacks on the U.S. Embassy in April 1983, and on the U.S. and French military barracks in October, which killed 265 marines and 58 French soldiers, persuaded Reagan to withdraw U.S. forces from Lebanon.[62]

Israeli forces subsequently withdrew from Beirut but remained in southern Lebanon. Syrian troops occupied the eastern part of the country, and Lebanon descended into a violent, destructive, multi-sided civil war, which continued for the rest of the 1980s.

It is sometimes said by political scientists that democracies do not wage war against other democracies. But the invasion of Lebanon by Israel, of an Arab democracy by a Jewish democracy, resulted in the destruction of democracy in Lebanon for a generation. A 1990 peace agreement among various parties in Lebanon greatly reduced the violence and brought an end to open civil war, but it did not reestablish the kind of democracy that existed before the migrations, invasions, and civil wars of the 1970s and 1980s.[63]

INTIFADA I

By the mid-1980s, Israel had persuaded Egypt to retire from the Arab-Israeli conflict, forced the PLO out of Jordan and then Lebanon, and established buffers in Lebanon and Syria to secure its borders against the Syrian army and residual Arab militias. But while it had secured their borders against external threats, Israeli officials had done little to address the problems facing Palestinians in the occupied territories.

Palestinians in the West Bank and Gaza could not claim citizenship in Israel, vote in its elections, or serve in its armed forces. Instead, they were subject to military rule, not civil law, and could be arrested without cause, jailed indefinitely, and deported at will. They were denied due legal process and subject to collective punishment for infractions by individuals, the most dramatic being the destruction of houses belonging to relatives or friends of individuals charged with violations of military law. Palestinians in the occupied territories were treated, under Israeli law, as illegal, undocumented immigrants. Israeli officials maintained that Palestinians were Jordanian or Egyptian citizens living illegally in Israel. They were, however, allowed to work in Israel, but under conditions that approximated illegal immigrant workers in the United States or illegal workers in China (see chapter 7). Not surprisingly, Palestinians grew increasingly restive.[64]

Between 1948 and 1967, Palestinians had relied on Arab states to advance their cause. Then, after the destruction of Arab armies in the 1967 war, they turned to the PLO to fight on their behalf. But after the PLO was forced from Lebanon, Palestinians in the occupied territories were left to their own devices.

After four Palestinian laborers were killed in a traffic accident in Gaza on December 8, 1987, mourners stoned an Israeli army compound, a development that triggered riots throughout the occupied territories. The revolt by unarmed, but stone-throwing civilians against heavily armed Israeli soldiers and settlers became known as the *intifada*, "an Arabic word referring to the shivering of someone in fever, or the shaking of a dog with fleas."[65]

Palestinians conducted strikes and organized demonstrations, refused to pay taxes, boycotted Israeli goods, and encouraged police and tax collectors to resign. Israeli forces responded by imposing collective punishments, such as destroying homes, arresting and detaining demonstrators, deporting others to neighboring countries, curbing the press, and for a period, shooting at demonstrators and rock throwers with live ammunition.[66] In January 1988, Israeli defense minister Itzhak Rabin decreed that live ammunition not be used and ordered soldiers to attack demonstrators with batons and tear gas instead. "Gentlemen, start using your hands or clubs and simply beat the demonstrators in order to restore order," Rabin said, instructions that became known as a "break their bones" policy.[67]

Because the *intifada* at first took place without direction or assistance from the PLO, other indigenous military groups emerged, Hamas among them. Hamas, an acronym meaning "zeal" in Arabic, described itself as an offshoot of Egypt's Muslim Brotherhood, which was responsible for the assassination of Sadat.[68]

During the first eighteen months, 50,000 Palestinians were arrested. By 1993, 1,145 Palestinian civilians had been killed, tens of thousands wounded, 1,473 homes demolished, and 413 people deported. In the same period, 160 Israeli soldiers and civilians had been killed, and hundreds injured.[69]

The *intifada* ground on for six years, longer than all five Arab-Israeli wars combined, largely because the Palestinians demanded that the PLO represent them in negotiations and the Israelis refused to recognize the PLO as a legitimate negotiating partner. Finally, in 1993, secret negotiations between Israelis and PLO representatives in Oslo, Norway, resulted in an agreement that ended the *intifada* and started a peace process designed to address the Israeli-Palestinian conflict.

THE OSLO PEACE ACCORDS, 1993

Under the Oslo Accords, the PLO agreed to recognize Israel's right to exist as a state and Israel agreed to recognize the PLO as the political representative of the Palestinians in the West Bank and Gaza. Both sides agreed that Israel would transfer some of its administrative powers in the occupied territories to a Palestinian Authority, pending a negotiated settlement of other outstanding issues.[70] Essentially, Israeli officials proceeded with promises they had first made at Camp David in 1978, agreements that were neglected and allowed to lapse for fifteen years.

On September 13, 1993, Israeli prime minister Itzhak Rabin and PLO chairman Yasir Arafat met to ratify the agreement on the White House lawn and, encouraged by President Bill Clinton, shook hands for the first time.

A Palestinian Authority was soon established in the occupied territories, and Arafat returned to run it on July 1, 1994. Palestinians could, for the first time, vote for representatives to local government.

But while the Oslo Accords brought an end to the *intifada* and promised to bring peace, finally, to Israelis and Palestinians, the peace process was undermined by three developments over the next several years.

First, Israel continued to build Jewish settlements in the West Bank and Gaza. Many of them were designed to accommodate some of the nearly one million Jews who immigrated to Israel from Russia after the collapse of the Soviet Union.[71] Overall, the number of Jewish settlers in the West Bank, not including the 130,000 around Jerusalem, grew from 80,000 in 1990 to 140,000 in 1998.[72] Palestinians resented continued Jewish immigration and the financial assistance and military security provided to them.

Second, the Israeli government implemented a series of "closure" policies designed to restrict Palestinian migration into Israel proper and to prevent terrorist attacks by splinter groups like Hamas. This meant that the Palestinian Authority was treated much like the municipal government in a mid-sized American city, where the governor had declared a state of emergency and had deployed the National Guard to enforce martial law. Israeli roadblocks and security checkpoints prevented Palestinian workers from traveling to Israel to work and restricted the movement of people or goods between the West Bank and Gaza.[73] These policies contributed to unemployment (35 percent of the Palestinian labor force depended on income from unskilled jobs in Israel), falling wages, and declining standards of living.[74] These economic losses were compounded by the loss of remittance income from Palestinians working in the Gulf states. Between 300,000 and 400,000 Palestinians working in the Gulf were forced to leave at the end of the 1990 Gulf War because Palestinian leaders had supported Iraq during the war. Many took refuge in Jordan.[75] As a result of falling income and Israeli closure policies, "real per capita GNP declined by 37 percent in the West Bank and Gaza [from 1992 to 1996], falling from $2,700 to $1,700."[76] By 1996, nearly 15 percent of the Palestinian population lived on less than $500 a year, a level comparable to poor workers in China (see chapter 7).[77]

Third, ongoing violence by splinter groups corroded the peace process. Arafat and the Palestinian Authority were either unwilling or unable to curb terrorist attacks—such as the suicide bomb that killed twenty-one Israelis in Beit Lid—by militant groups like Hamas and Islamic Jihad. The Israelis were unable to curb Jewish settler attacks on Palestinians, such as the massacre of twenty-nine worshipers at a mosque in Hebron by a Jewish settler.[78] Violence triggered retributions, and retribution revenge. But the most devastating blow to peace was the assassination of Israel's prime minister Itzhak Rabin, who was murdered on November 4, 1995, by an Israeli militant opposed to the Oslo Accords.[79]

Rabin's death led to the election in Israel of a conservative prime minister, Benjamin Netanyahu, who had opposed the Oslo Accords and did what he could to prevent or slow its implementation.[80] By the late 1990s, the peace process had ground to a halt.

The election in 1999 of Ehud Barak, a decorated war veteran, briefly revived the peace process. At President Clinton's invitation, Barak agreed to meet with Arafat at Camp David to hammer out a final agreement. But the talks broke down. Arafat, constrained by Palestinian militants opposed to any concessions with Israel, would not agree to compromise and reach an agreement. Barak, constrained by conservatives and settlers led by former general Ariel Sharon, could not offer concessions that might clinch an agreement.[81] The Oslo peace process collapsed. One year later, a second *intifada* began.

INTIFADA II

On September 28, 2000, Ariel Sharon, accompanied by members of the conservative Israeli opposition and 1,000 police officers, made a visit to the Al-Aqsa Mosque in Jerusalem, one of Islam's most important shrines. Palestinians viewed it as a provocation, which is what Sharon intended, and riots erupted throughout the West Bank and Gaza. The second *intifada* was under way.[82]

Sharon immediately took advantage of the riots to advance his political position. Running on promises to crack down on the *intifada*, which he had helped to ignite, Sharon was elected prime minister in February 2001. The violence then escalated rapidly: Palestinian suicide bombers attacked

Israeli targets; Israeli jet fighters attacked the Palestinian Authority and helicopters strafed demonstrators and militants in the occupied territories.[83] On the eve of 9/11, Israelis and Palestinians were once again at war, fighting over problems first created by partition in 1948.

NOTES

1. Marc Charney, "Arab and Israeli: The Roots of the Conflict," *New York Times*, 28 February 1988.

2. Michael J. Cohen, *The Origins and Evolution of the Arab-Zionist Conflict* (Berkeley: University of California Press, 1987), 90–93.

3. Amos Perlmutter, *Israel: The Partitioned State* (New York: Scribner's, 1985), 55.

4. Cohen, *Origins and Evolution*, 154.

5. Cohen, *Origins and Evolution*, 9, 11.

6. Walter Laqueur, *A History of Zionism* (New York: Schocken, 1976), 577.

7. J. C. Hurewitz, *The Struggle for Palestine* (New York: Schocken, 1976), 296.

8. Cohen, *Origins and Evolution*, 126–27.

9. Hashim S. H. Behbahani, *The Soviet Union and Arab Nationalism, 1917–1966* (London: KPI, 1986), 58–59.

10. Ritchie Ovendale, *The Origins of the Arab-Israeli Wars*, 4th ed. (Harlow, U.K.: Pearson Education, 2004), 135.

11. Ovendale, *Origins of the Arab-Israeli Wars*, 139.

12. Avram S. Bornstein, *Crossing the Green Line between the West Bank and Israel* (Philadelphia: University of Pennsylvania Press, 2002), 40.

13. Edward Said and Christopher Hitchens, *Blaming the Victims* (London: Verso, 1988), 74; Ovendale, *Origins of the Arab-Israeli Wars*, 140.

14. Said and Hitchens, *Blaming the Victims*, 75.

15. Benny Morris, *The Birth of the Palestinian Refugee Problem, 1947–1949* (Cambridge: Cambridge University Press, 1988), 155–69.

16. Morris, *Birth of the Palestinian Refugee Problem*, 141.

17. Morris, *Birth of the Palestinian Refugee Problem*, 174.

18. Morris, *Birth of the Palestinian Refugee Problem*, 240.

19. Morris, *Birth of the Palestinian Refugee Problem*, 255.

20. R. F. Holland, *European Decolonization, 1918–1981* (New York: St. Martin's, 1985), 121.

21. Morris, *Birth of the Palestinian Refugee Problem*, 297–98.

22. Ovendale, *Origins of the Arab-Israeli Wars*, 139.

23. Diane B. Kunz, *The Economic Diplomacy of the Suez Crisis* (Chapel Hill: University of North Carolina Press, 1991), 44–45, 64.

24. Kunz, *Economic Diplomacy*, 73, 75, 76.

25. Kunz, *Economic Diplomacy*, 73, 75, 76.

26. Walter LaFeber, *America, Russia and the Cold War, 1945–1984* (New York: Knopf, 1985), 185.

27. Tom Hartman and J. Mitchell, *A World Atlas of Military History, 1945–1984* (New York: Da Capo, 1985), 11.

28. William Roger Louis and Roger Owen, eds., *Suez 1956: The Crisis and Its Consequences* (Oxford: Clarendon, 1989), xv–xvi.

29. Yaacov Bar-Simon-Tov, *Israel, the Superpowers and the War in the Middle East* (New York: Praeger, 1987), 50.

30. Stephen Ambrose, *Eisenhower: The President* (New York: Simon & Schuster, 1984), 361.

31. John C. Campbell, "The Soviet Union, the United States, and the Twin Crisis of Hungary and Suez," in *Suez 1956: The Crisis and Its Consequences*, ed. William Roger Louis and Roger Owen (Oxford: Clarendon, 1989), 246.

32. Bar-Simon-Tov, *Israel, the Superpowers and the War*, 60–61.

33. Robert R. Bowie, "Eisenhower, Dulles, and the Suez Crisis," in *Suez 1956: The Crisis and Its Consequences*, ed. William Roger Louis and Roger Owen (Oxford: Clarendon, 1989), 215; Kunz, *Economic Diplomacy*, 171.

34. McGeorge Bundy, *Danger and Survival* (New York: Random House, 1988), 474.

35. Leonard S. Spector, *Going Nuclear* (Cambridge, Mass.: Ballinger, 1987), 131.

36. Mitchell Reiss, *Without the Bomb* (New York: Columbia University Press, 1988), 45–46, 140, 148.

37. Reiss, *Without the Bomb*, 146.

38. Thomas L. Friedman, "Israel and the Bomb: Megatons of Ambiguity," *New York Times*, 9 November 1986.

39. Ovendale, *Origins of the Arab-Israeli Wars*, 204–8.

40. Ovendale, *Origins of the Arab-Israeli Wars*, 208.

41. Ovendale, *Origins of the Arab-Israeli Wars*, 9, 12.

42. Judith Miller, "Yasir Arafat, Palestinian Leader and Mideast Provocateur, Is Dead at 75," *New York Times*, 12 November 2004.

43. Geoffrey Aronson, *Israel, Palestinians and the Intifada: Creating Facts on the West Bank* (London: Keegan Paul, 1990), 12.

44. Aronson, *Israel, Palestinians*, 88.

45. Aronson, *Israel, Palestinians*, 19.

46. Bornstein, *Crossing the Green Line*, 44.

47. Bornstein, *Crossing the Green Line*, 44.

48. Ovendale, *Origins of the Arab-Israeli Wars*, 214–15.

49. Miller, "Yasir Arafat."

50. Miller, "Yasir Arafat."

51. Miller, "Yasir Arafat."

52. Bornstein, *Crossing the Green Line*, 217–18.

53. Ovendale, *Origins of the Arab-Israeli Wars*, 223–24.

54. Ovendale, *Origins of the Arab-Israeli Wars*, 232.

55. Ovendale, *Origins of the Arab-Israeli Wars*, 236.

56. Ovendale, *Origins of the Arab-Israeli Wars*, 238–39.

57. Ovendale, *Origins of the Arab-Israeli Wars*, 232–33.

58. Ovendale, *Origins of the Arab-Israeli Wars*, 233.

59. Ovendale, *Origins of the Arab-Israeli Wars*, 244.

60. Ovendale, *Origins of the Arab-Israeli Wars*, 245.

61. Ovendale, *Origins of the Arab-Israeli Wars*, 246; Ahron Bregman, *Israel's Wars, 1947–93* (London: Routledge, 2000), 115.

62. Ovendale, *Origins of the Arab-Israeli Wars*, 247.

63. Muhammad Muslih, "The Shift in Palestinian Thinking," *Current History* (January 1992), 29–33.

64. Sara Roy, "The Gaza Strip: Past, Present, and Future," *Current History* (February 1994), 67.

65. Ovendale, *Origins of the Arab-Israeli Wars*, 258.

66. Ovendale, *Origins of the Arab-Israeli Wars*, 260.

67. Bregman, *Israel's Wars*, 129; Baruch Kimmerling, "The Power-Oriented Settlement: PLO-Israel—The Road to the Oslo Agreement and Back?" in *The PLO and Israel: From Armed Conflict to Political Solution, 1964–1994*, ed. Avraham Sela and Moshe Ma'oz (New York: St. Martin's, 1997), 236.

68. Ovendale, *Origins of the Arab-Israeli Wars*, 261; Bregman, *Israel's Wars*, 126–27.

69. Bregman, *Israel's Wars*, 131.

70. Michael C. Hudson, "The Clinton Administration and the Middle East: Squandering the Inheritance?" *Current History* (January 1994); Ovendale, *Origins of the Arab-Israeli Wars*, 282–84.

71. Ovendale, *Origins of the Arab-Israeli Wars*, 276; Muslih, "The Shift in Palestinian Thinking," 22.

72. Don Peretz, "Israel since the Persian Gulf War," *Current History* (January 1992), 36; Ovendale, *Origins of the Arab-Israeli Wars*, 275–76.

73. Sara Roy, "The Palestinian Economy after Oslo," *Current History* (January 1998), 21.

74. Muhammad Muslih, "Jericho and Its Meaning: A New Strategy for the Palestinians," *Current History* (February 1994), 74; Roy, "The Gaza Strip," 68–69; Roy, "The Palestinian Economy," 21–22.

75. Don Peretz, "The Palestinians since the Gulf War," *Current History* (January 1993), 32.

76. Roy, "The Palestinian Economy," 23.

77. Roy, "The Palestinian Economy," 24.

78. Ovendale, *Origins of the Arab-Israeli Wars*, 284.

79. Ovendale, *Origins of the Arab-Israeli Wars*, 287.

80. Ovendale, *Origins of the Arab-Israeli Wars*, 289–90.

81. Ovendale, *Origins of the Arab-Israeli Wars*, 304–6.

82. Ovendale, *Origins of the Arab-Israeli Wars*, 300.

83. Ovendale, *Origins of the Arab-Israeli Wars*, 309.

10

❊

Revolution and War in Iran and Iraq: 1978–1980

Events in 1947–1948 shaped political developments in India and Palestine for the next fifty years. The years 1978–1980 were just as important for Afghanistan, Iran, and Iraq. In 1978, secular communist revolutionaries overthrew the government in Afghanistan and, in 1979, a revolutionary coalition composed of secular and religious movements deposed the shah of Iran. Both revolutions were quickly followed by invasion and war. In 1979, the Soviet Union invaded Afghanistan and, in 1980, Iraq invaded Iran. These invasions led to long, costly, and bitter wars.

Both wars ended in 1988. But the wars' end did not bring an end to war. In Afghanistan, a multi-sided civil war erupted after Soviet troops withdrew. This civil war continued for another ten years and led to the emergence of a Taliban government by the end of the 1990s. It was to be one of the Taliban's allies—Al Qaeda—that would attack the United States on 9/11.

In the Persian Gulf, the end of the First Iraq-Iran War was followed, two years later, by a Second Gulf War, which began when Iraq invaded Kuwait in 1990. A coalition led by the United States forced Iraq out of Kuwait and destroyed its armies on the battlefield, but did not occupy Iraq or force Saddam Hussein's dictatorship from power. The end of the Second Gulf War created conditions that would contribute to a Third Gulf War after 9/11.

Although 1978–1980 was an important watershed in both Afghanistan and the Persian Gulf, their stories are best told separately. We will look first at revolution and war in the Persian Gulf. In this case, oil played an extremely important role in developments there. The 1973 oil embargo, which was triggered by the fourth Arab-Israeli, Yom Kippur War (see chapter 9), led to rising world oil prices. Rising oil prices in the 1970s contributed directly to revolution in Iran and its invasion by Iraq. But oil prices fell in the 1980s. Falling oil prices contributed to the end of the First Gulf War and, ironically, also to the beginning of the Second Gulf War in 1990. To appreciate these developments, we will now examine the role that oil played in revolution and war in the Persian Gulf.

OIL, REVOLUTION, AND WAR IN THE GULF

During the 1973 Yom Kippur War, Arab members of OPEC cut back oil production and restricted its sale.[1] The embargo reduced world oil supplies by 25 percent, from 20.7 to 15.9 million barrels per day.[2] Falling supplies led to rising prices. Within a few months, oil prices increased from $2.90 to $11.65 per barrel.[3]

Rising oil prices had important global and regional consequences. In global terms, they led to inflation and stagnation in rich countries, a development that would lead to a battle against inflation in the 1980s (see chapter 3). In poor countries, rising prices led to heavy borrowing and, in the 1980s, to a debt crisis (see chapter 4). The debt crisis, in turn, contributed to democratization in many countries (see chapter 6).

In regional terms, the oil embargo and rising prices contributed to revolution in Iran, competition for control of oil supplies in the region, and the invasion of Iran by Iraq in 1980.

The 1973 oil embargo was itself the product of developments long in the making. OPEC was founded by oil ministers from Venezuela and Saudi Arabia in 1960, after Western oil companies unilaterally cut the prices they paid to governments in oil-producing countries.[4] At their first meeting in Baghdad, oil ministers from Venezuela, Iran, Iraq, Kuwait, and Saudi Arabia decided to cooperate in an effort to raise their oil revenues. But they had little success during the 1960s.

OPEC first tried to embargo oil during the 1967 Six Days' War (see chapter 9). Arab oil producers reduced their output by 1.5 million barrels per day. But the United States, with help from Indonesia and Iran, was able to increase production, make up the difference, and meet the world's demand for oil.[5] As a result, the first oil embargo failed.

Despite OPEC's efforts, oil prices fell slowly during the 1970s. Then, in 1971, President Richard Nixon devalued the dollar (see chapter 2). This reduced the real price of oil because producers were paid in dollars for their oil.

Faced with a decline in oil prices, and a fall in the value of dollars they received for their oil, OPEC members became even more determined to alter a system that, in their view, short-changed them. They began to demand higher prices from oil companies, threatening to nationalize oil fields if they refused. Their bargaining position had improved, in part because many of the thirty-year leases they had signed with Western oil companies in the 1930s and 1940s were expiring, forcing the oil companies to renegotiate their contracts with oil producers. As a result, oil prices increased modestly, from $2.18 in 1971 to $2.90 in 1973.[6]

The 1973 Yom Kippur War provided oil producers with an opportunity to raise prices dramatically. Arab oil producers were able to prepare their embargo in advance of war because Egyptian president Anwar Sadat informed them of his plan to launch a surprise attack on Israel (see chapter 9). So when the war broke out, they moved quickly to cut oil supplies.

This time, the embargo worked. Unlike the previous attempt in 1967, the 1973 embargo succeeded because the Arab states made a bigger reduction in oil supplies (down 5 million barrels a day, not just 1.5) and because the United States and its allies could no longer make up the difference. U.S. oil production had peaked, at 11.3 million barrels per day, in 1970.[7] By 1973, world demand for oil had grown and the United States no longer had any surplus capacity.[8] So when Arab producers cut back, oil prices shot up.

Rising oil prices greatly increased the revenues of oil-producing countries. It made governments in the Middle East and elsewhere rich. But while one would expect rising oil revenues to solve economic problems, in some cases it created problems, as it did in Iran. There, rising prices contributed to inflation, which led to revolution in Iran by the end of the decade.

INFLATION AND REVOLUTION IN IRAN

In 1973, Iran was ruled by the shah, a dictator who imagined himself king. In fact, he had inherited the title from his father, Reza Khan, who took power in 1925 and crowned himself king. During World War II, British and Soviet troops invaded Iran to secure oil supplies and keep them out of Axis hands. They forced the shah to abdicate power to his son, Mohammed Reza.

During the 1950s, the shah ceded power to parliament. But a coup sponsored by British and U.S. intelligence agencies restored him to power in 1953. In the 1960s, the shah introduced a series of agricultural reforms that he called the "White Revolution." But reform drove poor rural farmers off the land and into the cities, without substantially increasing food production. This was a problem because Iran's population was growing rapidly in this period, increasing from twenty million in 1956 to thirty-three million in 1976.[9] Because food production could not keep pace with the growing population, Iran had to import more and more food, a costly proposition for the government.[10] Moreover, world food prices rose sharply in the 1970s, largely as a result of poor Soviet grain harvests. This increased the cost of imported food and contributed to inflation in Iran.

The failure of agricultural reform created a second set of problems. Poor farmers who were forced off the land by reform migrated into Iran's cities. But the government was unprepared to house them or provide them with jobs. Iranian industry was backward and did not create many new jobs. As rural people crowded into the cities, housing rents rose sharply and unemployment soared. By the late 1970s, urban households paid as much as 60–70 percent of their income on rent.[11] In Tehran, rents increased fifteenfold between 1960 and 1975, then tripled between 1975 and 1976.[12]

The government also saw its oil revenues increase dramatically during the 1970s. Oil revenues rose from $5.6 billion in 1973 to $20.5 billion in 1977.[13] The dictatorship might have used this money to invest in industry to provide jobs, or in agriculture to increase food production. Instead, the regime spent it on the military. Between 1972 and 1976, the shah purchased $10 billion worth of arms from the United States and billions more from Britain.[14] The shah spent lavishly on the military because he wanted to replace Great Britain as the major military power in the region—Britain had withdrawn its military forces from the Gulf in 1971. Iran also used oil revenues to import food to feed its growing population, and to purchase the manufactured and consumer goods that domestic industry could not produce. So, despite rising oil revenues, the regime spent more than it earned and, by the late 1970s, began running trade deficits, a remarkable achievement for an oil-rich country.[15] Meanwhile, rising oil prices, combined with rising food prices and rents, created high, double-digit rates of inflation in Iran.

As already noted (see chapter 3), inflation is a discriminatory economic process. In Iran, it disadvantaged rural peasants, urban workers, small business owners, students, and clerics. Government employees and the rich used their positions to keep up with inflation, even get ahead. The result was a widening gap between rich and poor. While oil revenues enriched the elites, the inflation associated with rising oil prices, food prices, and rents eroded standards of living for the vast majority of Iranians.

Falling real wages led first to strikes by workers in the oil fields in 1978.[16] They were soon joined by students opposed to the regime, then by urban workers and shopkeepers. Massive demonstrations were met by government violence, which led to new, bigger protests. Oil production ground to a halt. By December 25, 1978, urban residents had to stand in line for cooking kerosene—this in a country rich with oil. As Iranian oil production fell, world oil prices rose sharply. It was as if Iran had imposed a unilateral embargo by cutting off its own supply of oil to the world.

On January 16, 1979, the shah, ill with cancer, appeared at the airport and announced he was leaving on a vacation.[17] He departed and never returned. Two weeks later, Ayatollah Khomeini arrived in Tehran from exile in Paris and appointed a new government.[18] The revolution in Iran had begun.

It is important to note that the Iranian revolution was the result of efforts by a coalition of groups, secular and Islamic. Urban workers and students tended to support secular unions and communist parties; rural farmers, recent migrants to the cities, and shopkeepers supported Muslim clerics. They joined together to welcome Khomeini, a leading figure in the anti-shah opposition since he was exiled in 1964. But while these movements were briefly joined in the opposition to the shah in late 1978 and early 1979, they would soon fight with each other for control of the government after his departure. As it happened, Iraq's invasion of Iran would disadvantage secular revolutionary groups and shift the balance of power to Islamic revolutionaries allied with Khomeini.

COMPETITION IN THE GULF

The 1979 revolution reduced oil production in Iran. This forced up world oil prices, which tripled from $11 to $30 a barrel in the next two years. But while OPEC countries were united in the determination to raise prices and welcomed this bonanza, they disagreed with each other about how high prices should go. Basically, they divided into two camps: price "doves" and price "hawks."[19] Saudi Arabia, Kuwait, and the United Arab Emirates preferred relatively low prices. These doves were willing to settle for lower prices than the hawks because they did not want high prices to cripple the world economy (and kill the golden goose), because they had small populations (Kuwait had only one million people in 1980), so a little revenue could go a long way, and because they had large supplies that would last a long time, which allowed them to take the long view. By contrast, the price hawks—Iraq and Iran—wanted prices to rise as high as possible. They felt they had long been cheated by the West and were indifferent to any economic distress high prices might cause the "imperial" powers. They also had large populations (Iraq had eleven million and Iran thirty-three million in 1976) and needed oil revenues to finance economic development (a lot of money only goes a little way), and because their supplies would not last very long at high rates of production.

In this dispute, Saudi Arabia generally won because it was the biggest producer with the largest reserves. And because Saudi Arabia was closely allied politically with the United States, it was willing to use its dominant position in OPEC to keep prices lower than the price hawks wanted. This greatly annoyed leaders in Iraq and Iran. As Saudi Arabia's oil minister, Sheik Zaki Yamani, said of the price hawks in OPEC: "They're too greedy, they're too greedy. They'll pay for it."[20]

It was in this context, the dispute among OPEC members over price levels, that the First Gulf War emerged. In Iraq, Saddam Hussein decided that if he invaded Iran and seized its oil, he could corner the market in the Gulf, wrest control of OPEC from Saudi Arabia, and drive up prices. His greed would cost him dearly. But it would also impose serious costs on Iran, on Kuwait, and on Saudi Arabia and the rest of OPEC.

IRAQ AND THE BAATH DICTATORSHIP

In July 1979, just six months after the fall of the shah in Iran, Saddam Hussein assumed control of the Baath Party in Iraq. He did so by elbowing aside President Ahmad Hassan Bakr, executing twenty-two party leaders who had opposed his ascent, and purging the party of remaining dissidents.[21]

The Baath Party (*Baath* means "renaissance" or "awakening" in Arabic) originated in Syria and spread to Iraq in the 1950s.[22] But the Baath Party in Iraq broke with its counterpart in Syria. In Iraq, the Baath Party first took power in 1963, but it was soon forced out of office. It recaptured power in 1968.[23]

As a quasi-socialist party, the Baath regime used rising oil revenues in the 1970s in a manner that contrasted sharply with the shah's in Iran. The regime cut taxes; raised the minimum wage; subsidized foodstuffs; and invested in education and social services, housing and infrastructural development.[24] The government also provided large-scale employment in the bureaucracy and the military. The number of civilian employees in government grew from 400,000 to nearly 700,000 between 1972 and 1978, and the army grew substantially during this period.[25] Because the dictatorship wanted to remain in power, one-fifth of all government employees were assigned to its internal security forces.[26] Moreover, the regime recruited one to two million party members to serve as its representatives and agents in shops, factories, and schools.[27] As a result of these policies, per capita income in Iraq tripled between 1974 and 1980, and the regime secured substantial political support among sections of the population—primarily the large number of government employees.

But while the regime used oil revenues to provide public benefits, most of the government jobs were distributed to members of the Sunni minority, who lived in the triangle (called the Sunni Triangle) west and north of

Baghdad. The Baath Party had its origins there and most of its leaders were drawn from this group. So Bakr and later Hussein lavished aid on relatives and fellow Sunni to purchase their political support. This development would have important consequences after the U.S. invasion of Iraq in 2003, as we will see.

Of course, these policies antagonized other groups in Iraq. Kurdish rebels in the North opposed the regime and fought against it during the 1960s and early 1970s. But the Baath regime was able to suppress the Kurdish rebellion in 1975, after the shah of Iran agreed to discontinue Iranian aid to the Kurds in exchange for Iraqi concessions on disputed territory along the Shatt al Arab waterway.[28]

Excluded from any real power under the regime, the Shiite majority also mounted protests against it. But the Shiite opposition was forcibly suppressed by a regime willing to arrest, torture, and kill dissidents.

When Hussein came to power in the summer of 1979, Iran was in chaos. The Iranian military had been weakened by purges—hundreds of officers had been arrested or shot, terms of enlistment had been cut, and equipment maintenance had been neglected—oil production was at a standstill, its economy was in tatters, and the government was in turmoil as secular and religious factions struggled for power.[29]

Moreover, Iran was a country without a friend in the world. Soviet leaders viewed Islamic revolutionaries as a threat to secular revolutionaries, whom they supported, and objected to Iranian support for mujahideen forces fighting the Soviets in Afghanistan (see chapter 11). U.S. leaders, who had supported the shah, were dismayed by the revolutionaries who overthrew him and, after the U.S. embassy was seized and American hostages taken in November 1979, they were furious at Iran for its violation of international law. Leaders in Saudi Arabia and Kuwait, meanwhile, worried that Iran would try to export its revolution to their countries or use its military to menace them, particularly in the absence of any superpower military protection in the Gulf.

Iran's internal weakness, coupled with universal hostility toward the regime, provided Hussein with a golden opportunity. He calculated that Iraq could invade Iran without incurring superpower opposition, secure economic aid from Arab oil states fearful of Iran, capture Iranian oil, and put Iraq in a position to dictate world oil prices. He could then use oil revenues to promote economic development, expand the military, and finance Iraq's development of nuclear weapons.

NUCLEAR PROLIFERATION

In 1976, Iraq acquired the technology and training necessary to develop nuclear weapons from France and, in 1980, purchased weapons-grade nuclear fuel.[30] The regime was determined to develop nuclear weapons in response to the Israeli nuclear program, which grew out of the 1956 Suez War (see chapter 9). But the emergence of an Iraqi nuclear program would alarm not just Israel, but also Iran, which began to develop nuclear weapons in the 1980s, and the United States.

During his first year in power, Hussein prepared his army and secured political support and financial aid from Saudi Arabia and Kuwait for an invasion of Iran.[31] On September 21, 1980, Hussein struck, sending his army into the oil-rich region of Iran, where 90 percent of its oil reserves were located, triggering the onset of the First Gulf War.[32]

As Hussein expected, the Iranian army fell back before the onslaught, leaders in the United States and the Soviet Union turned their backs on Iraqi aggression, even though it was an obvious violation of the UN Charter, and the UN Security Council passed only a mild resolution urging both sides "to refrain immediately from any further use of force." But they did not call for a ceasefire or demand that Iraq withdraw its forces from Iran.[33] Iraq's Arab allies, meanwhile, provided war matériel and no-interest loans to assist the Iraqi war effort. Kuwait provided $6 billion in aid to Iraq during the first year of the war, while other Gulf states kicked in another $4 billion.[34]

THE FIRST GULF WAR

At the beginning of the war, the Iraqi air force tried to destroy the Iranian air force on the ground, much as the Israelis had done in the 1967 Six Days' War (see chapter 9). But unlike the Israelis, they failed in their mission. The Iraqi army then advanced into Iran, but did so slowly, giving Iranian forces time to rally. When they did, they fought Iraqi columns to a halt. The Iraqi army was eventually driven out of Iran, but Iranian troops made little progress once inside Iraq, and the war bogged down in a savage stalemate for the next eight years.

To break the stalemate, Hussein repeatedly tried to escalate the fighting. In 1983 and again in 1986, Iraqi forces used chemical weapons on the battlefield.[35] In 1984, Iraq launched air strikes and missile attacks on undefended civilian populations in Iranian cities.[36] Then, after a moratorium, Iraq renewed its attacks in 1985.[37] Iraq also targeted Iranian oil terminals and attacked ships carrying Iranian oil.[38] The Iranians eventually retaliated by attacking ships belonging to Kuwait, Iraq's ally in the war.[39] Kuwaiti officials then appealed to both the United States and the Soviet Union to protect their vessels in the Gulf, and the United States in 1986 sent naval forces to protect them, thereby indirectly assisting Iraq in its "tanker war." It was in this context, in 1987, that Iraqi planes "mistakenly" attacked the USS *Stark*, killing thirty-seven.[40]

Iraq's invasion of Iran was a violation of the UN Charter, and its uses of chemical weapons were violations of international law. But these violations did not prompt the United States, the Soviet Union, or the international community to take action against Hussein. Instead, they registered mild protest or, after the *Stark* was attacked, accepted Hussein's explanation and apology. Hussein took this to mean that he could do what he pleased without risk of censure, a view that persuaded him to take aggressive new initiatives in subsequent years.

"Nuclear" War

While the superpowers were prepared to let Hussein conduct war on his own initiative, regional powers were not ready to let Iraq develop nuclear weapons. The Iranian air force first struck Iraq's nuclear facility at Darbandi Khan shortly after the war began, damaging it.[41] Then in June 1981, Israel, which was not a party to the conflict but was nonetheless alarmed by Iraq's nuclear program, attacked and destroyed the Iraqi nuclear reactor.[42]

Although nuclear weapons were not used in either raid, the attacks can be considered a "nuclear" war insofar as they were designed to preempt the development of nuclear weapons by Iraq. Nuclear facilities in Iraq would again be attacked during the Second Gulf War in 1990, and then cited by U.S. officials as a reason to begin the Third Gulf War in 2003, though postwar investigations revealed that Iraq had not, in fact, developed nuclear weapons or made significant progress toward acquiring them.

But while Iraqi nuclear facilities were destroyed, the Iranians began developing nuclear weapons of their own in 1982.[43] This would become a problem for the United States and other countries in the region after 9/11.

The Persian Gulf War had other consequences in Iran. The war led to the ouster of moderates and secular political factions in the revolutionary coalition and to the consolidation of power by radical Islamic forces under Khomeini. Moderate president Abu al-Hassan Bani-Sadr was forced from office, and he fled into exile in June 1981. This was followed by the massacre of 10,000 secular, leftist students and teachers, and widespread arrests and executions.[44] A violent campaign by secular mujahideen to assassinate Islamic leaders, including the president and prime minister, triggered civil war. Islamic forces eventually suppressed the mujahideen, arresting and executing many of them.[45]

Gender and Revolution

The ascent of Islamic revolutionaries in Iran, as elsewhere in the region (particularly Afghanistan), pushed women to the margins of political power and out of the labor force. Before the revolution, in 1976, the percentage of women in the labor force stood at 13.8 percent.[46] This rate was far above Pakistan (3.7 percent), and a little below Egypt (18.7 percent) or Iraq (19 percent).[47] But by 1986, the percentage of women in Iran's labor force had dropped by one-third, to only 8.9 percent.[48]

The political and economic marginalization of women in Iran was accompanied by cultural oppression: the forcible removal of unveiled women from public life—women were publicly flogged for "indecency" and failure to conform with male-imposed dress codes—and the passage of laws that restricted their legal rights, including measures that required two women to provide testimony in court, but allowed only "one honest man" to testify in court.[49]

Iran-Contra

Although U.S. officials generally sided with Iraq during the war, they also provided covert aid to Iran. In 1986, investigators revealed that officials in the Reagan administration had secretly agreed to sell U.S. weapons and spare parts to Iran (with assistance from Israel) in return for Iranian help in securing the release of U.S. hostages in Lebanon. This, of course, violated the Reagan administration's public policy of not negotiating with hostage-taking terrorists and of supporting a regime that had itself earlier taken American hostages from the U.S. embassy in Tehran.[50]

It was also discovered that this was part of a larger, more complicated scheme to evade U.S. law. Money from the sale of U.S. weapons to Iran was used to finance contra rebels, an anti-communist militia fighting to overthrow the Sandinista regime in Nicaragua. Using Iranian money to fund the contras violated U.S. law because Congress had passed legislation prohibiting the government from providing aid to the contras. This scheme, which violated both U.S. policy principles and law, forced the resignation of low-level officials who participated in the plan, but other responsible officials avoided indictment and jail.[51]

WAR AND OIL

The First Gulf War had important consequences for oil, though not what one might expect. The battle for control of oil in the Gulf eventually rebounded against Iraq and Iran and forced them to end the fighting.

Initially, revolution in Iran and war with Iraq drove oil prices up to historic highs. By 1981, oil was selling for $34 a barrel.[52] But high prices encouraged the development of new supplies, particularly in the North Sea and Mexico. Producers in these regions were not members of OPEC and did not have to observe its quota system. So new oil flowed into the market. Moreover, high oil prices encouraged consumers to conserve energy. They purchased higher mileage cars, installed more energy-efficient refrigeration and heating systems in offices and homes, and turned thermostats down. World oil consumption fell from fifty-one million barrels a day in 1979 to forty-five million barrels a day in 1983. So by 1983, consumers were daily saving as much oil as Saudi Arabia produced each day.[53]

High oil prices also encouraged OPEC members to cheat. Because they wanted the revenues from high prices, many OPEC countries pumped more oil than their quotas allowed. Widespread cheating forced Saudi Arabia to cut back its production to keep prices from falling in the face of growing supplies. But despite the Saudis' efforts, oil prices fell from $34 to $29 a barrel in 1983, the first time prices had fallen in a decade.[54]

Because Saudi Arabia kept cutting back its production to sustain prices, its revenues fell from $119 billion in 1981 to only $26 billion in 1985.[55] To the astonishment of Saudi officials, the United Kingdom earned more from the sale of North Sea oil than Saudi Arabia.[56]

By November 1985, Saudi officials were tired of propping up oil prices, at their own expense, for everyone else. Unwilling to keep playing the sap, the government announced it would increase production to recapture its market share and let prices fall. Saudi Arabia became a dove with a vengeance. Prices quickly plummeted from near $30 a barrel to only $10.[57]

Moreover, the falling price of oil was soon accompanied by the falling value of the dollar. Recall that in 1985, the United States and its G-7 partners agreed in the Plaza Accords to devalue the dollar by half over the next three years (see chapter 2). So the real price of oil fell twice—first because its price in dollars fell from $30 to $10, second because the value of the dollar itself declined by half.

In fact, prices fell so low that Vice President George Bush traveled to Saudi Arabia in 1986 and asked the Saudis to cut production and raise prices. He made this remarkable request, which would have been extremely unpopular with consumers and voters in the United States if it had become known, because oil producers in Texas could not make money at $10 a barrel. They have thicker oil, deeper wells, and higher costs than Saudi producers, so low prices were driving many of them out of business. The Saudis agreed to Bush's request and helped raise prices back to $16–$18 a barrel, a level at which Texas producers could profit.[58]

Falling oil prices reduced revenues for oil-producing countries around the world. But this development was particularly painful for Iraq and Iran, which desperately needed oil revenues to continue fighting the war. Fortunately, falling oil prices crippled their ability to wage war and forced them to negotiate an end to the war. Ironically, Hussein's attempt to corner oil supplies, which was supposed to increase prices and revenues, led instead to falling prices, which crippled his ability to wage war and capture the supplies he coveted. Falling prices and the costs of war wiped out the $34 billion in savings Iraq had accumulated before the war and left it deeply in debt to creditors in Kuwait.[59]

Iran was impoverished, too. Iran's oil receipts fell from $20 billion in 1983 to only $7 billion in 1988.[60] By 1988, the revolutionary regime faced a large trade deficit and found itself $12 billion in debt.[61] Iran could no longer wage war effectively or hope to break the stalemate at the front. So Khomeini accepted peace terms, which returned both sides to prewar borders, even though he confessed that his acceptance of the UN resolution ending the war "was more deadly for me than taking poison. I submitted myself to God's will and drank this for His satisfaction."[62] After drinking the poison of defeat, he died one year later.

The First Gulf War, which ended on August 20, 1988, killed one million people, wounded two million more, and forced one million people from their homes. It cost both countries more than $1 trillion and drove both parties deep into debt.[63]

THE SECOND GULF WAR

War's end did not bring an end to war in the Gulf. In its aftermath, Iraq was saddled with substantial debt, much of it owed to Kuwait. Leaders there refused to let Iraq off the hook and insisted that Hussein repay Iraq's debts, estimated to be about $50 billion.[64] Low oil prices, which fluctuated between $13 and $19 a barrel in the late 1980s, made it difficult for Hussein's regime to repay its debts, refurbish its army, or rebuild its economy, which had been ground down by years of war.[65]

But Hussein still wanted to capture oil supplies that would enable him to corner the market and dictate higher prices. So he decided to invade Kuwait. By seizing Kuwait's oil, Iraq could "control 20 percent of OPEC production and 25 percent of world oil reserves . . . [and] would be the dominant power in the Persian Gulf, well equipped to resume his war with Iran."[66] It would also enable Iraq to erase its debts to Kuwait. This could be more easily accomplished than war with Iran, Hussein calculated, because Kuwait's army was really just a police force. So on August 2, 1990, just two years after the end of the First Gulf War, Hussein sent Iraqi tanks down the six-lane highway to Kuwait City, starting the Second Gulf War (known in the United States as Operation Desert Storm).[67]

But while the superpowers, the United Nations, and governments in the region did not object to Iraq's invasion of Iran, they did object, furiously, to Iraq's invasion of Kuwait. They did so for several reasons that Hussein did not anticipate or appreciate.

U.S. officials objected because Kuwait, unlike Iran, was a close ally. Kuwait was a U.S. ally in part because U.S. oil firms had long obtained oil concessions in Kuwait (but not in Iran or Iraq, which had been a British preserve), and because Kuwait, along with Saudi Arabia, worked as a dove to keep prices relatively low in OPEC. Soviet leaders, who had provided some military assistance to Iraq during the war, were no longer interested in supporting dictatorships outside the Soviet Union. They had already withdrawn from Afghanistan and abandoned communist regimes in Eastern Europe. They no longer viewed Iran as a serious threat and so did not need Iraq as a counter. Members of the United Nations viewed the Iraqi invasion as a serious violation of the UN Charter. They had not invoked the charter during the First Gulf War because they viewed Iranian hostage taking as a violation of international law. But the invasion of Kuwait, a small country, by a much larger neighbor, triggered a different response. Finally, Saudi Arabia now viewed Iraq, its erstwhile ally, as a military threat, and saw the invasion as an attempt to corner oil supplies and displace Saudi leadership in OPEC.

Because no one supported Iraq, the United States quickly organized a multinational military alliance, with the consent of the Soviet Union and China, to dislodge Iraq from Kuwait. It took some months to organize a response, but in early 1991, U.S. and coalition forces began conducting air raids in Iraq. After a month-long bombing campaign, U.S. and coalition troops forced Iraq from Kuwait and destroyed the Iraqi army on desert battlefields in just three days.

Despite its battlefield success, U.S. forces did not attempt to capture Baghdad or overthrow Hussein's Baathist regime, largely because U.S. leaders did not want to take responsibility for the political chaos and economic difficulties that would follow in its wake. Secretary of Defense Dick Cheney explained it this way:

> What kind of government [would take Hussein's place]? Would it be a Sunni government or Shi'a government or a Kurdish government or a Ba'athist regime? Or maybe we want to bring in some Islamic fundamentalists? How long would we have had to stay in Baghdad to keep that government in place? What would happen to the government once U.S. forces withdrew? How many casualties should the United States accept in that effort to try to create clarity and stability in a situation that is inherently unstable?[68]

Ironically, as vice president, Cheney would later support invasion and occupation in Iraq.

Hussein stayed in power and used the remnants of his army to suppress a Shiite uprising in the South. The Bush administration and the United Nations settled for a containment policy that restricted the regime's ability to use troops to destroy Kurdish rebels in the North, and its ability to use oil revenues to rebuild its army or its economy. Nonetheless, Hussein used the resources available to him to reward his political supporters in the Sunni Triangle, punish his enemies, and tighten his grip on power for another decade.

The Second Gulf War led briefly to higher oil prices. But they quickly fell back to prewar levels and remained low throughout the 1990s, a development that contributed to the growing use of low-mileage pickup trucks, minivans, and sports utility vehicles in the United States, and the expansion of car fleets around the world.

Throughout this period, oil played a contradictory role. Rising oil prices in the 1970s contributed to revolution and war in the Persian Gulf. Falling oil prices in the 1980s forced an end to one war, but also led to the start of another.

NOTES

1. Benjamin Shwadram, *Middle East Oil Crises since 1973* (Boulder, Colo.: Westview, 1986), 43–44.

2. Daniel Yergin, *The Prize: The Epic Quest for Oil, Money, and Power* (New York: Simon and Schuster, 1991), 614; Shwadram, *Middle East Oil Crises*, 58.

3. Yergin, *The Prize*, 625.

4. Shwadram, *Middle East Oil Crises*, 17; Stephen Pelletière, *Iraq and the International Oil System: Why America Went to War in the Gulf* (Westport, Conn.: Praeger, 2001), 139.

5. Yergin, *The Prize*, 557.

6. Yergin, *The Prize*, 625.

7. Yergin, *The Prize*, 567.

8. Yergin, *The Prize*, 594, 614.

9. Fred Halliday, *Iran: Dictatorship and Development* (Harmondsworth, U.K.: Penguin, 1979), 10.

10. Mansoor Moaddel, *Class, Politics and Ideology in the Iranian Revolution* (New York: Columbia University Press, 1993), 77, 85–86.

11. Halliday, *Iran*, 190.

12. Halliday, *Iran*, 190.

13. Halliday, *Iran*, 143.

14. Pelletière, *Iraq and the International Oil System*, 155; Nikki R. Keddie, *Roots of Revolution: An Interpretative History of Modern Iran* (New Haven, Conn.: Yale University Press, 1981), 176.

15. Halliday, *Iran*, 160.

16. W. Thom Workman, *The Social Origins of the Iran-Iraq War* (Boulder, Colo.: Lynne Rienner, 1994), 51; Moaddel, *Class, Politics and Ideology*, 128.

17. Yergin, *The Prize*, 682.

18. Yergin, *The Prize*, 683.

19. Yergin, *The Prize*, 637; Shwadram, *Middle East Oil Crises*, 54.

20. Yergin, *The Prize*, 705.

21. Marion Farouk-Sluglett and Peter Sluglett, *Iraq since 1958: From Revolution to Dictatorship* (London: KPI, 1987), 208–9; Dilip Hero, *The Longest War: The Iran-Iraq Military Conflict* (New York: Routledge, 1991), 30.

22. Judith S. Yaphe, "Reclaiming Iraq from the Baathists," *Current History* (January 2004), 13; Farouk-Sluglett and Sluglett, *Iraq since 1958*, 90.

23. Hero, *The Longest War*, 29.

24. Farouk-Sluglett and Sluglett, *Iraq since 1958*, 179–80.

25. Farouk-Sluglett and Sluglett, *Iraq since 1958*, 248.

26. Hero, *The Longest War*, 21; Pelletière, *Iraq and the International Oil System*, 169.

27. Farouk-Sluglett and Sluglett, *Iraq since 1958*, 184; Yaphe, "Reclaiming Iraq," 13.

28. Stephen C. Pelletière, *The Iran-Iraq War: Chaos in a Vacuum* (New York: Praeger, 1992), 9; Workman, *The Social Origins of the Iran-Iraq War*, 71.

29. Pelletière, *The Iran-Iraq War*, 35.

30. Edgar O'Ballance, *The Gulf War* (London: Brassey's Defense Publisher, 1988), 26.

31. Hero, *The Longest War*, 38.

32. Yergin, *The Prize*, 710.

33. Elaine Sciolino, *The Outlaw State: Saddam Hussein's Quest for Power and the Gulf Crisis* (New York: Wiley, 1991), 109.

34. Hero, *The Longest War*, 77.

35. O'Ballance, *The Gulf War*, 149–50, 179; Pelletière, *Iraq and the International Oil System*, 205.

36. O'Ballance, *The Gulf War*, 153; Pelletière, *The Iran-Iraq War*, 109.

37. O'Ballance, *The Gulf War*, 169–70.

38. O'Ballance, *The Gulf War*, 154.

39. O'Ballance, *The Gulf War*, 156.

40. Pelletière, *Iraq and the International Oil System*, 185; Pelletière, *The Iran-Iraq War*, 128.

41. Pelletière, *The Iran-Iraq War*, 36.

42. Pelletière, *The Iran-Iraq War*, 11; O'Ballance, *The Gulf War*, 76; Hero, *The Longest War*, 74.

43. O'Ballance, *The Gulf War*, 90–91.

44. Ahmad Asraf, "Charisma, Theocracy, and Men of Power in Post-Revolutionary Iran," in *The Politics of Social Transformation in Afghanistan, Iran, and Pakistan*, ed. Myron Weiner and Ali Banuazizi (Syracuse, N.Y.: Syracuse University Press, 1994), 116, 118.

45. O'Ballance, *The Gulf War*, 110–12.

46. Haideh Moghissi, "Public Life and Women's Resistance," in *Iran after the Revolution: Crisis of an Islamic State*, ed. Saeed Rahnema and Sohrab Behad (London: I. B. Tauris, 1996), 253.

47. Workman, *Social Origins of the Iran-Iraq War*, 77; Valentine M. Moghadam, "Gender Inequality in Iran," in *The Politics of Social Transformation in Afghanistan,*

Iran, and Pakistan, ed. Myron Weiner and Ali Banuazizi (Syracuse, N.Y.: Syracuse University Press, 1994), 408.

48. Moghissi, "Public Life and Women's Resistance," 253.

49. Asraf, "Charisma, Theocracy, and Men of Power," 137; Moaddel, *Class, Politics and Ideology*, 263; Moghissi, "Public Life and Women's Resistance," 255.

50. Hero, *The Longest War*, 215–25.

51. Pelletière, *Iraq and the International Oil System*, 197–202.

52. Yergin, *The Prize*, 720.

53. Yergin, *The Prize*, 718; Shwadram, *Middle East Oil Crises*, 184.

54. Yergin, *The Prize*, 720.

55. Yergin, *The Prize*, 747.

56. Yergin, *The Prize*, 748.

57. Yergin, *The Prize*, 750.

58. Yergin, *The Prize*, 756–60; Pelletière, *Iraq and the International Oil System*, 182; Abbas Alnasrawi, "Economic Devastation, Underdevelopment and Outlook," in *Iraq since the Gulf War: Prospects for Democracy*, ed. Fran Hazelton (London: Zed Books, 1994), 76.

59. Alnasrawi, "Economic Devastation," 73; Farouk-Sluglett and Sluglett, *Iraq since 1958*, 265.

60. Vahid F. Nowshirvani and Patrick Clawson, "The State and Social Equity in Postrevolutionary Iran," in *The Politics of Social Transformation in Afghanistan, Iran, and Pakistan*, ed. Myron Weiner and Ali Banuazizi (Syracuse, N.Y.: Syracuse University Press, 1994), 235–36.

61. Sohrab Behad, "The Post-Revolutionary Economic Crisis," in *Iran after the Revolution: Crisis of an Islamic State*, ed. Saeed Rahnema and Sohrab Behad (London: I. B. Tauris, 1996), 110, 112.

62. Hero, *The Longest War*, 243.

63. Workman, *The Social Origins of the Iran-Iraq War*, 1.

64. Hero, *The Longest War*, 250.

65. Alnasrawi, "Economic Devastation," 76.

66. Yergin, *The Prize*, 772.

67. Yergin, *The Prize*, 770.

68. Adeed Dawisha, "The United States in the Middle East: The Gulf War and Its Aftermath," *Current History* (January 1992), 2.

11

❀

Revolution and
War in Afghanistan

In Afghanistan, as in Iran, revolution in 1978–1979 led to invasion and war. The war in Afghanistan, as in the Gulf, dragged on until 1988. But in Afghanistan, the end of war led to a civil war that continued for thirteen more years. During the latter stages of this protracted civil war, in the late 1990s, the Taliban seized the capital and took control of most of the country. Then, on September 11, 2001, the Taliban's Al Qaeda ally launched attacks on the United States. The United States then joined with anti-Taliban forces still fighting in the North and invaded Afghanistan, overthrowing the Taliban and forcing its supporters and Al Qaeda allies to flee into the mountains and seek refuge in Pakistan's rugged borderlands.

Wars in Afghanistan were the product of superpower conflict and regional competition for power in the region. The Soviet Union wanted to secure power for communist revolutionaries in Afghanistan, and the United States worked to prevent this by supporting indigenous, anticommunist rebels—the mujahideen—in their fight with domestic communists and Soviet invaders. During the war, the mujahideen, who were made up of different ethnic groups and political factions, were united in their effort to force the withdrawal of Soviet forces and overthrow the communist regime in Kabul. But when the Soviets withdrew in 1989, different mujahideen groups, each supported by different countries in the region, began fighting among themselves, which allowed the communist regime to remain in power for several years after the Soviets departed. Because neighboring states—Pakistan, Iran, and Saudi Arabia—wanted their mujahideen allies to gain power, they encouraged them to continue fighting, leading to a long, indecisive civil war.

In the mid-1990s, the Taliban, which was supported by Pakistan, seized Kabul, defeated mujahideen supported by Iran and Uzbekistan, and took over much of the country. The Taliban regime imposed strict Islamic law and forced women out of public life. They also allowed Al Qaeda, led by Osama bin Laden, to set up training camps where recruits from around the region were trained to fight on behalf of Islamic movements in Afghanistan and Kashmir, and conduct terrorist campaigns against targets in Saudi Arabia, Western Europe, and the United States. It is important to note that the Taliban, which assisted and protected Al Qaeda and bin Laden, was supported and financed by Pakistan, which has long been the principal ally of the United States in the region. Pakistan supported the Taliban and Al Qaeda because it wanted to establish a friendly government in Afghanistan and deploy Islamic and Al Qaeda fighters in Kashmir, where Pakistan supported a Muslim insurgency against the Indian government (see chapter 8). To appreciate these developments, it is important to return to 1978–1979, when revolution and war in Afghanistan began.

COMMUNIST REVOLUTION AND SOVIET INVASION

Afghanistan, a country about the size of France, was one of the few countries never colonized by European or Asian empires. During the 1960s and early 1970s, the Muslim monarchy received money from both the Soviet Union and the United States, which regarded Afghanistan as a neutral buffer state.[1] In 1973, Mohammed Daoud Khan, a cousin of King Mohammed Zahir Shah, deposed the king and established a republic. But his suppression of domestic communist parties persuaded them to make a revolution of their own. In 1978, the Communist Party seized power in a coup, killed Daoud, and established a secular, communist state. But the communists who took power encountered two serious problems from the outset. Their seizure of power started a fight between factions of the Communist Party, and it triggered a revolt by anticommunist groups in the army, the bureaucracy, and ethnic groups around the country.

The Communist Party that took power in 1978 actually consisted of two separate factions that had only recently joined together. The Khalq or "Masses" faction consisted primarily of ethnic Pashtuns based in the military.[2] The Parcham or "Banner" faction consisted of ethnic Tajiks based in the government bureaucracy.[3] These two factions were bitter rivals until 1977, when Iranian and Soviet communists brokered an agreement that merged them into a single party.[4] With Soviet help, the party immediately began planning a coup. According to Babrak Karmal, leader of the Parcham faction, "Russia wanted that there should be a revolution [in Afghanistan]."[5]

In 1978, the Communist Party launched its revolt. The new government was headed by Hafizullah Amin, leader of the Khalq faction, and the Parcham faction was assigned a minor role in government. Once in power, Amin purged the party and arrested and killed dissidents, both communist and noncommunist. His brief reign of terror resulted in the death of some 50,000 people.[6] The regime imposed unpopular land reforms and passed anti-Islamic legislation, which antagonized the population and triggered mutiny and desertion by noncommunist groups in the army, who then took up arms against the dictatorship.[7]

By 1979, the regime was fighting on two fronts: against the Parcham members of the party and against mujahideen defectors and insurgents. At this point, Soviet leaders decided to step in to end the murderous faction fighting and defeat the growing anticommunist insurgency. So on December 24, 1979, Soviet agents assassinated Amin, installed Babrak Karmal, the pro-Soviet leader of the Parcham faction, as head of the regime, and rushed Soviet troops into Afghanistan to protect the government and combat the insurgency. The regime could count on only a few Afghan troops because most of the army had already deserted, many of them forming the nucleus of mujahideen militias.[8]

Soviet leader Leonid Brezhnev defended the assassination of Amin and invasion of Afghanistan, arguing that "to have acted otherwise would have meant leaving Afghanistan a prey to imperialism . . . to watch passively [the creation] of a serious danger to the security of the Soviet State."[9]

The Soviet invasion ignited a wider war with mujahideen insurgents, who were supported by the United States, Pakistan, and Saudi Arabia. The United States provided billions in aid and arms to the mujahideen during the next eight years. U.S. officials did so because they regarded the Soviet invasion as a violation of Afghanistan's neutrality as a buffer state and because they wanted to repay the Soviets for earlier helping communist forces fight Americans during the war in Vietnam. Saudi officials provided billions of dollars in aid because they wanted to support Islamic, anticommunist rebels. Pakistani leaders did so because they wanted to consolidate their alliance with the United States, a position enhanced by their newfound status as a front-line, anticommunist state, and obtain economic assistance from the United States. (During the 1980s, Pakistan became the third largest recipient of U.S. aid, after Israel and Egypt.[10]) The Pakistanis also wanted to establish a pro-Pakistani, Islamic government in Afghanistan, one that might support Pakistan in its struggles with India over the Kashmir. Together, these countries gave about $1 billion a year in financial aid and about $5 billion in arms during the war to the mujahideen, providing fighters in Afghanistan with more weapons than India and Pakistan combined.[11]

Pakistan, meanwhile, forced different mujahideen groups into a wartime alliance, providing them with military training and intelligence, and directed the war effort from secure bases in Pakistan.[12] Pakistani officials also welcomed and enlisted fighters from Saudi Arabia, Egypt, and other Arab countries in the Middle East, Osama bin Laden among them, to assist the mujahideen.

Soviet forces in Afghanistan grew from 7,000 on the eve of the 1979 invasion to 115,000 troops by 1984.[13] But the large Soviet deployment failed to crush the insurgents. Counterinsurgency experts estimated that Soviet military strength was "far short of the minimum (about 500,000) . . . necessary to mount a serious military challenge to the resistance."[14] Instead of relying on ground troops, the Soviets used their air force to combat the insurgency. But Soviet air power was neutralized after U.S. officials provided advanced surface-to-air Stinger missiles to the mujahideen.

The war resulted in the death of one million Afghans, forced four to five million Afghans to flee and seek refuge in Pakistan and Iran, and displaced another 1.5 million people inside the country.[15] Altogether, about one-quarter of the population became refugees.[16]

As the war dragged on, Soviet casualties mounted—15,000 Soviet troops were killed during the war and 37,000 were wounded.[17] In 1986, Mikhail Gorbachev, who had become the new leader of the Soviet Union one year earlier (see chapter 6), admitted that "counterrevolutionaries have turned Afghanistan into a bleeding wound."[18] So he replaced Afghan leader Babrak Karmal with Mohammed Najibullah, decided to withdraw Soviet troops in 1988, and began conducting negotiations to end the war.[19] But because the United States and its allies did not take Soviet initiatives seriously, negotiations dragged on until February 8, 1988, when Gorbachev announced that he would withdraw all Soviet forces when a UN-sponsored agreement was reached. Negotiations were then rapidly concluded, and on April 14, parties to the conflict signed an agreement providing for a full Soviet withdrawal by February 15, 1989.[20]

The UN agreement provided for a Soviet withdrawal in return for an end to outside aid to the mujahideen.[21] But when the Soviets withdrew, the United States and its allies continued to supply arms and aid to the mujahideen. The Soviets then continued supplying the communist regime in Kabul with money and food, which the government used to purchase the continued support of its army and the population in Kabul.[22] As a result, fighting continued, now as a civil war.

Contrary to all expectations, Najibullah's regime stayed in power until 1992, largely because its mujahideen opponents began to quarrel and refused to coordinate their efforts. Cooperation became increasingly difficult because regional powers now armed and supported different mujahideen factions inside Afghanistan. Iran supported Shiite forces, Pakistan supported Pashtuns, Uzbekistan supported Uzbeks, while Tajik forces, lacking a foreign benefactor, fought alone.[23]

When Najibullah's regime finally fell, no one faction was strong enough to take charge and none of the factions were willing to share power or let another dominate a coalition government. So the central government created by the communists dissolved, and a multisided civil war continued, as factions battled for control of Kabul and the countryside. In the early 1990s, mujahideen battles for control of Kabul resulted in the death of 50,000 people, mostly civilians, and reduced much of the city to rubble.[24]

RISE OF THE TALIBAN AND AL QAEDA

In 1994, a new group known as the Taliban, or "Religious Students," joined the fighting. They captured Kabul in 1996 and by 1999 ruled most of the country. Only a small area controlled by the Northern Alliance remained outside their control.

During the war against Soviet occupation, millions of Afghan refugees fled into the borderlands of western Pakistan. There, representatives of the Deobandi movement established Islamic religious schools, or madrassas, for male youths, providing the only education available to refugees.[25] Deobandi teachers offered an extremely conservative form of religious instruction, which originated in nineteenth-century India and rejected "all forms of itihad, the use of reason to create innovations in sharia [Islamic law] in response to new conditions."[26]

In these religious schools, Deobandi teachers trained a generation of young Afghan males, who came of age in the 1980s and early 1990s. After the communist regime fell, many returned to Afghanistan, to southern cities like Kandahar, and began organizing a political-military organization, consisting primarily of Pashtuns, under the leadership of Mullah Muhammad Omar. In 1994, they took up arms to end the internecine fighting and establish a regime based on strict adherence to Deobandi religious tenets.

The Taliban, as they became known, were supported in their campaign by Pakistan, which provided aid and military advisors to the movement; by Saudi Arabia, which provided money; and by opium trafficking, which provided them with money to purchase arms and food. They were also assisted by Osama bin Laden, who contributed money, arms, and soldiers drawn from countries outside Afghanistan and trained by his organization, which he called Al Qaeda, or "Military Base."[27]

THE TALIBAN, WOMEN, AND OPIUM

When the Taliban defeated their mujahideen rivals or won them over with bribes, they imposed strict Islamic laws, which were extremely hostile to women. Taliban soldiers literally drove women from public life, prohibiting women from even appearing in public—whether to work, shop, visit relatives, or seek medical care—unless fully covered or veiled and accompanied by a male relative. For the many women who lost husbands and sons in Afghanistan's long wars, or had become separated from male relatives, these laws were catastrophic, making it nearly impossible for them to work, feed their families, secure an education, or survive. Women who broke these dictates were publicly flogged by Taliban vigilantes.[28]

Under Taliban rule, opium production flourished, after having been severely reduced by the wars. By 2000, areas under Taliban rule produced 97 percent of Afghanistan's crop and about 75 percent of the world's total crop.[29] Although the Taliban banned opium cultivation in 2000–2001, they continued to sell from huge opium stockpiles and profited from the higher prices they commanded. Most experts think that the ban was essentially an embargo designed to force up opium prices.[30]

Not only did the Taliban oppress women and profit from opium, they also allied with bin Laden and allowed Al Qaeda to train terrorists for wars in Kashmir, where they joined Pakistani and Kashmiri insurgents fighting against India (see chapter 8). "It is also true that some [Taliban and Al Qaeda] Afghans are fighting against the Indian occupation forces in Kashmir," Mullah Omar admitted in 1998, "but these Afghans have gone on their own."[31]

Bin Laden, one of fifty-seven children sired by his Yemeni father, joined the mujahideen during the war against Soviet occupation and used money from his family's construction company to build mujahideen bases in Pakistan and, later, the Khost tunnel complex in the Afghan mountains.[32] He organized Al Qaeda in 1989 and later allied it with the Taliban. Under its protection, he trained terrorists to launch attacks on U.S. targets outside Afghanistan—the 1998 bombing of U.S. embassies in Kenya and Tanzania, which killed 220 people, being the most deadly.[33]

There are three ironies here. The first is that the Taliban and its Al Qaeda ally were supported, financed, and assisted by Pakistan, America's principal ally in the region. Pakistani officials supported the Taliban and deployed Al Qaeda fighters in Kashmir because they wanted a friendly regime in Afghanistan and allies in their conflict with India. In a sense, we have come full circle. The origins of the attack on 9/11—which was conducted by Al Qaeda, supported by the Taliban, and assisted by Pakistan, a major U.S. ally—can be traced back to the partition of India and to the problems related to partition in the subcontinent, Kashmir chief among them.

There is no evidence that Pakistani intelligence officers knew of or encouraged Al Qaeda to attack the United States on 9/11. But they certainly turned a blind eye to Al Qaeda's agenda, which bin Laden made quite explicit. In 1998, three years before 9/11, Al Qaeda issued the following call to Muslims around the world:

> The ruling [fatwa] to kill the Americans and their allies—civilians and military—is an individual duty for every Muslim who can do it in any country in which it is possible to do. . . . We—with God's help—call on every Muslim who believes in God and wishes to be rewarded to comply with God's order to kill Americans and plunder their money wherever and whenever they find it.[34]

The second irony is that Iran, one of America's main enemies since the revolution in 1979, was the only country in the region to support the Northern Alliance, the mujahideen still fighting against the Taliban/Al Qaeda in Afghanistan during the late 1990s and early 2000s.[35] Iran supported the Northern Alliance because Taliban theology was extremely hostile to the Shiite ideology of Iranian revolutionaries and because Taliban forces had attacked and murdered members of the Iranian consulate in Afghanistan. After 9/11, the United States would turn to the Northern Alliance, which survived because of Iranian assistance, and, with its help, defeat the Taliban and Al Qaeda in Afghanistan. Iran, a longtime foe of the United States, assisted the United States in the aftermath of 9/11; while Pakistan, a longtime U.S. ally, indirectly contributed to attacks on the United States.

The third irony is that Iran and Pakistan, though rivals in Afghanistan, were allies in the proliferation of nuclear weapons. Pakistani officials, having developed nuclear weapons to confront India, secretly transferred nuclear designs to Iran during the 1990s, helping Iran develop nuclear weapons that might deter both Israeli and Iraqi military threats.[36] This development, of course, has to be seen in the context both of conflicts related to partition in Palestine (see chapter 9), and of revolution and wars in the Persian Gulf (see chapter 10).

The attacks of 9/11 can be seen as the product of complex but related developments across the Middle East, developments related to partition in India and Palestine, revolution and war in the Gulf, and revolution and war in Afghanistan.

NOTES

1. Thomas Barfield, "The Afghan Morass," *Current History* (January 1996), 39; Barnett R. Rubin, *The Search for Peace in Afghanistan: From Buffer State to Failed State* (New Haven, Conn.: Yale University Press, 1995), 22.

2. Ralph H. Magnus and Eden Naby, *Afghanistan: Mullah, Marx, and Mujahid* (Boulder, Colo.: Westview, 2002), 59.

3. Magnus and Naby, *Afghanistan*, 111; Barfield, "Afghan Morass," 39.

4. Magnus and Naby, *Afghanistan*, 105.

5. Magnus and Naby, *Afghanistan*, 121.

6. Anthony Arnold, "The Ephemeral Elite: The Failure of Socialist Afghanistan," in *The Politics of Social Transformation in Afghanistan, Iran and Pakistan*, ed. Myron Weiner and Ali Banuazizi (Syracuse, N.Y.: Syracuse University Press, 1994), 42.

7. Arnold, "Ephemeral Elite," 40–41, 46–47.

8. Arnold, "Ephemeral Elite," 47; Barnett R. Rubin, *The Fragmentation of Afghanistan: State Formation and Collapse in the International System*, 2nd ed. (New Haven, Conn.: Yale University Press, 2002), 120.

9. Peter W. Rodman, *More Precious than Peace: The Cold War and the Struggle for the Third World* (New York: Scribner, 1994), 205.

10. Shahid Javed Burk, "The State and the Political Economy of Redistribution in Pakistan," in *The Politics of Social Transformation in Afghanistan, Iran and Pakistan*, ed. Myron Weiner and Ali Banuazizi (Syracuse, N.Y.: Syracuse University Press, 1994), 280.

11. Rubin, *Fragmentation of Afghanistan*, 196; Burk, "The State and the Political Economy of Redistribution," 280.

12. Rubin, *Fragmentation of Afghanistan*, 201.

13. Rubin, *Search for Peace in Afghanistan*, 29, 63; Rodman, *More Precious than Peace*, 206.

14. Rubin, *Search for Peace in Afghanistan*, 63.

15. Barfield, "Afghan Morass," 40.

16. Rodman, *More Precious than Peace*, 315.

17. Rodman, *More Precious than Peace*, 315.

18. Rodman, *More Precious than Peace*, 327–28.

19. Rubin, *Search for Peace in Afghanistan*, 8, 68, 74–75.

20. Rubin, *Search for Peace in Afghanistan*, 7, 84, 91.

21. Fred Halliday, *Two Hours That Shook the World: September 11, 2001; Causes and Consequences* (London: Saqi Books, 2002), 37–38.

22. Rubin, *Fragmentation of Afghanistan*, 147, 170, 265.

23. Rubin, *Fragmentation of Afghanistan*, 273.

24. Larry P. Goodson, *Afghanistan's Endless War: State Failure, Regional Politics, and the Rise of the Taliban* (Seattle: University of Washington Press, 2001), 75.

25. Goodson, *Afghanistan's Endless War*, 76; Barnett R. Rubin, "Afghanistan under the Taliban," *Current History* (February 1999), 80.

26. Rubin, "Afghanistan under the Taliban," 82.

27. Rubin, *Fragmentation of Afghanistan*, xvi; Ahmed Rashid, *Taliban: Militant Islam, Oil and Fundamentalism in Central Asia* (New Haven, Conn.: Yale University Press, 2000), 132.

28. Goodson, *Afghanistan's Endless War*, 162; Rubin, *Fragmentation of Afghanistan*, xvii; Rubin, "Afghanistan under the Taliban," 79–80.

29. Rubin, *Fragmentation of Afghanistan*, xxiii.

30. Rubin, *Fragmentation of Afghanistan*, xxiii.

31. Rashid, *Taliban*, 186.
32. Rashid, *Taliban*, 131–32.
33. Rashid, *Taliban*, 134–35.
34. Halliday, *Two Hours That Shook the World*, 218–19.
35. Rubin, "Afghanistan under the Taliban," 86.
36. Douglas Jehl, "C.I.A. Says Pakistanis Gave Iran Nuclear Aid," *New York Times*, 24 November 2004, 2.

12

☙

The Aftermath of 9/11

T he attacks on September 11, 2001, had important consequences for people in the United States and around the world.

In global terms, 9/11 had a number of important political and economic consequences. First, 9/11 joined together separate problems across the Middle East: conflicts related to partition in India and Palestine and revolutions and wars in the Gulf and Afghanistan. Before 9/11, these conflicts were treated as distinct problems. Today, they are seen as inextricably linked.[1] Because they are now understood as global problems, it could be said that 9/11 contributed to the "globalization" of conflict in the Middle East.

Second, in regional terms, 9/11 prompted U.S. invasions of Afghanistan and Iraq. The Bush administration's "war on terror" also led to the intensification of already existing conflicts between Israelis and Palestinians in the West Bank and Gaza, and between Indians, Pakistanis, and Muslim insurgents in the Kashmir. Moreover, government officials in Sri Lanka, Russia, Colombia, and the Philippines used the war on terror to justify a crackdown on domestic insurgencies that had begun years earlier.

Third, in economic terms, 9/11 and the wars that followed contributed to recession and, more recently, a new housing crisis. To appreciate the immediate and possible long-term consequences of these developments, we will examine each of them briefly.

INVASION OF AFGHANISTAN

After 9/11, the United States moved quickly to attack Al Qaeda and its Taliban sponsors in Afghanistan. On October 7, 2001, U.S. forces began bombing Al Qaeda–Taliban targets, forged an alliance with mujahideen fighters in the Northern Alliance, and then conducted a joint military campaign to overthrow the Taliban and destroy or capture members of Al Qaeda. Kabul fell to U.S. forces and their allies on November 13, Kandahar on December 6.[2] The United States and its allies then established an interim government under the leadership of Hamid Karzai, who was subsequently elected president in democratic elections conducted in the fall of 2004. This was a remarkable development, given the fractious character of politics in Afghanistan.

Although the U.S. invasion resulted in the quick overthrow of the Taliban regime and, after a time, the creation of a civilian, democratic government in Kabul, fighting between U.S. and government forces and the remnants of Taliban and Al Qaeda militias has continued. The government's authority does not extend effectively outside Kabul, where mujahideen warlords and their militia hold sway.

Afghanistan was so wrecked and impoverished by war during the last twenty years that it is "no longer even listed in the tables of the World Development Report published yearly by the United Nations because it has no national institutions capable of compiling such data."[3]

Given its economic distress, it is not surprising that opium production increased after the Taliban's fall, so that "87 percent of the world's opium is produced in Afghanistan," and opium now provides 60 percent of the country's gross domestic product.[4]

Osama bin Laden and other Al Qaeda leaders slipped away and have eluded determined efforts to capture or kill them. Many of their supporters fled to Pakistan, where they found protection amid a sympathetic populace. There they have launched attacks against Pakistan's military government, which turned against them and now supports the U.S. war against their former partners.

The events of 9/11 created a particular problem for Pakistan's government, which had supported the Taliban in Afghanistan and Muslim insurgents, assisted by Al Qaeda, in Kashmir.[5] Pakistani officials were forced to abandon their allies in Afghanistan, but were reluctant to do so in Kashmir. And by abandoning the Taliban, they antagonized Pakistanis and Afghan refugees who supported them, and the religious institutions—madrassas—that nurtured them. This has led to attacks on the government by Taliban-like forces in Pakistan.

INVASION AND OCCUPATION IN IRAQ

After conducting a successful campaign against the Taliban in Afghanistan, the Bush administration moved to confront Saddam Hussein's Baathist regime in Iraq, launching an invasion of Iraq in April 2003. Before the invasion, President Bush and administration officials argued that Hussein had supported the Al Qaeda attacks on 9/11, had developed chemical and biological weapons of mass destruction, and had worked to acquire nuclear weapons, which threatened other countries in the Gulf and posed an imminent threat to U.S. interests in the region. U.S. officials argued that Hussein's brutal dictatorship was unpopular and that U.S. forces would be welcomed by Iraqis as liberators.

But U.S. officials misrepresented the various rationales for war and were mistaken about its consequences. In fact, Hussein's regime did not have ties with Al Qaeda and did not assist the attacks of 9/11. It no longer possessed chemical or biological weapons, and had not developed any significant nuclear capability. Although the administration argued that the invasion was prompted by the events of 9/11, U.S. officials had actually begun preparing for war in the summer before 9/11.

In a matter of weeks, U.S. and allied forces, chiefly British, destroyed and dissolved Iraq's conventional army, deposed the Baathist regime, and occupied the country. But several problems quickly emerged, and the invasion became, in President Bush's words, a "catastrophic success."[6]

First, U.S. and allied forces were not prepared, or numerous enough, to prevent widespread looting, chaos, and criminal violence across the country, which resulted in the destruction of government offices, private industry, schools, and public infrastructure (electricity, water, and sewage treatment). Administration officials had insisted that the war could be fought and won with a relatively small army (140,000 troops) and that the bulk of it would soon be withdrawn, leaving only 30,000 troops to occupy Iraq by September 2003.[7]

But while the army was large enough to defeat conventional armies in the field, it was not sufficient to police the large civilian population in the absence of a functioning Iraqi bureaucracy. "We never had enough troops on the ground," U.S. ambassador L. Paul Bremer III, the chief administrator in Iraq after the war, later admitted.[8]

How many U.S. troops would have been needed? The Pentagon estimated that if it used "the same ratio of peacekeepers to population as it had in Kosovo, the United States would have to station 480,000 troops in Iraq. If Bosnia was used as the benchmark, 364,000 troops would be needed."[9] The Bush administration decided to make do with less than half of this number.

Second, the U.S. occupation authority ordered the demobilization of Iraqi military and security services, and closed the Iraqi Defense Ministry, putting an estimated 450,000 Iraqi soldiers and police out of work.[10] Moreover, they ordered a purge of Baath Party members from public institutions, from positions of authority in government bureaucracies, schools, hospitals, and businesses. This resulted in the loss of jobs for tens of thousands more.

Recall that Hussein and the Baath Party provided some 450,000 government jobs to supporters and another 450,000 jobs in the military and security forces, filling most of these positions with members of the Sunni community west of Baghdad—the so-called Sunni Triangle. The decision to terminate government jobs for Baath Party members and supporters triggered widespread protest by this group and laid the ground for the insurgency that soon emerged. Although it soon took on a religious character, the insurgency was, at bottom, an economic revolt by unemployed supporters of the previous regime. Using weapons collected from the battlefield or seized from arsenals during the looting, Sunni insurgents began attacking U.S. and allied forces; civilian contractors working with the occupation government; nonprofit aid agencies, including the United Nations; and Iraqis—Shiites and Kurds—who opposed them. Recall that the Baath regime fought to suppress a revolt by Kurds in the 1960s and early 1970s and suppressed Shiite uprisings after the Second Gulf War.

The Sunni insurrection against U.S. occupation authority and the civil war among different Iraqi groups—Sunni, Shiite, and Kurd—intensified after conventional warfare ended. U.S. forces have been forced to retake cities first captured during the conventional war. And more U.S. troops died during the insurrection than were killed in conventional fighting during the war.

Third, the war may contribute to the breakup or partition of Iraq. The war exacerbated conflicts among Sunni, Shiite, and Kurdish populations, and much of the current fighting can be characterized as a three-way struggle for political power in post-Hussein Iraq. If this conflict persists, parties to the conflict may advance the partition of the country along ethnic-religious lines. This has already occurred, to some extent, with Kurds demanding and receiving promises granting them considerable political autonomy in the North. Sunni insurgents are fighting, in part, to protect the political power they had under Hussein, while the Shiite majority wants to claim power, which was long denied, in a united country. Some political analysts have urged partition as a way to solve ethnic-religious conflicts in Iraq.[11] Partition in Iraq might take the form that it did in India, with the creation of separate, sovereign states; the form it did in Palestine, with the creation of one sovereign state and another, nonsovereign entity (the West Bank and Gaza); or the form it more recently took in Bosnia, where Serb and Bosnian enclaves are minimally joined under the authority of an outside administrator supported by "peacekeeping" forces from other countries. In any event, recall that partition in other settings did not solve the problems its sponsors imagined, but instead created conflicts that resulted in ongoing wars and insurgencies.

In the fall of 2007, the Bush administration organized a "surge" of U.S. forces in Iraq to combat the insurgency and curb the violence associated with multi-sided civil war. During the next twelve months, the surge reduced violence in Iraq, though it did not defeat the insurgents or end the civil war. It is not clear why the Bush administration waited so long to increase the number of troops devoted to the fighting. But it is clear that the surge cannot be sustained, and U.S. commanders have said they will have to reduce the number of troops devoted to the surge in early 2009. After that, security may again deteriorate, unless the Iraqi government can deploy effective Iraqi forces to replace U.S. and coalition forces and organize an effective administration of the country on its own.

Fourth, the war and insurrection crippled Iraqi oil production. Before the war, Iraq produced about two million barrels a day. It could have produced more, but UN sanctions against Iraq, which were supported by the United States, restricted the production and sale of oil. Although the seizure of Iraqi oil was not used by the Bush administration as a rationale for war, oil did figure prominently in their postwar plans. The Bush administration wanted to wage a quick, decisive war, relying on a relatively small number of troops (see above), in part to minimize damage to Iraqi oil facilities. Once the war was over, they hoped to increase Iraqi oil production substantially, from two to four, perhaps even six million barrels a day.

Increased Iraqi oil production would provide the money necessary to repay Iraq's debts (which dated from the First Gulf War; see chapter 10) and rebuild infrastructure damaged by the fighting. It would also increase world oil supplies and help keep prices low. To some extent, increased oil production would transform Iraq from a price "hawk" to a price "dove." During the first two Gulf wars, Hussein had tried to corner supplies in Iran and then in Kuwait, and use control of their supplies to drive prices up. During the Third Gulf War, the Bush administration hoped to do just the opposite: gain control of Iraqi oil supplies, increase production, and drive prices down.

But the Bush administration's strategy quickly foundered. Although Iraqi oil facilities were not damaged by the war, the looting, chaos, and then insurgent attacks on oil pipelines reduced oil production by 95 percent, and U.S. contractors were forced to import oil to meet domestic demand in Iraq. Oil production slowly recovered, but experts do not expect it to increase substantially anytime soon.[12]

The decline of oil production in Iraq had important domestic and global consequences. First, it meant that Iraqi oil revenues could not be used to repay debt or finance reconstruction and development. So the United States has had to provide billions of dollars in aid to Iraq to rebuild the country. U.S. officials finally persuaded creditors in Europe, Russia, and

Kuwait to write off many of their loans to Iraq, but that did not provide new funding for reconstruction or development. A 2002 study found that the war in Iraq "could cost the United States as much as $1.9 trillion including lost economic output" over a ten-year period.[13]

Second, the decline of Iraqi oil production came at a time when world supplies were already tight. Political and economic turmoil in Venezuela, Russia, and Nigeria reduced world oil supplies during this period. Inadequate refinery capacity in the United States made it difficult to meet growing demand in the United States. (Refineries are designed to use certain kinds of oil. So if the supply of one kind is restricted, say from Venezuela, it is difficult to switch quickly to oil from a different source.) And rising demand for oil in China has helped increase world demand (see chapter 7). As a result, oil prices went up, not down, climbing from $30 a barrel before the war to more than $50 a barrel in the fall of 2004 and $140 a barrel in 2008.

Rising oil prices have contributed both to inflation and stagnation (see chapter 3). High prices have already contributed to an economic slowdown in countries like India and Japan, which are heavily dependent on imported oil. If rising oil prices are combined with rising food prices, which might climb as a result of growing demand in China (see chapter 7), then the kind of problems encountered during the 1970s might return. But rising prices might also speed the development of alternative fuel technologies and energy conservation, which would probably be a good thing. The particular consequences depend on whether high prices persist or whether they retreat. What is clear, however, is that the world's dependency on oil from the Gulf will only increase, because three-fifths of all proven oil reserves are to be found there. This means that developments in Iraq will no doubt play an important role in determining oil prices in the future.

THE WAR ON TERROR

After 9/11, government officials in many countries announced they would support the Bush administration's call for a global war on terror. Officials in some countries used the war on terror to justify a crackdown on domestic rebels, not foreign terrorists, appealing to the United States to support their efforts. So, for example, Russian officials announced a new campaign against Muslim insurgents in Chechnya; Colombian officials broke off talks and renewed fighting with Marxist insurgents in the countryside; Filipino officials requested U.S. military assistance so they could defeat Muslim rebels; and Sri Lankan officials broke off negotiations and hardened their stance against Tamil separatists.

It is important to note that the rebels described by these governments as "terrorists" had no connection with Al Qaeda and that their revolts had grown out of domestic quarrels with their governments, which had begun years before 9/11. These old, domestic insurgencies were described, after 9/11, as part of a new kind of global terrorism. Before 9/11, the governments waging war against these groups were subjected to considerable criticism, by U.S. officials among others, for violating human rights and rejecting peaceful, negotiated solutions to conflict. After 9/11, they were encouraged, by the United States and others, to pursue their counter-insurgency campaigns to a successful conclusion. As a result, 9/11 led to an intensification of domestic conflict and civil war in many settings. This was particularly true in India and in Israel and the occupied territories, where governments used 9/11 to justify suppressing ongoing insurgencies or *intifadas*.

INDIA

Since 1989, the government in India has been battling Muslim insurgents in Kashmir, a region where disputes with Pakistan triggered several wars (see chapter 8). After 9/11, Indian officials increased military efforts to crush the insurgents, arguing that they were linked to Al Qaeda and supported by Pakistan. This was, in fact, true. They hoped that the events of 9/11 and the advent of a war on terror would weaken U.S. support for Pakistan, make it more difficult for Pakistan to aid rebels in Kashmir, win sympathy for India's claim to Kashmir, and persuade U.S. officials to adopt a more friendly diplomatic attitude toward India.

Attacks by Kashmiri insurgents on parliaments in Kashmir and in India further intensified the conflict in the fall of 2001. India mobilized troops along its border with Pakistan. Pakistan did the same, setting the stage for a fourth Indo-Pakistani war.[14] Because both countries now possess nuclear weapons, U.S. officials moved quickly to defuse the crisis. Although they backed down, Indian officials achieved some of their objectives. They got U.S. officials to restrain Pakistan and view Indian claims more favorably, an important shift in the diplomatic balance of power in the region. The insurgency in Kashmir continued, but it was more isolated, in diplomatic and military terms, than it was before 9/11.

ISRAEL

As in Kashmir, the insurgency or *intifada* in Israeli-occupied territories began before 9/11 (see chapter 9). But Israeli prime minister Ariel Sharon used 9/11 as an opportunity to intensify his government's military efforts to suppress Palestinian rebellion in the West Bank and Gaza, undermine the Palestinian Authority and its leader, Yasir Arafat, and win U.S. support for Israel's "new" war on "terror."[15]

In March 2002, the Israeli army reoccupied cities and refugee camps in the West Bank and besieged Arafat at his compound in Ramallah.[16] The Israeli government "closed" much of the occupied territories and accelerated construction of a 24-foot-tall, 400-mile-long, concrete security barrier. This huge wall, four times longer than the Berlin Wall, which serpentines through the West Bank, will totally surround many Palestinian towns, put 14.5 percent of West Bank land, "some of it the most fertile," onto the "Israeli" side, and disrupt communication and travel for Palestinians divided by the barrier.[17] In the West Bank, the construction of the wall and the re-enforcement of long-standing occupation policies resulted in a situation that approximates a prison, a walled ghetto, an American Indian reservation, and a South African Bantustan under apartheid.

Since 9/11, fighting has intensified. The Israeli air force has routinely used jet fighters to launch air strikes in the occupied territories; Palestinian militants have regularly used suicide bombers, homemade missiles, and small arms to attack Israelis in the occupied territories and in Israel. More than 100 Israelis died in attacks in Israel between March and May 2002; 497 Palestinians were killed in the same period.[18] As part of its military campaign in this period, "the Israeli government demolished 1,601 Palestinian homes and damaged 14,436 more, affecting 96,100 people," according to the U.S. Agency for International Development.[19] And the World Bank estimated that 13,000 acres of agricultural land in the West Bank and Gaza were destroyed or seized.[20]

As a result of these developments, economic conditions in the West Bank and Gaza deteriorated significantly. The World Bank estimated that the Palestinian economy fell by half since 2000, when the second *intifada* began, and 50 percent of the 2.2 million Palestinians in the West Bank live below the poverty line, 68 percent in Gaza, which is home to 1.3 million people.[21]

Before 9/11, the Bush administration had announced its support for a Palestinian state, the first U.S. administration to do so. But after 9/11, it retreated from this position, criticized Palestinians for obstructing peace, and muted previous criticisms of Israel's occupation policies. With the death of Palestinian leader Yasir Arafat in November 2004, U.S. and Israeli leaders indicated that they might resume the peace process, but on Israel's terms. As a result, the Palestinian negotiating position has weakened. As Ritchie Ovendale, a British expert on the Arab-Israeli wars, observed, "In 1947, the United Nations vote had awarded the Palestinians 47 percent of their 'homeland'; the Oslo Accords of 1993, with the offer of the West Bank and Gaza, offered 22 percent of the 'homeland'; at Camp David in 2000 the Palestinians were offered 80 percent of the 22 percent of the 100 percent of their original homeland."[22] And if the 14.5 percent of West Bank lands that were placed on the Israeli side of the new wall are taken off the bargaining table, as many expect, the Palestinians would be allowed to negotiate for even less. Given these circumstances, which deepen animosity and intensify conflict, it would not be surprising if the Israeli-Palestinian conflict became even more obdurate and intractable.

Taken together, 9/11 had important political and military consequences. It led to wars in Afghanistan and Iraq. In Afghanistan, invasion brought an end to civil war; in Iraq, it triggered civil war. Moreover, 9/11 and the global war on terror intensified conflicts between governments and domestic insurgencies in a number of countries. In India, conflict very nearly led to a fourth Indo-Pakistani war, while in Israel, it deepened conflict, shredded what remained of the Oslo peace process, and impoverished Palestinians living under Israeli military rule.

RECESSION AND 9/11

The events of 9/11 also had important economic repercussions in the United States and around the world. In the United States, the economy had slowed and stock prices had fallen—largely a result of corporate scandals and the collapse of the dot-com boom—prior to 9/11. Recession in the United States slowed economic growth in Western Europe and prevented the Japanese economy, already in the doldrums, from recovering (see chapter 1).

Although the events of 9/11 did not trigger recession, they deepened and prolonged it. To prevent the recession from deepening, Alan Greenspan and the Federal Reserve dramatically lowered interest rates. The Fed's policy made it possible for banks to offer low-interest loans to low-income households who had previously been unable to borrow money and purchase homes, either because they had little savings (and could not make a substantial down payment on a home) or poor credit histories. These "sub-prime" loans substantially increased the demand for houses and helped increase housing prices across the country.

The housing boom provided jobs for domestic and immigrant workers in the construction industry (see chapter 5), which reduced unemployment rates, and increased the rate of home ownership, which had been falling. The rising price of homes increased property-tax assessments and provided new revenue for local governments and schools. During the mid-2000s, the housing boom helped the United States weather the economic storms associated with 9/11. But in 2008, the widespread practice of making sub-prime loans triggered a major financial crisis.

When they made sub-prime loans to low-income households, bank managers offered mortgages that had low initial interest rates. But the interest rate on many loans could rise sharply if the homeowner paid the mortgage bill late or missed a payment; if the Federal Reserve raised interest rates on an adjustable-rate mortgage; or if the loan was structured to charge a low rate of interest when people were starting out and then increase or "balloon" after a few years, when the household earned more money and could afford to pay a higher rate. The problem was that rising interest rates and bigger mortgage payments caused a "debt crisis" for many households, much as they did for Latin American countries during the 1980s (see chapter 4). As a result, many borrowers were forced to default on their loans. But widespread foreclosure sales increased the number of homes on the market. The supply of homes outstripped demand, housing prices fell dramatically, and the housing "bubble" burst.

For banks, falling prices meant they could not recover the value of their loans, and they lost money. In earlier years, this would have led to the collapse of banks that made bad loans, as it did in the 1980s when the savings and loans went under (see chapter 3). But in recent years, large financial institutions—major commercial banks, investment banks and brokerage houses, and mortgage institutions—had purchased sub-prime mortgages from local banks because they were profitable and because housing prices were rising (so if a loan went bad, the bank could recover its costs). But as housing prices fell, the financial firms that had purchased these loans could not recover their costs; they lost money and were threatened with ruin. The bankruptcy of large firms could have wrecked the U.S. economy and caused financial havoc throughout the world. But the Fed moved quickly to reorganize money-losing banks and brokers and provide them with the financial resources and guarantees they needed to stay in business. Still, the collapse of the housing boom brought the construction industry to a standstill, which resulted in rising unemployment, and the fall in home prices reduced revenue for cities, school districts, and states. Both of these developments contributed to an economic recession in the United States.

But two other problems remain that could also affect the economic climate: U.S. trade deficits and U.S. budget deficits.

The United States has been running trade deficits since 1971 (see chapter 2). But these deficits increased dramatically during the 1990s, largely as a result of increased trade with China (see chapter 7). As a result, the value of the dollar has fallen sharply in relation to the euro, but only a little against the yen, and not at all against the Chinese yuan, which is set at a fixed rate by the Chinese dictatorship.

Asian governments have prevented the dollar from falling in relation to the yen and the yuan by purchasing U.S. Treasuries, amounting to $900 billion in November 2004.[23] If the dollar devalued against the yen and the yuan, Japanese and Chinese businesses would find it more difficult to sell their goods in the United States, a development that would undermine economic growth and increase unemployment in both countries. But because the dollar has devalued against the euro, the real value of the Japanese and Chinese hoard of U.S. Treasuries has declined. In late 2004, their purchases of U.S. Treasuries slowed and there was some indication that they might sell off U.S. bonds or start buying euros. If they did, this could lead to a rapid and large devaluation of the dollar.

A dollar devaluation in Asia would be good for U.S. businesses that compete with firms based in China and Japan (see chapter 7). A devaluation would increase U.S. exports and, perhaps, reduce U.S. trade deficits. Of course, it would hurt businesses based in Asia, slow economic growth, and increase unemployment there.

Under normal circumstances, a dollar devaluation would probably help the U.S. economy. But remember that the Bush administration cut taxes and started running huge and growing budget deficits. If governments in China and Japan stopped buying U.S. Treasuries, which are used to finance U.S. budget deficits (money from the sale of treasuries is used to pay the U.S. government's bills), or started buying euros instead, the Federal Reserve would have to raise U.S. interest rates to persuade public and private investors to purchase U.S. Treasuries so the government could pay its bills.[24] But rising interest rates would slow economic growth and increase unemployment in the United States, much as occurred under Paul Volcker in the early 1980s (see chapter 3). A recession, in turn, would reduce tax revenues (government tax revenues fall when the economy slows and unemployment rises) and increase budget deficits. This would increase the amount of money the government has to raise, pressuring the Fed to raise interest rates even higher. Higher interest rates would slow the economy and increase unemployment even further.

In global terms, a serious recession in the United States would slow economic growth in Asia and Western Europe, where businesses have come to depend on the sale of goods to U.S. consumers. To prevent this, the government could raise taxes and/or curb spending, and cut budget deficits, as the Clinton administration did in the 1990s. This approach is anathema to most politicians in either party. But the way these problems are handled may well determine the global economic climate in coming years.

NOTES

1. Myron Weiner and Ali Banuazizi, "Introduction," in *The Politics of Social Transformation in Afghanistan, Iran, and Pakistan*, ed. Myron Weiner and Ali Banuazizi (Syracuse, N.Y.: Syracuse University Press, 1994), 1.

2. Barnett R. Rubin, *The Fragmentation of Afghanistan: State Formation and Collapse in the International System*, 2nd ed. (New Haven, Conn.: Yale University Press, 2002), xxxii.

3. Barnett R. Rubin, "A Blueprint for Afghanistan," *Current History* (April 2002), 153.

4. Carlotta Gall, "Afghan Poppy Growing Reaches Record Level, U.N. Says," *New York Times*, 19 November 2004.

5. Hooman Peimani, *Falling Terrorism and Rising Conflicts: The Afghan "Contribution" to Polarization and Confrontation in West and South Asia* (Westport, Conn.: Praeger, 2003), 71.

6. Michael R. Gordon, "The Strategy to Secure Iraq Did Not Foresee a 2nd War," *New York Times*, 19 October 2004.

7. Gordon, "Strategy to Secure Iraq."

8. Elisabeth Bumiller and Jodi Wilgoven, "Bremer Critique Raises Furor," *New York Times*, 6 October 2004.

9. Gordon, "Strategy to Secure Iraq."

10. Judith S. Yaphe, "Reclaiming Iraq from the Baathists," *Current History* (January 2004), 13; Michael R. Gordon, "Debate Lingering on Decision to Dissolve the Iraqi Military," *New York Times*, 21 October 2004.

11. Edward Wong, "Making Compromises to Keep a Country Whole," *New York Times*, 4 January 2004.

12. Adriana Lins de Albuquerque, Michael O'Hanlon, and Amy Unikewicz, "The State of Iraq: An Update," *New York Times*, 26 November 2004.

13. Anna Bernasek, "Counting the Hidden Costs of War," *New York Times*, 24 October 2004.

14. Alexander Evans, "India, Pakistan, and the Prospect of War," *Current History* (April 2002), 160–64; Peimani, *Falling Terrorism and Rising Conflicts*, 74.

15. Ritchie Ovendale, *The Origins of the Arab-Israeli Wars*, 4th ed. (Harlow, U.K.: Pearson Education, 2004), 312–13.

16. Ovendale, *The Origins of the Arab-Israeli Wars*, 315.

17. Greg Myre, "U.N. Estimates Israeli Barrier Will Disrupt Lives of 600,000," *New York Times*, 12 November 2003.

18. Ovendale, *The Origins of the Arab-Israeli Wars*, 317.

19. Sara Roy, "The Palestinian State: Division and Despair," *Current History* (January 2004), 35.

20. Roy, "The Palestinian State," 35.

21. Judith Miller, "Yasir Arafat, Palestinian Leader and Mideast Provocateur, Is Dead at 75," *New York Times*, 12 November 2004; Roy, "The Palestinian State," 32.

22. Ovendale, *The Origins of the Arab-Israeli Wars*, 322.

23. Edmund L. Andrews, "Foreign Interest Appears to Flag as Dollar Falls," *New York Times*, 27 November 2004.

24. Andrews, "Foreign Interest Appears to Flag."

13

❧

Global Climate Change

On June 23, 1988, NASA scientist James Hansen told Congress he was "99 percent certain" that global warming had begun. "It is time to stop waffling so much and say that the evidence is pretty strong that the greenhouse effect is here," he argued.[1]

Scientists first warned of global warming a century ago. Nobel laureate scientist Svante Arrhenius argued in 1896 that increasing levels of carbon dioxide in the atmosphere would raise its temperature.[2] Since then, the amount of carbon dioxide in the atmosphere has increased about 27 percent, from 280 parts per million to 356 parts per million, as have other important heat-trapping gases such as methane, chlorofluorocarbons (CFCs), and nitrous oxide.[3] Hansen and others have argued that these gases, which were largely a by-product of human activity, retained heat from the sun, creating a "greenhouse effect" that has warmed the planet. The Earth's temperature is about 1 degree Fahrenheit hotter today than it was a century ago.[4] And the UN's Intergovernmental Panel on Climate Change (IPCC), a scientific task force first assembled in 1988 to assess the problem, predicts that global temperatures will increase between 1.4 and 6.3 degrees during the next century unless drastic steps are taken to curb greenhouse gas emissions.[5] If this occurs, the IPCC reported, "the rate of change is likely to be greater than that which has occurred on Earth any time since the end of the last ice age."[6]

Rapidly rising temperatures could create serious problems for people in different settings, scientists argue. Rising temperatures could melt polar ice—as it is, the North Pole is free of ice certain times of the year—and raise sea levels, inundating islands and low-lying coastal plains where millions live. A 1-meter rise would flood deltas on the Nile, Po, Ganges, Mekong, and Mississippi rivers, displacing millions of people and swamping the croplands now used to feed them.[7] Higher sea levels could drown coral reefs, destroying the fish and ruining the livelihood of people who depend on reefs in the Caribbean and the Pacific.[8] Warmer water could also increase the strength of hurricanes and typhoons, causing greater damage for people living along their path in the Western Atlantic and Western Pacific.[9] The insurance industry is particularly concerned about this prospect because windstorms caused $46 billion in losses between 1987 and 1993.[10]

Higher temperatures could also disrupt agriculture. While farmers in northern latitudes—North America and northern Europe and Asia—could benefit from higher temperatures, longer growing seasons, and higher levels of carbon dioxide (which plants use to grow), even modest increases could devastate farmers in tropical zones in Asia, Africa, and Latin America. Rice yields decline significantly if daytime temperatures exceed 95 degrees, and in many Asian countries, temperatures are already near this limit.[11] One group of scientists predicted that cereal prices could increase between 25 and 150 percent by the year 2060, a development that would cause hunger and starvation for between 60 million and 350 million poor people, most of them in the tropics.[12]

The prospect that global warming could have its most serious impact on people in relatively poor and populous tropic countries is ironic given the fact that most of the gases believed to contribute to global warming are generated by small populations in relatively rich northern countries, particularly the United States. In 1960, the United States produced one-third of all global carbon dioxide emissions and about 20 percent in 1987; its per capita annual production of 5.03 tons was nearly five times the world per capita average of 1.08 tons, but thirteen times that of Brazil and 167 times that of Zaire.[13] This irony is not lost on poor countries, which are now being asked to reduce or defer fossil fuel consumption as a way to mitigate the effects of global warming. Brazil, for example, has rejected calls to reduce burning of tropical woods in the Amazon basin. President José Sarney argued that the industrial nations were conducting "an insidious, cruel and untruthful campaign" against Brazil to distract attention from their own large-scale pollution, acid rain, and "fantastic nuclear arsenal" that threaten life.[14] Chinese officials maintain that they need to consume vast quantities of coal to provide electricity for China's growing economy (see chapter 7). At present, "the average Chinese consumes less than 650 kilowatt hours [annually], barely enough to burn a 75-watt bulb year round . . . and 120 million rural Chinese . . . live without electricity in their homes or villages."[15] But if it does expand coal-fired electricity production to light villages, run refrigerators, and power industry, as the government plans, China would overtake the United States as the world's largest producer of carbon dioxide sometime during this century.[16]

Because the environmental and social consequences of rising temperatures are not uniform, but unevenly distributed, it might be best to describe these developments as products of "global climate change" rather than "global warming," which suggests that temperature change will be experienced by people in much the same way everywhere.[17] Moreover, it might be better to use the phrase "global climate change" given the scientific uncertainties about the process and the debate surrounding claims that the world is warming significantly and that human activity is largely to blame.

Although Senator Al Gore, later vice president, asserted that "there is no longer any significant disagreement in the scientific community that the greenhouse effect is real and already occurring," global climate change remains the subject of scientific debate. While a majority of scientists share Gore's view that the greenhouse effect is demonstrable, a minority disagree, making two kinds of objections to the conclusions reached by the IPCC.[18]

First, some scientists argue that the Earth may not be warming much, if at all. S. Fred Singer, former director of the U.S. weather satellite program, notes that satellite data collected between 1979 and 1994 showed "no appreciable recent warming."[19] Other scientists agree that modest warming, about 1 degree, has occurred in the past one hundred years, but argue that this is well within the range of natural temperature change. They note that global temperatures in the 1980s, when some of the hottest years of the past century were recorded, were slightly lower than those recorded during the Middle Ages, some five hundred years ago, "when Scandinavians grew grain near the Arctic Circle," and more than 2 degrees cooler than global temperatures six thousand years ago.[20]

Second, some scientists argue that there is little proof that higher temperatures are the product of human activity and increased atmospheric concentrations of greenhouse gases. They note that global temperatures actually declined between 1940 and 1970, data that greenhouse proponents do not dispute, at a time when industry expanded and world carbon dioxide emissions tripled.[21] If human activity and atmospheric concentrations of carbon dioxide and other waste gases were closely related to temperature change, then global temperatures should have increased during this period, they maintain. The fact that they did not suggests that temperature change may be a result of natural fluctuation rather than a product of human activity.[22]

Because some scientists believe that the relation between human activity and temperature change is weak, they express little confidence in the computer models that use rising carbon dioxide levels to predict higher temperatures in the next century.[23] "I do not accept the model results [used by the IPCC and others] as evidence," Richard Lindzen of MIT argues. Trusting them, he says, "is like trusting a ouija board."[24] Lindzen and others argue that computer models of climate change do not adequately account for natural variation or environmental processes—increased cloud cover, water vapor—that may mitigate warming tendencies. "I don't think we've made the case yet" that serious climate change is now occurring, he argues.[25]

This kind of skepticism finds some support in the IPCC report, which conceded in 1992 that "it is not possible at this time to attribute all, or even a large part, of the observed global-mean warming to the enhanced greenhouse effect [the extra warming attributable to human-produced gases] on the basis of the observable data currently available."[26] More recently, in 1995, the IPCC expressed greater confidence in the relation, writing that the warming of the past century "is unlikely to be entirely due to natural causes and that a pattern of climatic response to human activity is identifiable in the climatological record."[27] But criticism still persists. Lindzen argues that because computer models do not estimate natural variability accurately, there is "no basis yet for saying that a human influence on the climate has been detected."[28]

At issue is whether drastic steps should be taken to reduce the emission of greenhouse gases. If human activity does, in fact, contribute to global warming, as a majority of scientists believe, then their call for a 50 to 60 percent cut in greenhouse gas emissions, effectively returning emissions to 1950 levels, should be heeded. But if climate change is not a product of human activity, then efforts to reduce greenhouse gas emissions would be both ineffective and socially disruptive, particularly for poor countries that are trying to use carbon sources to promote economic growth. "Poverty is already a worse killer than any foreseeable environmental distress [associated with global warming]," Lawrence Summers, former chief economist of the World Bank, has argued. "Nobody should kid themselves that they are doing Bangladesh a favor when they worry about global warming."[29]

Although scientific uncertainty makes global climate change a difficult issue, it is nonetheless possible to develop a constructive approach. If steps were taken to solve environmental problems that are serious in their own right, problems that may also contribute to global climate change, then people could address real as well as potential problems, a strategy that would minimize both social and environmental risks. There are, for example, sound environmental and social reasons to reduce energy consumption and car use and slow deforestation. Because these activities also release vast quantities of carbon dioxide, efforts to curb the consumption of fossil fuels and wood might also reduce global warming. (The carbon dioxide released by these activities accounts for about half of all greenhouse gases.) The same is true for other activities that produce other greenhouse gases. A reduction of world cattle herds would reduce hunger and deforestation, and also curb emissions of methane, which makes up about 18 percent of all greenhouse gases.[30] The ban on CFCs, which was scheduled to take effect in 2000, will slow destruction of the ozone layer, about which there is no serious scientific dispute, and reduce its contribution (about 14 percent) to global climate change. And if nitrogen fertilizer use were curbed, the problems associated with groundwater pollution could be addressed and nitrous oxide levels in the atmosphere (about 6 percent of the total) could be reduced. However, in the case of nitrous oxide, fertilizer reductions could adversely affect global food supplies and contribute to hunger, which suggests that efforts to curb fertilizer use should be approached with great caution.

In this context, it is important not just to curb human activities that adversely affect the environment and may contribute to global climate change, but to do so in ways that are equitable and sensible. Any effort to address environmental problems should take into account the social impact of proposed solutions. With these issues in mind, let us look at pressing environmental problems that might be addressed, both because they contribute to known environmental and social problems, and because they may contribute to global warming. We will look first at human activities that produce carbon dioxide, the main greenhouse gas, and then at activities that produce methane, CFCs, and nitrous oxide.

POWER, FORESTS, AND CARS (CARBON DIOXIDE)

In the early 1970s, long before global warming became a concern, environmentalists urged energy conservation. They argued that the profligate use of coal and oil to generate electricity for home and industry contributed to serious environmental and economic problems.[31] (Coal and oil are used to generate more than one-half of the electricity around the world.) Because coal often contains sulfur, combustion in power plants contributes acid rain, which ruins forests near and far.[32] Although oil burns cleaner than coal, the production of oil creates toxic waste that is hard to discard safely, and the transport of oil across oceans frequently results in spills that harm marine life.

The use of oil to produce electricity or to fuel cars also contributes to serious economic problems, environmentalists argue. Countries that import oil spend vast sums to purchase it, contributing to trade deficits and sometimes debt. For the United States, "oil imports alone have accounted for three-fourths of [its] trade deficit since 1970," and the United States has paid $1 trillion between 1970 and 1990 for imported oil.[33] For countries without substantial domestic supplies, the economic costs can be higher. During the 1970s, many Latin American countries borrowed heavily and fell deeply into debt to pay for oil imports (see chapter 4). Moreover, competition for control of oil supplies has recently led to two wars in the Persian Gulf—the 1980–1990 Iraq-Iran War and the 1990–1991 Gulf War—which forced consumer countries to increase defense spending and intervene militarily to protect supplies in the region (see chapters 10 and 12). "Even before Iraq invaded Kuwait, U.S. forces earmarked for gulf deployment were costing taxpayers around $50 billion a year—yearly $100 a barrel for oil imported from the Persian Gulf."[34] The war to dislodge Iraqi troops cost tens of billions more.

Energy conservation can solve many of these environmental and economic problems. During the 1970s, when oil prices rose, it became cost effective to curb energy use by persuading consumers to conserve and by introducing new technologies to produce and use electricity and fossil fuels more efficiently. Between 1973 and 1986, these steps—lowering thermostats, insulating homes, driving less, and improving gas mileage in cars—enabled the U.S. economy to grow without expanding energy use, a development that helped the environment and resulted in $150 billion in annual energy savings.[35] But as oil prices fell in the mid-1980s, many conservation measures were abandoned and technological improvements were deferred, resulting in growing levels of energy consumption. Energy consumption could again be curbed through consumer conservation programs and the introduction of new technologies—fuel cells, photovoltaic cells, computerized electricity distribution systems, gas-turbine engines, and solar-hydro-geothermal technologies—to create what engineers call a "nega-watt revolution": producing more energy with less.[36] Some economists have urged U.S. officials to pursue energy-efficient policies in part because they would make U.S. industry more competitive with businesses in Europe and Japan, where ongoing conservation practices have lowered the real cost of energy for domestic industry.[37] If the rich countries developed and introduced energy-saving technologies, their cost would decline and they would become more affordable for poor countries. And because the power sector accounts for one-third of global carbon dioxide emissions, energy conservation would also help avert potential warming.

Deforestation

During the early 1980s, environmentalists recognized that tropical deforestation had become a problem. Attention was first drawn to Brazil, where 33 percent of the world's tropical forests are located (58 percent are in Latin America, 23 percent in Southeast Asia and the Pacific, and 19 percent in Africa), and where deforestation was rapidly accelerating.[38]

In 1981, the United Nations estimated that 7.3 million hectares of tropical forest were burned or cut every year, and some environmental groups argued that as much as 11.3 million hectares were lost annually.[39] These figures were revised downward in the early 1990s, to about 1.5 million hectares.[40] More recent studies indicate that deforestation has accelerated again, particularly in the Amazon.[41] But whatever the precise rate, deforestation remains a serious and growing problem, especially if the vast temperate Siberian forest (the size of the continental United States) is opened to intensive cutting, which experts expect to occur as a result of the breakup of the Soviet Union and the entry of foreign companies into the Russian economy.[42]

Environmentalists have identified four problems with tropical deforestation. First, deforestation in the tropics, where rainfall is heavy—a one-hour downpour in Ghana can dump more rain than showers in London can deliver in a month—leads to rapid runoff, triggering floods and eroding soils.[43] Runaway forest soils fill streams with sediments, kill fish, and clog reservoirs behind dams, shortening their life span and reducing their ability to irrigate fields and generate electricity.[44]

Second, tropical woods do not easily recover from deforestation. Some scientists estimate that it takes tropical forests between 150 and 1,000 years to recover fully from a clearcut, in contrast to temperate woods, which can be cut more often on a sustainable basis.[45] Because tropical forests do not easily recover, deforestation is typically associated with the loss of plant and animal species. Again, scientists debate the rate of extinction, or argue that tropical rain forests are not the only or most important reservoirs of plant and animal species (one scientist argues that savannas may be more important for mammal species), but the fact is not disputed that deforestation leads to extinction in species-rich, tropical forest environments.[46] If Latin American forests were reduced to 50 percent of their original size, scientists estimate that 15 percent of forest plant species and 12 percent of bird species would be lost.[47] The problem with species loss is that it undermines the genetic base for cultivated plants and domestic species that are used by people throughout the world, and for pharmacology, which relies on biological resources for human drugs.[48] For example, scientists recently discovered a primitive corn species in a tiny corner of a Mexican rain forest that was threatened with destruction. Because the surviving corn is a perennial, resistant to a variety of plant viruses, and hardy in cold and elevated climates, the plant holds great promise for plant geneticists who hope to infuse modern corn with some of the rain forest corn's genetic properties.[49] The irony is that species loss is reducing genetic resources just when genetic engineering technologies are being deployed to make effective use of them.[50]

Third, deforestation adversely affects indigenous Indian populations, particularly in Latin American and Pacific island forests. The destruction of rain forests and exposure to outside populations have reduced their number. In 1994, one entire village of Jaguapure Indians threatened to commit suicide unless seized lands were returned, and suicide rates among remnant tribes are high.[51] The opening of forests and the introduction of long-isolated cultures have resulted in the import and export of disease. River blindness and malaria have been taken into the forests, while scientists think that dengue fever, the Ebola virus, and perhaps AIDS have been brought out of disrupted rain forest environments.[52]

Deforestation also adversely affects other populations. In India and Africa, where there are few residual forest tribes, deforestation has increased fuel wood costs and forced poor people to forage farther and longer for fuel to cook meals.[53]

Fourth, deforestation accelerates carbon dioxide emissions. Although undisturbed tropical forests emit carbon dioxide on balance, forest clearing and burning have greatly increased net carbon dioxide emissions, adding "perhaps 1 to 2.6 billion tons of carbon dioxide to the atmosphere annually, or between 20 and 50 percent as much as the burning of fossil fuels."[54] It matters, of course, how forests are cut. Selective cutting for valuable hardwoods produces less destruction and fewer emissions than conversion to tropical plantations. And these uses do less to destroy forest cover or release carbon dioxide than clear-cutting or burning, as occurred on a grand scale in 1976, when Volkswagen set 25,000 hectares of Amazonian forest afire to clear land for a cattle ranch.[55]

Although there are good reasons to curb deforestation in the tropics, and to prevent deforestation in temperate Siberian woods, different social solutions are needed because the economic causes of deforestation vary regionally.

In Latin America, the primary cause of deforestation is cattle ranching, from which beef is exported to the fast-food hamburger industry in the United States.[56] Central American ranchers, for example, export 85 to 95 percent of their beef to the United States.[57] The migration of poor farmers into the forests, where they engage in subsistence agriculture, also contributes to deforestation in Latin America. Governments in the region typically provide subsidies to cattle ranchers and assist the settlement of migrant farmers. They do so to promote beef exports that can be used to repay debt—deforestation accelerated after the onset of the debt crisis— and to provide land for small farmers displaced by the introduction of large-scale Green Revolution agriculture in the fertile valleys.[58]

In this context, deforestation might be slowed if the United States curbed consumption or imports of Latin American beef and if governments in the region introduced land reform to provide land to poor farmers and ease pressure on marginal agricultural environments. They might also encourage sustainable forestry—rubber tapping, nut harvesting—for indigenous and some settler groups already living in the woods. And northern countries would probably have to extend substantial debt relief and economic aid to persuade governments in the region to conserve forests rather than use them as a source of export earnings or a way to absorb displaced rural populations.

The causes of deforestation in Southeast Asia are rather different. In Thailand, Malaysia, Indonesia, Papua New Guinea, and the Philippines, most tropical hardwoods are cut and exported to Japan, which imports 53 percent of the world's tropical hardwoods for use in construction and paper manufacture.[59] Taiwan and South Korea are also big tropical timber importers. Although timber exports are the main cause of deforestation in this region, conversion to tropical plantations—rubber, palm, and coffee—and the migration of subsistence farmers into the woods, which in Indonesia is directed by a government transmigration program designed to reduce population density on Java, also contribute to deforestation.[60]

In this context, deforestation might be slowed if Japan and the other tropical timber importers curbed hardwood consumption—much of the imported plywood is used once for concrete-building forms and then simply discarded—and if the Indonesian government emphasized birth control policies rather than transmigration programs. Thailand has banned the export of tropical timber, and other countries have banned the export of raw logs, in an effort to protect and manage their forests more effectively, though new free trade agreements may undermine their ability to use export controls to protect their resources. And while conversion of forests to plantations contributes to species loss, it is less damaging than conversion to pasture or agriculture and may represent a kind of compromise for countries needing export earnings.

In India and tropical Africa, the use of forests for fuel wood and subsistence agriculture is the main cause of deforestation, though countries in West Africa export some timber to European consumers.[61] The destruction of forests for fuel woods is particularly acute where population densities are high. In this context, birth control programs to reduce population density; the introduction of agro-forestry programs, where villages are given resources to plant fast-growing trees for fuel; the provision of more efficient stoves to use fuel more effectively; and in some cases, the provision of cheap fuel alternatives, like kerosene, can help reduce pressure on the forests. These steps can improve conditions for poor women, who may spend hours searching for fuel woods or may switch to faster-cooking but less nutritious foods where fuel is scarce or too expensive.[62] While these programs have proven effective, they are also costly for governments with few resources, which means that rich countries would have to provide economic aid if they were to be adopted widely.

Gender and Forests

In the South, deforestation generally hurts women because they are responsible in most poor, rural households for gathering firewood to cook food and warm family dwellings. Reforestation projects would mean that fuel supplies would be more plentiful and easier to obtain, which would save women considerable time, energy, and expense. In the North, as we have seen, deforestation generally hurts men because the logging, transport, milling, and pulping industries employ men, not women. Reforestation would provide employment for men, not women, though the growing reliance on technology and machinery in these industries means that reforestation could occur without significantly increasing job opportunities for men.

Automobiles

At the first Earth Day rallies in 1970, students sometimes bashed or buried cars, which they saw as symbols of environmental pollution and waste. Although smog and oil spills were then recognized as problems associated with cars, the explosive growth of the world car fleet and the widespread use of automobiles have since brought a host of other car-related environmental and social problems into sharp relief.

In 1950, there were fifty million cars worldwide, 75 percent of them in the United States. This number doubled by 1960, doubled again by 1970, and doubled again by 1990, an eightfold increase to more than 400 million cars.[63] And experts predicted that the world car fleet would reach 600 million by 2010, with much of the growth occurring in the former communist countries.[64]

The exploding car population intensifies and widens the impact of cars on the environment. But perhaps more important is the extensive use of cars. By 1990, U.S. owners drove their cars two trillion miles every year, the equivalent of a round-trip from Earth to Pluto every day of the year.[65]

The practice of driving interplanetary distances results in the consumption of vast quantities of oil, about six billion barrels a year. As has already been noted, the production, transportation, and use of oil are associated with a whole set of environmental, economic, and political problems. When burned to power a car, the oil is converted into a complex set of waste gases, carbon dioxide among them.[66] The average car produces its weight in carbon dioxide every year; the world car fleet generates about 14 percent of all carbon dioxide emissions.[67] But carbon dioxide emissions were long neglected because it is not a toxic gas, like most of the other one thousand assorted pollutants created by internal combustion engines.[68] Early efforts to reduce auto pollution focused on lead in gasoline, benzene, carbon monoxide, nitrogen oxides, unburned hydrocarbons, aldehydes, particulates, and trace metals because they contributed to smog and were known to have adverse effects on human health. More recent efforts have identified the release of CFCs from auto air conditioners (the number one source of CFC emissions in the United States) and carbon dioxide as problematic, even though they are not toxic to humans. Recently, more attention has been given to tiny particles of black carbon soot, which may be responsible for 50,000 to 60,000 deaths annually in the United States, affecting primarily young children with respiratory problems.[69] Carbon soot is also thought to make an important contribution to global warming, though the scientific evaluation of soot has only recently begun.

While cars release toxic pollutants, nontoxic gases, and particulates that adversely affect human health or the environment, they are also responsible for considerable death and injury. Worldwide, 265,000 are killed every year in auto accidents, and ten million more are injured.[70] Death rates are highest in poor countries, where cars share roads with pedestrians, bikes, and animals, and lower in rich countries where governments spend heavily to segregate traffic, police roads, improve auto safety, and, increasingly, prosecute drunk drivers.[71] But even in rich countries, cars are prodigious killers. In the United States, cars kill about 50,000 annually, and have killed three million during the century since the car was invented. Nearly twice as many Americans have died on the highway as on the battlefield in all of this country's wars since 1776.[72]

In social and economic terms, the growth of car fleets has persuaded governments to build highways and transform cities to accommodate them, developments that have contributed to the deterioration of urban neighborhoods and downtown businesses, fueled urban sprawl, and consumed rural farms near cities. Because mass transit systems become more expensive and less efficient as cities sprawl—mass transit carries only as many people in the United States today as it did in 1900—people rely more heavily on the automobile for transport, at considerable personal expense.[73] A study by the Hertz Corporation reported that Americans devote 15 percent of their income to automobile transportation, and the average male spends 1,600 hours a year in his car.[74] The government, too, spends vast sums supporting private transportation, particularly on road construction, maintenance, and police services. The California Department of Transportation reported that the state spent $2,500 more for each vehicle on the road than it received from car owners in taxes and fees.[75] The cost to taxpayers of government automobile subsidies is only now

becoming apparent, largely because the interstate highways and bridges
(which were built in the 1950s) are now due for repair, at staggering cost.

During the 1970s, some of the problems associated with car use were
addressed by government policy and market forces. Clean air legislation
required cars to get better mileage and emit fewer pollutants. The elimi-
nation of lead from gasoline was particularly important, though U.S.
manufacturers began exporting tetraethyl lead to other countries after the
U.S. ban took effect. The introduction of safety belts, crash standards, and
speed limits reduced fatalities 6 percent, even though the car population
grew 50 percent.[76] At the same time, rapidly rising oil prices spurred technological innovation, resulting in smaller, more fuel-efficient cars, many
of them from Japan, and reduced consumer use.

But many of the gains made during the 1970s eroded during the 1980s.
The growing car population undermined technological improvements.
"All of the progress we are making through technology is being eaten up
by growth," one official of the California Air Resources Board reported.[77]
The Reagan, Bush, and Clinton administrations cut mass transit systems,
reduced fuel-efficiency standards, and increased speed limits. These re-
laxed government policies undercut previous environmental and safety
gains. After 1985, falling oil prices encouraged consumers to increase car
use. In the past fifteen years, consumers have purchased trucks, Jeeps,
and minivans in increasing numbers, which is a problem because these
vehicles get poor gas mileage, lowering the average fuel efficiency of the
U.S. fleet. According to the Environmental Protection Agency, some forty
million Americans live in cities that do not meet federal clean air standards, despite two decades of pollution-reduction efforts.[78]

Many environmentalists and government officials have proposed higher
taxes, tickets, and tolls as a way to curb auto use in the United States. They
note that a 50-cent-a-gallon increase in the gas tax would produce $55 bil-
lion in annual revenue and discourage car use.[79] Although the Clinton administration raised the gas tax by only 4 cents, state officials around the
country have increased registration fees, tickets, and tolls to raise money for
strapped state budgets. In California, state officials raised registration fees
and slapped a 40 percent penalty on drivers who paid one day late, while
cities hiked parking tickets and tolls to finance budgets and reduce conges-
tion. San Francisco, for example, increased its parking ticket revenues from
$5.5 million in 1990 to $42 million in 1993, a sevenfold increase.[80]

The problem with this approach is that it is an inefficient and unfair way
to reduce car use. The rising cost of fuel, fees, taxes, and tolls is felt first by
low-income drivers, discouraging them from hitting the road. But this is not
an efficient way to curb use because poor people own few cars and drive
them sparingly. Families with annual incomes over $35,000 own three times
as many cars and drive them three times as often and three times as far as

families earning under $10,000.[81] The increasing popularity of gas-guzzling cars, vans, and trucks among well-to-do drivers means they consume more than three times as much gas. Higher costs are inefficient because they do little to curb use by people who have the greatest impact. They are also unfair because, in a country where mass transit systems are inadequate, poor people need cars to get to work just as much as wealthy families do. There are also regional inequalities. In Wyoming, the average commuter pays $243 a year in gas taxes, the average New Yorker only $91.[82] An across-the-board tax increase would have a greater impact on drivers in western states, where commutes are long, than on drivers in the East.

Gender and Automobiles

The rising cost of car ownership in the United States is socially unfair not only because it disadvantages poor people, but also because it adversely affects poor women and single mothers. As a result of welfare reform in 1996, poor women who received government benefits were assigned jobs while looking for permanent employment. But the rising cost of car ownership made it difficult for poor women to own cars, making it hard for them to find or keep jobs that were far from home or could not easily be reached by public transit. It was even harder for women trying to get to work and child care providers. This problem has been exacerbated by the suburbanization of towns and industries, which has located jobs in places that require workers to own cars. So poor women and single mothers without cars have found it extremely difficult to keep assigned jobs or find permanent employment.

There is, of course, a way to curb car use that is efficient and equitable: gas rationing. But it is such an anathema to consumers and politicians—despite demonstrable success during World War II—that it does not yet figure in any political discussion of car use.

Increasing car ownership and use in the United States and around the world is the source of serious environmental and social problems, global climate change among them. There are good reasons to reduce car use. But the automobile is so deeply embedded in the economic, social, and psychological life of the wealthy countries, particularly the United States, that it will be difficult to curb car ownership or use.

SWAMPS, RICE, AND CATTLE (METHANE)

After carbon dioxide, methane is the most important greenhouse gas, representing about 18 percent of all the gases with climate-changing potential.[83] Most of the methane in the atmosphere, about 65 percent, is emitted from natural processes, from anaerobic fermentation in wetlands, peat bogs, and swamps.[84] Human activity is responsible for about 35 percent of all methane emissions. Rice cultivation accounts for about 20 percent of this, and animal husbandry contributes about 15 percent.[85]

Although the use of synthetic fertilizer could lower methane emissions from rice cultivation, it would be difficult to reduce methane from this source without jeopardizing rice production, which feeds so much of the world.[86] But while it would be difficult to reduce rice production, there are sound human and environmental reasons to reduce animal herds, particularly of cattle. Cattle release about 80 million metric tons of methane into the atmosphere, considerably more than other domestic animals— ten times more than sheep, fifty times more than pigs.[87] Moreover, the expansion of the world cattle herd, which increased from 500 million to 1.2 billion between 1950 and 1990, has contributed to two important human and environmental problems.[88]

First, cattle consume vast quantities of grain that might otherwise feed a hungry and growing world population. Cattle consume about one-third of the world's grain, nearly 70 percent of the grain grown in the United States.[89] In recent years, feed grain for cattle has increasingly replaced grain grown for human consumption in poor countries, which has reduced the supply and increased the price of staple foods for hungry people.[90] Many countries, particularly in Latin America, have increased cattle production to supply beef for the U.S. market and earn money that can be used to repay debt. But the expansion of cattle ranching, with its resulting "protein flight," has reduced the amount of land devoted to agriculture and has led to rural job loss because ranching employs very few workers (one worker for every 47.6 hectares, compared with one worker for every 2.9 hectares in agriculture).[91] Cattle ranching has thereby contributed to rising prices and falling incomes for rural families in many countries, one reason why "one-third of rural families [in Mexico] never eat meat or eggs and 59 percent never drink milk."[92]

While the expansion of the world cattle herd contributes to hunger in poor countries, it also contributes to obesity and disease in wealthy countries where consumers eat vast quantities of beef.

Between 1945 and 1976, per capita beef consumption in the United States grew from 71 pounds to 129 pounds annually.[93] Nearly 40 percent of this was consumed as hamburgers, a postwar phenomenon associated first with the spread of backyard barbecue grills (ground pork falls apart on a grate, while ground beef sticks together) and later with the spread of fast-food hamburger franchises.[94]

The growing consumption of beef protein created a diet heavy in fat: 37 percent of calories in U.S. diets comes from fat.[95] As a result, obesity has become a serious problem. More than thirty-four million Americans are overweight, according to the Centers for Disease Control and Prevention, and the number of women considered obese rose from 13.3 percent to 17.7 percent between 1960 and 1980.[96] More important, fat-heavy diets and obesity contribute to disease and death. The U.S. surgeon general estimated that 1.5 million deaths in 1987 were related to dietary factors and said that diets high in saturated fat and cholesterol contributed to the high incidence of heart attack, colon and breast cancer, and stroke in America.[97]

Second, the world's cattle herds contribute to a series of environmental problems. The expansion of cattle herds in tropical regions has led to extensive deforestation, particularly in Latin America, and increased carbon dioxide emissions. Outside rain forest settings, cattle grazing on desert fringe areas or on marginal lands can lead to desertification, as it has in the Sahel of Africa, or to the degradation of grasslands.[98] Cattle also consume large quantities of water, far more than other domesticated animals, and can drain water resources in arid regions.[99] On pasture land, cattle waste can foul streams ("an ungrazed part of a stream in Montana produced 268 more trout than did a grazed part of the same stream") while waste from feed lots can contaminate groundwater supplies ("the organic waste generated by a 10,000-head feed lot is equivalent to the human waste generated in a city of 110,000 people"[100]).

After increasing for many years, beef consumption in the wealthy countries fell in the past twenty-five years. From a high of 129 pounds per capita in 1976, U.S. beef consumption fell to 78.2 pounds by 1983, back to World War II levels.[101] This rapid decline was due to two developments: rising grain prices during the 1970s and changing diets.[102] As Americans became more concerned about diet and health in the 1970s and 1980s, consumers reduced their purchase of beef, which was growing more expensive because feed prices had soared, and increasingly turned to chicken, fish, and pasta. As a result of falling consumer demand, the growth of the world's cattle herds slowed and then stabilized at about 1.25 billion head by 1990.[103]

While this was a welcome development in human and environmental terms, further cattle herd reductions would be feasible and beneficial. A reduction in dietary fat in the United States, say from 37 to 30 percent, a level most dietary scientists believe is necessary, would reduce beef consumption by another 20 percent.[104] And if fat made up only 14 percent of American diets, a level that some scientists believe is optimal because it would greatly reduce health risks, beef consumption would fall and the world cattle herd could shrink dramatically, releasing large amounts of grain for human consumption. "If the 130 million metric tons of grain that are fed yearly to U.S. livestock were consumed directly as human food, about 400 million people—1.7 times larger than the U.S. population—could be sustained for one year," Cornell scientist David Pimentel estimated.[105] However, even scientists who argue that "we are basically a vegetarian species and should be eating a wide variety of plant food and minimizing our intake of animal food" do not argue for a completely vegetarian diet or an agriculture without animal husbandry.[106] Rather, they argue that consumption of animal protein should be reduced in wealthy countries, that consumers should rely on other animals that produce protein more efficiently (chickens and pigs), and that cattle herds should be reduced but not eliminated so they can continue to provide milk, cheese, traction, and manure in many agricultural systems.

COOLANTS AND SPRAY CANS (CFCs)

In 1985, British scientists discovered that a set of gases called chlorofluorocarbons (CFCs) were responsible for depleting ozone in the atmosphere.[107] CFCs were first invented by DuPont scientist Thomas Midgley in 1930.[108] Because CFCs were nontoxic, nonflammable, noncorrosive, and inert, they soon found widespread application as coolants in refrigerators and air conditioners, propellants in spray cans, and blowing agents in plastic and styrofoam. Later, related gases called halons found use in fire extinguishers: "Used in Army tanks, in which engine and munition fires can spread like lightning, [automatic] halon fire extinguishers . . . can snuff out a raging gasoline fire in thousandths of a second."[109]

In 1974, scientists warned that while CFCs were beneficial in many respects, when they were released into the atmosphere, sunlight would detach chlorine from the rest of the molecule, and that chlorine would then attack and deplete ozone.[110] If this occurred, researchers warned, CFCs could weaken the ozone layer, the planet's protection against harmful ultraviolet radiation from the sun. They predicted that higher radiation levels would increase the incidence of skin cancer, particularly among Caucasians; make cataracts more common; and suppress the human immune system, which fights off viruses, tumors, and other infectious diseases.[111] Increased radiation would also harm plants, particularly crops like soybeans, and affect marine organisms in waters up to 100 feet deep.[112] Scientists noted, however, that the effects would be hard to predict because radiation levels around the globe would increase unevenly, and because places with heavy cloud cover would block much of the radiation, whereas areas with direct sunlight would receive heavier doses.[113]

Scientists found evidence in 1985 that CFC levels and ozone depletion were closely related and discovered serious ozone depletion over Antarctica. This discovery of an "ozone hole . . . larger than the United States and taller than Mount Everest" spurred efforts to reduce and eventually ban CFCs and other ozone-depleting gases, halons, and later methyl bromide.[114]

In 1987, twenty-four countries adopted the Montreal Protocol on Substances That Deplete the Ozone Layer, agreeing to freeze CFC production at 1986 levels, followed by a 20 percent reduction by 1993 and another 30 percent reduction in 1998.[115] During the next few years, as evidence mounted that ozone depletion was accelerating and becoming more serious, signatories to the Montreal Protocol took additional steps. In 1989, they agreed to eliminate all CFC production by the end of the century, and in 1992, moved the deadline up to 1996.[116] The rich countries also agreed to provide $500 million to poor countries to help them reduce their reliance on CFCs.[117] The Bush administration used provisions of the Clean Air Act to ban some substances not covered by the Montreal Protocol.[118] And in 1995, one hundred governments agreed to phase out methyl bromide, a powerful ozone-depleting pesticide used in agriculture.[119]

Governments acted with rare unanimity and considerable speed because the scientific evidence was not disputed (as it has been with global warming); because CFCs contributed to ozone depletion but also to global warming (accounting for about 14 percent of greenhouse gases); because ozone depletion was accelerating rapidly—ozone was being depleted twice as fast in 1992 as it had been in 1985—and because the particular properties of CFCs made delay damaging. CFCs have long atmospheric life spans: CFC 11 lasts 76 years in the upper atmosphere, CFC 12 for 139 years.[120] And because they are now stored in foams, refrigerators, and spray cans, they will be slowly released as these containers erode, pumping ozone-depleting gases into the atmosphere even after their manufacture has stopped. As a consequence, a CFC ban in the year 2000 would only reduce CFCs to 1995 levels in 2073. By moving up the deadline to 1996, governments thought CFC levels would fall back to 1995 levels much sooner, by 2053. These realities made for rare agreement. As U.S. delegate Richard Benedick explained, "We're seeing something completely unprecedented in the history of diplomacy. Politicians from every block and region of the world are setting aside politics to reach agreement on protecting the global environment."[121]

These developments produced real and immediate benefits and created fewer economic problems than many first expected. Scientists found that production and use of CFCs slowed even before the Montreal Protocol took effect and, as a result, that ozone depletion slowed dramatically. "Here is a beautiful case study of science and public policy working well," said NASA scientist James Elkins.[122] Moreover, the cost to industry and consumers of replacing CFCs with other technologies was less expensive than industry officials first predicted. The switch to other technologies even resulted in energy-saving designs, which could result in savings of up to $100 billion during the next eighty-five years.[123] While efforts to reduce CFCs have been successful, it is important to note that these efforts occurred not so much because they contributed to averting global warming, but because they contributed to averting other serious social and environmental problems.

SYNTHETIC FERTILIZER (NITROUS OXIDE)

Synthetic nitrogen fertilizers emit nitrous oxide, a gas associated with global climate change. There are also natural sources of nitrous oxide emissions in oceans and soils, but as with methane emissions, little can be done about them. About one-third of all nitrous oxide emissions are associated with human activity. Of these, synthetic fertilizer use is the most important source, though nitrous oxide is also produced during the manufacture of nylon.[124] The increasing use of nitrogen fertilizers—global fertilizer use grew from 14 million tons in 1950 to 121 million tons in 1984, and an estimated 140 million tons in 2009—contributes not only to global warming (about 6 percent of the total), but also to groundwater pollution and algae blooms, which deprive rivers, estuaries, and oceans of oxygen, killing fish and other marine life.[125]

But while there are problems associated with nitrogen fertilizers, global climate change among them, the use of fertilizer greatly increases world food production, which is essential for a growing population. As the World Watch Institute estimated, "Eliminating [synthetic fertilizer] use today would probably cut world food production by at least a third," an extremely serious problem for the "billion and a half people now fed with the additional food produced with chemical fertilizer."[126]

Under these circumstances it is difficult to imagine or suggest that synthetic fertilizer use should be dramatically reduced or eliminated as part of an effort to reduce global warming. There are, however, some steps that could be taken to reduce fertilizer use and nitrous oxide emissions on the margins.

Because natural gas is used to produce synthetic fertilizers like anhydrous ammonia, fertilizer prices rise and fall with energy prices. When energy prices rose in the 1970s, fertilizer use slowed somewhat, curbing pollution and nitrous oxide emissions.[127] Most agronomists also think that changed farm practices could reduce some fertilizer-related problems. They argue that farmers could apply fertilizers more carefully, an important consideration because their effectiveness is relatively ephemeral.[128] They urge farmers to use more natural manures for fertilizer—much of it now wasted in cattle, pig, and chicken feed lot systems, which do not apply waste to farmlands—because natural manures release less nitrous oxide and require less energy (and carbon dioxide emissions) to produce.[129] They also urge farmers to plant nitrogen-fixing crops like alfalfa along with nitrogen-depleting crops like corn to reduce synthetic-fertilizer inputs and also reduce farmer costs, an important consideration if small-scale farmers are to survive.[130]

Because these practices would likely produce only modest reductions in synthetic fertilizer use, people concerned about global climate change will probably have to devote their attention to curbing other greenhouse gases, which, in any event, play a more significant overall role.

POLITICAL AND ECONOMIC SOLUTIONS

The scientific case for global warming persuaded government officials around the world to take steps to reduce greenhouse gas emissions. In 1992, government representatives meeting in Rio de Janeiro agreed that the industrialized countries would try to reduce greenhouse gas emissions to 1990 levels by the year 2000. President George H. W. Bush signed the agreement, which provided that measures taken to reduce greenhouse gases were voluntary, not mandatory.[131]

In 1997, officials for 170 countries met in Kyoto, Japan, to hammer out a formal treaty that would set more ambitious goals and require signatory governments to take mandatory steps to reduce greenhouse emissions. At this meeting, representatives agreed to reduce emissions to 5 percent below 1990 levels by 2012, an approach endorsed by the Clinton administration.[132] The Kyoto Protocols, as the treaty was called, would take effect only after fifty-five countries, including countries responsible for 55 percent of all 1990 emissions, had signed the accord.

During the next four years, officials haggled over the details so that governments could sign the protocols and bring them into effect. But as negotiations neared their conclusion in 2001, President George W. Bush announced that the United States had reversed its position and would not endorse the treaty.[133] He argued that if the United States took steps required in the treaty, it would "have a negative economic impact, with layoffs of workers and price increases for consumers."[134] He also complained that the agreement was unfair because it required major reductions by the United States but none by poor, developing countries like China and India.[135]

Negotiators had agreed to exempt India and China because their per capita production of greenhouse gases was only a fraction of that in the United States, Western Europe, and Japan. In 1998, per capita carbon dioxide emissions in the United States (20.1 tons) were ten times bigger than per capita emissions in China (2.3 tons).[136] If India and China were required to reduce emissions, they might be unable to provide even minimal supplies of electricity or fuel for transportation and heating. Moreover, while China was not obligated by Kyoto to reduce its greenhouse emissions, it did so anyway, reducing emissions by 17 percent between 1995 and 2001, even while its economy grew 36 percent (see chapter 7).[137] "Even without undertaking binding commitments under an international agreement, China has nevertheless contributed substantially to reducing growth in greenhouse emissions," scientists at the Lawrence Berkeley National Laboratory reported.[138]

China was able to reduce emissions voluntarily because it ended government subsidies for fuel, which increased energy prices and reduced consumption, and because it promoted energy conservation, particularly in big cities where air pollution is a serious problem.[139] "We've done what we can to reduce emissions," Gao Feng, a Chinese official, explained. "But it's not fair to ask the developing countries to take the lead."[140]

After the Bush administration pulled out of negotiations in the summer of 2001, adoption of the treaty was put in serious jeopardy. But negotiations continued in Marrakesh, Morocco, during the fall.[141] Finally, in November 2001, representatives of 164 countries announced that they had reached an agreement that would meet the approval of 55 countries, including countries responsible for 55 percent of global emissions.[142] With the United States on the sidelines, this meant crafting language that persuaded the EU, Japan, and Russia to sign on.

Although U.S. officials refused to participate, the treaty may still help reduce U.S. emissions even without government permission. Private corporations are moving on their own to reduce energy consumption and greenhouse emissions for a variety of reasons. First, many businesses and public utilities are converting from energy sources that are high in carbon (coal) to energy supplies that are lower in carbon (oil and natural gas).[143] When burned, natural gas produces only one-third as much carbon dioxide as coal, two-thirds that of oil.[144] Corporations practice this kind of substitutionism to take advantage of new technologies and reduce costs. The "decarbonization" of the energy system in the United States is being driven largely by market considerations, not government policy.

Second, many transnational corporations (TNCs) based in the United States do business in Western Europe, Japan, and other countries that are signing on to the Kyoto Protocols. As a result, "many multinational companies plan to continue reducing emissions because they face strong pressure to do so in Europe and Japan, fear rising energy costs, or want to promote their products as being friendly to the environment."[145]

TNCs recognize that if they do not take steps, their ability to compete with corporations in Western Europe and Japan may erode. This is particularly clear to U.S. automakers, who recognize that they must develop new, more energy-efficient, higher-mileage automobile designs if they are going to compete with automakers in Western Europe and Japan.[146] In the same spirit, DuPont and Alcoa have announced plans to reduce substantially their greenhouse gas emissions. DuPont said it would cut emissions to 65 percent below 1990 levels by the year 2010, Alcoa to 25 percent below 1990 levels by 2010.[147] These cutbacks would be more dramatic and more rapid than reductions called for in the Kyoto Protocols. It may be that U.S. businesses are willing to do what government officials cannot contemplate. In this case, the globalization of business may have positive environmental and social benefits.

NOTES

1. Philip Shabecoff, "Global Warming Has Begun, Expert Tells Senate," *New York Times*, 24 June 1988.

2. Jacqueline Vaughn Switzer, *Environmental Politics: Domestic and Global Dimensions* (New York: St. Martin's Press, 1994), 269.

3. Boyce Rensberger, "As Earth Summit Nears, Consensus Still Lacking on Global Warming's Cause," *Washington Post*, 31 May 1992.

4. William K. Stevens, "Experts Confirm Human Role in Global Warming," *New York Times*, 10 September 1995.

5. Christopher Flavin, "Slowing Global Warming," in *State of the World 1990*, ed. Lester Brown (New York: Norton, 1990), 17.

6. Paul Kennedy, *Preparing for the 21st Century* (New York: Random House, 1993), 108; William K. Stevens, "Earlier Global Warming Harm Seen," *New York Times*, 17 October 1990; Jeremy Leggett, "The Nature of the Greenhouse Threat," in *Global Warming: The Greenpeace Report*, ed. Jeremy Leggett (Oxford: Oxford University Press, 1990), 2.

7. Kennedy, *Preparing for the 21st Century*, 110.

8. William K. Stevens, "Violent World of Corals Is Facing New Dangers," *New York Times*, 16 February 1993.

9. Bill McKibben, *The End of Nature* (New York: Random House, 1989), 95–96.

10. Jeremy Leggett, "Gone with the Winds," *World Paper* (April 1993), 13.

11. Kennedy, *Preparing for the 21st Century*, 111–12.

12. David E. Pitt, "Computer Vision of Global Warming: Hardest on Have-Nots," *New York Times*, 18 January 1994; Claire Pedrick, "A Moveable Feast: Climate, Bread and Butter," *World Paper* (April 1993), 11.

13. Flavin, "Slowing Global Warming," 19; Kennedy, *Preparing for the 21st Century*, 117.

14. Marlise Simons, "Brazil, Smarting from the Outcry over the Amazon, Charges Foreign Plot," *New York Times*, 23 March 1989.

15. Patrick E. Tyler, "China's Power Needs Exceed Investor Tolerance," *New York Times*, 7 November 1994.

16. Patrick E. Tyler, "China's Inevitable Dilemma: Coal Equals Growth," *New York Times*, 29 September 1995.

17. William K. Stevens, "In a Warming World, Who Comes Out Ahead?" *New York Times*, 5 February 1991.

18. Patrick J. Michaels, *Sound and Fury: The Science and Politics of Global Warming* (Washington, D.C.: Cato Institute, 1992), 3; William K. Stevens, "Global Warming: The Contrarian View," *New York Times*, 29 February 2000; Andrew C. Revkin, "Debate Rises over a Quick(er) Climate Fix," *New York Times*, 3 October 2000.

19. Fred S. Singer, "Global Climate Change: Fact and Fiction," in *Environment 93/94*, ed. John L. Allen (Guilford, Conn.: Dushkin, 1993), 186; Richard A. Kerr, "Is the World Warming or Not?" *Science* 267, 3 February 1995, 612; Michaels, *Sound and Fury*, 53.

20. Rensberger, "As Earth Summit Nears"; William K. Stevens, "In New Data on Climate Changes, Decades, Not Centuries Count," *New York Times*, 7 December 1993; William K. Stevens, "Climate Roller Coaster in Swedish Tree Rings," *New York Times*, 7 August 1990.

21. William K. Stevens, "With Climate Treaty Signed, All Say They'll Do Even More," *New York Times*, 13 June 1992; Singer, "Global Climate Change," 17; Lester Brown, "A False Sense of Security," in *State of the World 1985*, ed. Lester Brown (New York: Norton, 1985), 15.

22. Singer, "Global Climate Change," 186.

23. Flavin, "Slowing Global Warming," 17.

24. William K. Stevens, "A Skeptic Asks, Is It Getting Hotter, or Is It Just the Computer Model?" *New York Times*, 18 June 1996.

25. Stevens, "A Skeptic Asks."

26. Rensberger, "As Earth Summit Nears."

27. William K. Stevens, "Experts Confirm Human Role in Global Warming," *New York Times*, 10 September 1995.

28. Stevens, "Experts Confirm Human Role."

29. Sylvia Nasar, "Cooling the Globe Would Be Nice, but Saving Lives Now May Cost Less," *New York Times*, 31 May 1992.

30. Mick Kelly, "Halting Global Warming," in *Global Warming: The Greenpeace Report*, ed. Jeremy Leggett (Oxford: Oxford University Press, 1990), 86; Leggett, "The Nature of the Greenhouse Threat," 17.

31. Michael Renner, "Reinventing Transportation," in *State of the World 1994*, ed. Lester Brown (New York: Norton, 1994), 64.

32. Sandra Postel, "Protecting Forests," in *State of the World 1984*, ed. Lester Brown (New York: Norton, 1984), 82.

33. Joseph J. Romm and Amory B. Lovins, "Fueling a Competitive Economy," *Foreign Affairs* (Winter 1992–1993), 47.

34. Romm and Lovins, "Fueling a Competitive Economy," 49.

35. Romm and Lovins, "Fueling a Competitive Economy," 48.

36. Renner, "Reinventing Transportation," 69.

37. Romm and Lovins, "Fueling a Competitive Economy," 50.

38. Catherine Caufield, *Tropical Moist Forests* (London: Earthscan, 1982), 7; Norman Meyers, "Tropical Forests," in *Global Warming: The Greenpeace Report*, ed. Jeremy Leggett (Oxford: Oxford University Press, 1990), 377.

39. Caufield, *Tropical Moist Forests*, 1; Sandra Postel and Lori Heise, "Reforesting the Earth," in *State of the World 1988*, ed. Lester Brown (New York: Norton, 1988), 85.

40. Stephen Budiansky, "The Doomsday Myths," in *Environment 95/96*, ed. John L. Allen (Guilford, Conn.: Dushkin, 1995), 35; "Instant Trees," *The Economist*, 28 April 1990, 93.

41. Diane Jean Schemo, "Burning of Amazon Picks Up Pace, with Vast Areas Lost," *New York Times*, 12 September 1996.

42. William K. Stevens, "Experts Say Logging of Vast Siberian Forest Could Foster Warming," *New York Times*, 28 January 1992.

43. Caufield, *Tropical Moist Forests*, 10.

44. Fred Pearce, "Hit and Run in Sarawak," *New Scientist*, 12 May 1990, 47; A. Kent MacDougall, "Worldwide Costs Mount As Trees Fall," *Los Angeles Times*, 14 June 1987; Postel, "Protecting Forests," 84.

45. Edward C. Wolf, "Avoiding a Mass Extinction of Species," in *State of the World 1988*, ed. Lester Brown (New York: Norton, 1988), 110; Patrick Anderson, "The Myth of Sustainable Logging: The Case for a Ban on Tropical Timber Imports," *The Ecologist* 19 (September–October 1989), 166. Studies have shown that even temperate woods do not recover as easily or as fast as foresters have long assumed. Catherine Dold, "Study Casts Doubt on Belief in Self-Revival of Cleared Forests," *New York Times*, 1 September 1992.

46. Caufield, *Tropical Moist Forests*, 10; Budiansky, "The Doomsday Myths," 34; Charles Petit, "Scientist Argues against Focus on Rain Forests," *San Francisco Chronicle*, 21 February 1992.

47. Wolf, "Avoiding a Mass Extinction of Species," 103.

48. David Pimentel, Laura E. Armstrong, Christine A. Flass, Frederic W. Hopf, Ronald B. Landy, and Marcia H. Pimentel, "Interdependence of Food and Natural Resources," in *Food and Natural Resources*, ed. David Pimentel and Carl W. Hall (San Diego: Academic Press, 1989), 42.

49. Norman Meyers, "Loss of Biological Diversity and Its Potential Impact on Agriculture and Food Productivity," in *Food and Natural Resources*, ed. David Pimentel and Carl W. Hall (San Diego: Academic Press, 1989), 52–53.

50. Meyers, "Loss of Biological Diversity," 53.

51. Caufield, *Tropical Moist Forests*, 19.

52. Caufield, *Tropical Moist Forests*, 33; Richard Preston, "Crisis in the Hot Zone," *The New Yorker*, 26 October 1992, 62.

53. Postel, "Protecting Forests," 83; MacDougall, "Worldwide Costs Mount As Trees Fall."

54. Postel and Heise, "Reforesting the Earth," 94.

55. Caufield, *Tropical Moist Forests*, 37.

56. Postel, "Protecting Forests," 77; Singer, "Global Climate Change," 86; Paul Harrison, *The Third Revolution: Population, Environment and a Sustainable World* (London: Penguin, 1993), 95–96.

57. Caufield, *Tropical Moist Forests*, 34.

58. "How Brazil Subsidizes the Destruction of the Amazon," *The Economist*, 18 March 1989, 69; Harrison, *The Third Revolution*, 96; Caufield, *Tropical Moist Forests*, 24–25.

59. Adam Schwarz, "Timber Troubles," *Far Eastern Economic Review*, 6 April 1989, 86; Caufield, *Tropical Moist Forests*, 29.

60. Caufield, *Tropical Moist Forests*, 28; Postel, "Protecting Forests," 77.

61. Postel and Heise, "Reforesting the Earth," 88–89.

62. Postel and Heise, "Reforesting the Earth," 88.

63. Motor Vehicle Manufacturers Association, *World Motor Vehicle Data, 1988 Edition* (Detroit: Motor Vehicle Manufacturers Association, 1988); Alan Attshuler, *The Future of the Automobile: The Report of MIT's International Automobile Program* (Cambridge, Mass.: MIT Press, 1984), 2–3, 13; Lester Brown, Christopher Flavin, and Colin Norman, *Running on Empty: The Future of the Automobile in an Oil-Short World* (New York: Norton, 1979), 86.

64. Attshuler, *The Future of the Automobile*, 113.

65. Alexandra Allen, "The Auto's Assault on the Atmosphere," *Multinational Monitor* (January–February 1990), 23.

66. Attshuler, *The Future of the Automobile*, 4.

67. Christopher Flavin, "Slowing Global Warming," in *State of the World 1990*, ed. Lester Brown (New York: Norton, 1990), 23; Attshuler, *The Future of the Automobile*, 58; Greenpeace, *The Environmental Impact of the Car: A Greenpeace Report* (Seattle: Greenpeace, 1992), 15.

68. Greenpeace, *The Environmental Impact of the Car*, 19.

69. Philip J. Hilts, "Studies Say Soot Kills up to 60,000 in U.S. Each Year," *New York Times*, 19 July 1993; Revkin, "Debate Rises over a Quick(er) Climate Fix."

70. Greenpeace, *The Environmental Impact of the Car*, 48; Attshuler, *The Future of the Automobile*, 5.

71. Wolfgang Zuckerman, *End of the Road: From World Car Crisis to Sustainable Transportation* (Post Mills, Vt.: Chelsea Green, 1993), 134.

72. Robert Schaeffer, "Car Sick: Autos ad Nauseam," *Greenpeace Magazine* (May–June 1990), 15.

73. Schaeffer, "Car Sick."

74. Brown, Flavin, and Norman, *Running on Empty*, 17; Zuckerman, *End of the Road*, 85–86; Romm and Lovins, "Fueling a Competitive Economy," 86.

75. Zuckerman, *End of the Road*, 215.

76. Attshuler, *The Future of the Automobile*, 66–67, 70.

77. Ronald Brownstein, "Testing the Limits," *National Journal* (July 29, 1989), 1918.

78. Robert Reinhold, "Hard Times Dilute Enthusiasm for Clean Air Laws," *New York Times*, 25 November 1993; James Sterngold, "A Back-and-Forth Smog War," *New York Times*, 12 September 1996.

79. Matthew L. Wald, "50-Cents-a-Gallon Tax Could Buy a Whole Lot," *New York Times*, 18 October 1992.

80. Phillip Matier and Andrew Ross, "State Drives Up Car Costs," *San Francisco Chronicle*, 21 June 1993.

81. Motor Vehicle Manufacturers Association, *World Motor Vehicle Data, 1988 Edition*.

82. Wald, "50-Cents-a-Gallon-Tax."

83. Ann Ehrlich, "Agricultural Contributions to Global Warming," in *Global Warming: The Greenpeace Report*, ed. Jeremy Leggett (Oxford: Oxford University Press, 1990), 401.

84. Ehrlich, "Agricultural Contributions to Global Warming," 402–3.

85. Ehrlich, "Agricultural Contributions to Global Warming," 403.

86. Ehrlich, "Agricultural Contributions to Global Warming," 407.

87. Ehrlich, "Agricultural Contributions to Global Warming," 404; McKibben, *The End of Nature*, 15; Alan Thein Durning and Holly B. Brough, "Reforming the Livestock Economy," in *State of the World 1992*, ed. Lester Brown (New York: Norton, 1992), 74.

88. Durning and Brough, "Reforming the Livestock Economy," 68.

89. Durning and Brough, "Reforming the Livestock Economy," 69–70.

90. Durning and Brough, "Reforming the Livestock Economy," 76; Steven E. Sanderson, "The Emergence of the 'World Steer': International and Foreign Domination in Latin American Cattle Production," in *Food, the State and International Political Economy: Dilemmas of Developing Countries*, ed. F. LaMond Tullis and W. Ladd Hollist (Lincoln: University of Nebraska Press, 1986), 133–34, 139–40.

91. Mark Edelman, "From Costa Rican Pasture to North American Hamburger," in *Food and Evolution*, ed. Marvin Harris and Eric B. Ross (Philadelphia: Temple University Press, 1987), 553, 554–55; Sanderson, "The Emergence of the 'World Steer,'" 146.

92. Sanderson, "The Emergence of the 'World Steer,'" 129.

93. Jimmy M. Skaggs, *Prime Cut: Livestock Raising and Meatpacking in the United States, 1607–1983* (College Station: Texas A&M University Press, 1986), 166.

94. Jeremy Rifkin, *Beyond Beef: The Rise and Fall of the Cattle Culture* (New York: Dutton, 1992), 260, 264.

95. Durning and Brough, "Reforming the Livestock Economy," 74.

96. Rifkin, *Beyond Beef*, 166.

97. Rifkin, *Beyond Beef*, 171.

98. Durning and Brough, "Reforming the Livestock Economy," 72–73; Kennedy, *Preparing for the 21st Century*, 98–99.

99. Vashek Cervinka, "Water Use in Agriculture," in *Food, the State and International Political Economy: Dilemmas of Developing Countries*, ed. F. LaMond Tullis and W. Ladd Hollist (Lincoln: University of Nebraska Press, 1986), 148–49; Durning and Brough, "Reforming the Livestock Economy," 70–71.

100. Rifkin, *Beyond Beef*, 206, 221.

101. Rifkin, *Beyond Beef*, 206, 221.

102. Skaggs, *Prime Cut*, 181–82.

103. Ehrlich, "Agricultural Contributions to Global Warming," 405.

104. Durning and Brough, "Reforming the Livestock Economy," 81–82.

105. Pimentel et al., "Interdependence of Food and Natural Resources," 36; Rifkin, *Beyond Beef*, 161.

106. Rifkin, *Beyond Beef*, 73–74.

107. Cynthia Shea, "Protecting the Ozone Layer," in *State of the World 1989*, ed. Lester Brown (New York: Norton, 1989), 77–78.

108. Shea, "Protecting the Ozone Layer," 85.

109. Malcome W. Browne, "As Halon Ban Nears, Researchers Seek a New Miracle Firefighter," *New York Times*, 15 December 1992.

110. Shea, "Protecting the Ozone Layer," 78–79.

111. Shea, "Protecting the Ozone Layer," 82–83; Tom Wicker, "Bad News from Above," *New York Times*, 10 April 1991.

112. Shea, "Protecting the Ozone Layer," 83; Malcome W. Browne, "Broad Effort Under Way to Track Ozone Hole's Effects," *New York Times*, 6 January 1992.

113. William K. Stevens, "Clouds May Retard Ozone Depletion," *New York Times*, 21 November 1995.

114. Shea, "Protecting the Ozone Layer," 78.

115. Shea, "Protecting the Ozone Layer," 93.

116. Craig R. Whitney, "Banning Chemicals That May Harm Ozone," *New York Times*, 3 March 1989; David Perlman, "Scientists Discover Huge Increase in Threat to Ozone," *San Francisco Chronicle*, 4 February 1992; Craig R. Whitney, "80 Nations Favor Ban to Help Ozone," *New York Times*, 3 May 1989; William K. Stevens, "Threat to Ozone Hastens the Ban on Some Chemicals," *New York Times*, 26 November 1992.

117. Stevens, "Threat to Ozone Hastens the Ban on Some Chemicals"; Craig R. Whitney, "Industrial Countries to Aid Poorer Nations on Ozone," *New York Times*, 6 May 1989.

118. Keith Schneider, "Bush Orders End to Making of Ozone-Depleting Agents," *New York Times*, 13 February 1992.

119. William K. Stevens, "100 Nations Move to Save Ozone Shield," *New York Times*, 10 December 1995.

120. Shea, "Protecting the Ozone Layer," 88.

121. Malcome W. Browne, "Ozone Fading Fast, Thatcher Tells Experts," *New York Times*, 28 June 1990.

122. William K. Stevens, "Scientists Report an Easing in Ozone-Killing Chemicals," *New York Times*, 26 August 1993.

123. David Doniger and Alan Miller, "Fighting Global Warming Is Good for Business" (College Park: University of Maryland, Center for Global Change, circa 1990).

124. Ehrlich, "Agricultural Contributions to Global Warming," 410–13; Keith Bradsher and Andrew C. Revkin, "A Pre-Emptive Strike on Global Warming," *New York Times*, 15 May 2001.

125. William J. Hudson, "Population, Food and the Economy of Nations," in *Food and Natural Resources*, ed. David Pimentel and Carl W. Hall (San Diego: Academic Press, 1989), 201–3; Hilary F. French, "Clearing the Air," in *State of the World 1990*, ed. Lester Brown (New York: Norton, 1990), 106; Lester Brown, "Reducing Hunger," in *State of the World 1994*, ed. Lester Brown (New York: Norton, 1994), 29.

126. Brown, "Reducing Hunger"; Ehrlich, "Agricultural Contributions to Global Warming," 419.

127. Brown, "Reducing Hunger," 31.

128. David A. Andow and David P. Davis, "Agricultural Chemicals: Food and Environment," in *Food and Natural Resources*, ed. David Pimentel and Carl W. Hall (San Diego: Academic Press, 1989), 195.

129. Andow and Davis, "Agricultural Chemicals"; Ehrlich, "Agricultural Contributions to Global Warming," 414, 416.

130. Norman Meyers, "Loss of Biological Diversity and Its Potential Impact on Agriculture and Food Productivity," in *Food and Natural Resources*, ed. David Pimentel and Carl W. Hall (San Diego: Academic Press, 1989); Ehrlich, "Agricultural Contributions to Global Warming," 416.

131. William K. Stevens, "Greenhouse Gas Issue: Haggling over Fairness," *New York Times*, 30 November 1997.

132. Andrew C. Revkin, "U.S. Move Improves Chance for Global Warming Treaty," *New York Times*, 20 November 2000.

133. David E. Sanger, "Bush Will Continue to Oppose Kyoto Pact on Global Warming," *New York Times*, 12 June 2001.

134. Sanger, "Bush Will Continue to Oppose Kyoto Pact."

135. Sanger, "Bush Will Continue to Oppose Kyoto Pact."

136. Erik Eckholm, "China Said to Sharply Reduce Emissions of Carbon Dioxide," *New York Times*, 15 July 2001.

137. Eckholm, "China Said to Sharply Reduce Emissions."

138. Eckholm, "China Said to Sharply Reduce Emissions."

139. Eckholm, "China Said to Sharply Reduce Emissions."

140. Eckholm, "China Said to Sharply Reduce Emissions."

141. Andrew C. Revkin, "Climate Talks Come Down to Haggling Over Details," *New York Times*, 9 November 2001; Andrew C. Revkin, "Delegates Work Late on a Treaty to Battle Global Warming," *New York Times*, 10 November 2001.

142. Andrew C. Revkin, "Deals Break Impasse on Global Warming Treaty," *New York Times*, 11 November 2001.

143. William K. Stevens, "Moving Slowly toward Energy Free of Carbon," *New York Times*, 31 October 1999.

144. Stevens, "Moving Slowly toward Energy Free of Carbon."

145. Bradsher and Revkin, "A Pre-Emptive Strike on Global Warming."

146. Bradsher and Revkin, "A Pre-Emptive Strike on Global Warming."

147. Bradsher and Revkin, "A Pre-Emptive Strike on Global Warming."

Index

About the Author

For many years, **Robert K. Schaeffer** worked as a journalist and an editor for Friends of the Earth, *In These Times*, *Nuclear Times*, and *Greenpeace Magazine*. He is now professor of sociology at Kansas State University. He participates in Pugwash Conferences on Science and World Affairs, which won the Nobel Peace Prize in 1995. He is the editor of *War in the World-System* (1990); author of *Warpaths: The Politics of Partition* (1990), *Power to the People: Democratization around the World* (1997), and *Severed States: Dilemmas of Democracy in a Divided World* (1999); and coauthor, with Torry Dickinson, of *Fast Forward: Work, Gender, and Protest in a Changing World* (2001) and *Transformations: Feminist Pathways to Global Change* (2008).